AF772460

CALCULUS

A computer algebra approach

Solutions Manual

International Press Incorporated, Boston
P.O. Box 2872
Cambridge, MA 02238-2872

Copyright © 1996 by International Press Incorporated

All rights reserved. No part of this work can be reproduced in any form, electronic or mechanical, recording, or any information storage and data retrieval system, without specific authorization from the publisher. The publisher will, in most cases, grant reproduction for personal and classroom use without charge.

Anshel, Iris
Goldfeld, Dorian
Calculus: A computer algebra approach. Second edition. Solutions manual.

ISBN: 1-57146-039-X

Printed on acid free paper, in the United States of America

Table of Contents

Solutions for Chapter VIII

Solutions for Chapter IX

Solutions for Chapter X

Solutions for Chapter XI

Solutions for Chapter XII

Solutions for Chapter XIII

Solutions for Chapter XIV

Solutions for Chapter XV

Solutions for Chapter XVI

Solutions for Chapter XVII

Solutions for Chapter XVIII

Solutions for Chapter XIX

Solutions for Chapter XX

Solutions for Chapter XXI

Solutions for Chapter XXII

Solutions for Chapter XXIII

Solutions for Chapter XXIV

Solutions for Chapter XXV

Solutions for Chapter XXVI

Solutions for Chapter XXVII

Solutions for Chapter XXVIII

Solutions for Chapter XXIX

Solutions for Chapter XXX

Solutions for §1.1

(2) $f(100) = 1930001, f(5.7) = 143.956$ and $f(\pi) = -6.074677447$.

(3) $f(1) = 0.080605, f(\frac{\pi}{2}) = -16.264$ and $f(3) = -242.98$.

(4) $f(1) = 12.5958...$, $f(3) = 20.0028...$, and $f(\frac{5\pi}{2}) = 73.68502750680849...$

(6) $g(-3) = 9, g(-2) = 4, g(1) = 2, g(10) = 1024$ and $g\left(\frac{1}{2}\right) = 1.414213562$.

(7) Assuming $c(t)$ is a linear function of t we get $c(t) = \frac{5}{9}(t - 32)$.

(8) $d(t) = 50t$.

(10) $L(52) = 14204.29$ and $L(104) = 2.0176 \times 10^6$.

(11) $f(m) = 50 + 0.2m$.

Solutions for §1.2

(1) For $f(x) = \frac{1}{x^2-x+2}$ we see that for real x the denominator is never zero. Letting $y = f(x)$ we obtain $x = \frac{1}{2} \pm \frac{1}{2}\sqrt{\frac{4}{y} - 7}$ and for x to be real $\Rightarrow 0 \le \frac{4}{y} - 7 < \infty \Rightarrow 0 < y \le \frac{4}{7}$. So we see that Domain($f$)= $\{x \mid -\infty < x < \infty\}$ and Range(f)= $\{y \mid 0 < y \le \frac{4}{7}\}$.

(2) $x^2 + 2x + 1 = (x+1)^2$ is zero for $x = -1$ implying Domain $(f) = \{x \mid x \ne -1\}$ also for $y = f(x)$ we have $x = -1 \pm \frac{1}{\sqrt{y}} \Rightarrow y > 0$ and Range(f)= $\{y \mid 0 < y < \infty\}$.

(3) For $y = f(x) = \frac{x}{x^2+2x+2}$ we find (when x,$y \ne 0$) that $x = \frac{1}{2y}\left((1-2y) \pm \sqrt{1-4y-4y^2}\right) \Rightarrow$ $1 - 4y - 4y^2 \ge 0 \Rightarrow \frac{-1-\sqrt{2}}{2} \le y \le \frac{-1+\sqrt{2}}{2}$ and this is the range of $f(x)$. Domain(f) is all of the reals since $x^2 + 2x + 2 \ne 0$ for all x.

(4) For $y = f(x)$ we get $x = -1 \pm \sqrt{\frac{1}{y-4}} \Rightarrow 0 \le \frac{1}{y-4} \le \infty \Rightarrow 0 < y \le \frac{1}{4}$, this being Range($f$). Domain($f$) is the set of all real numbers.

(5) $f(x) = -\sqrt{x}$ then Domain(f)= $\{x \mid 0 \le x < \infty\}$ and Range(f)= $\{y \mid -\infty < y \le 0\}$.

(6) For $c(x) = \sqrt[3]{x}$, we see that the domain and range of $c(x)$ are all of the real numbers.

(7) $c_{19}(10^6) = 2.069138081$ and there is no restriction on the domain or range.

(8) Domain(f)= $\{x \mid -\infty < x < \infty\}$ and Range(f)= $\{y \mid 0 \le y < \infty\}$.

(9) Domain($|f|$)= $\{x \mid -\infty < x < \infty\}$ and Range($|f|$)= $\{y \mid 0 \le y < \infty\}$.

(11) Even though the CAS can represent and manipulate numbers such as π and $\sqrt{3}$ and can evaluate them to any required number of decimal places (within the systems time and memory limitations) it cannot in general decide whether or not a given real number is rational (of the form $\frac{a}{b}$). The function f therefore is uncomputable.

Solutions for §1.3

(2)

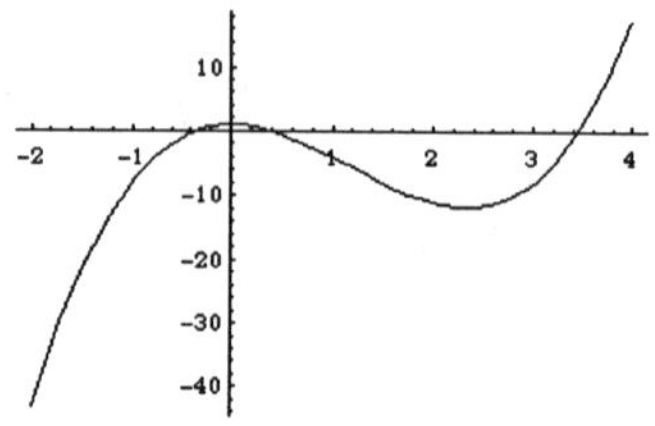

$$f(x) = 2x^3 - 7x^2 + 1.$$

(3)

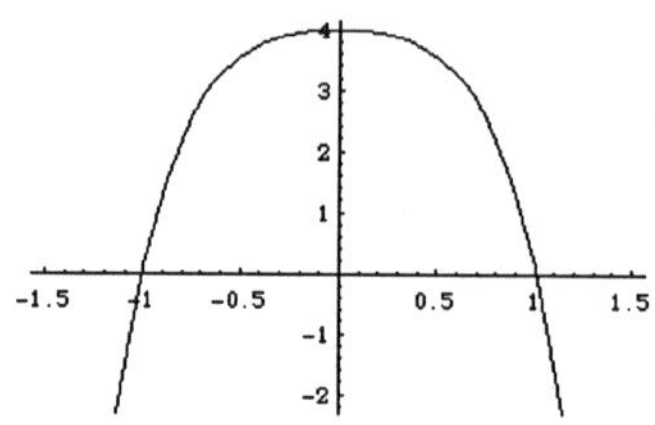

$$f(x) = 2\cos(x) - 3x^4 + 2.$$

(4)

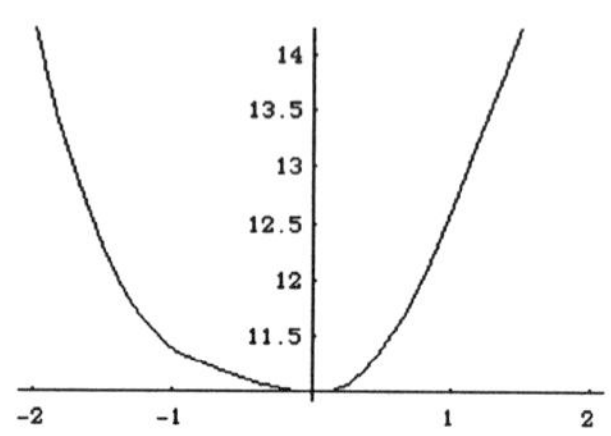

$$f(x) = \sin^3(x) + x^2 + 11.$$

(5)

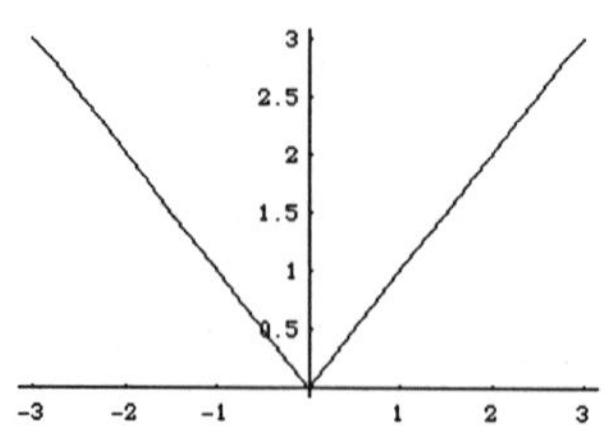

$$A(x) = |x|.$$

(6)

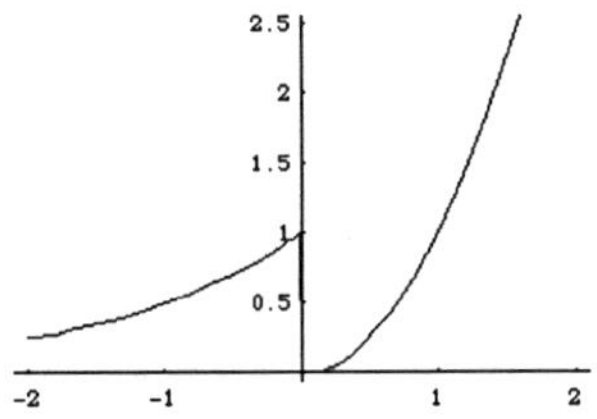

$$g(x) = \begin{cases} x^2 & \text{if } x \leq 0 \\ 2^x & \text{if } x > 0. \end{cases}$$

(7)

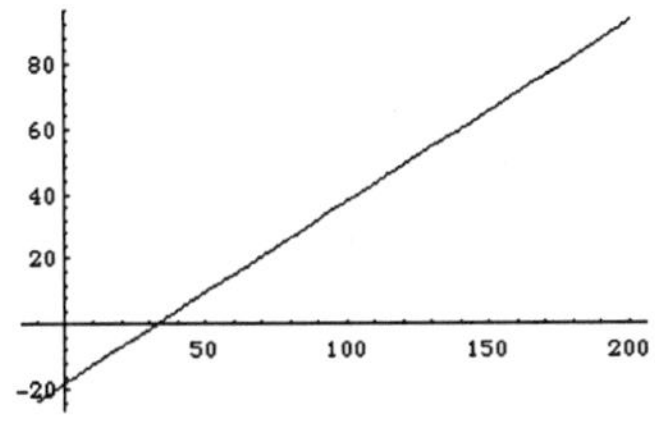

$$c(x) \;=\; \frac{5}{9}(x - 32).$$

(8)

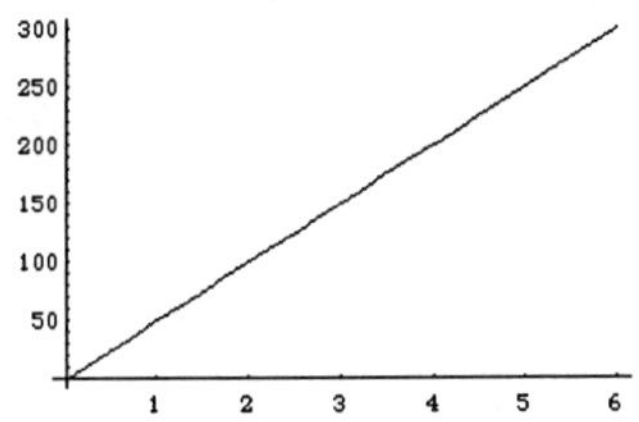

$$d(t) = 50t.$$

(9) Since we cannot visualize the words TRUE and FALSE, we replace them with the numbers 1 and -1 respectively. The graph is depicted below.

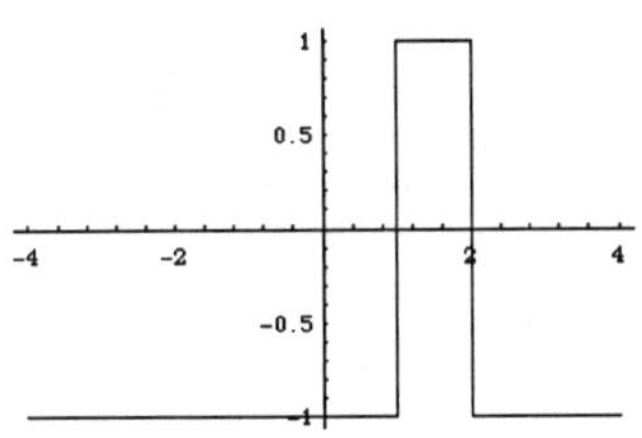

$$g(x) = \begin{cases} 1 & \text{if } \; 1 \leq x < 2 \\ -1 & \text{otherwise .} \end{cases}$$

(10),(11) The graphs of the functions $y = f(x+1), y = f(x+10)$ and $y = f(x-10)$ are simply the graph of $y = f(x)$ translated 1 and 10 units to the left and 10 units to the right, respectively. The graphs of $f(x) = x^2$ and $f(x) = x^3$ along with these translates are given below.

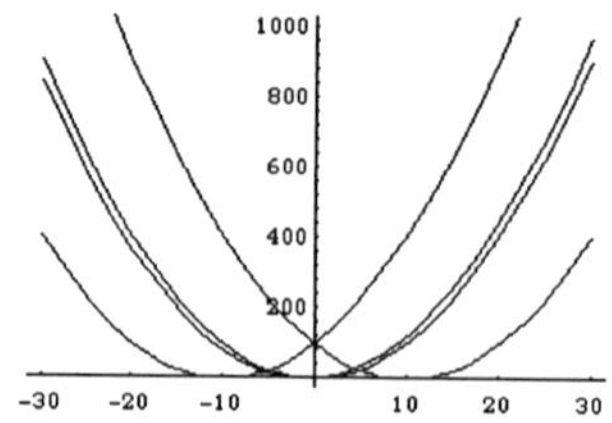

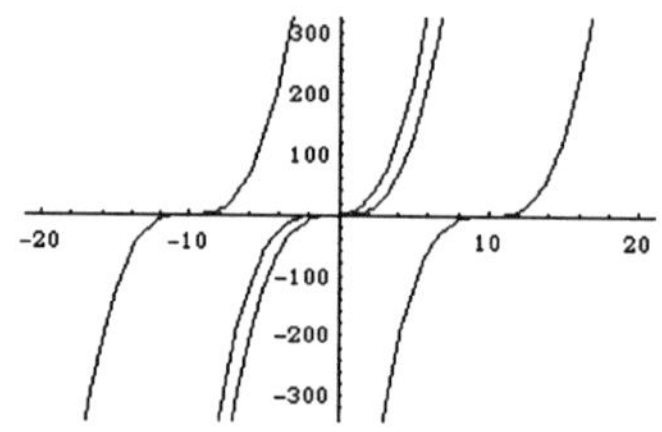

(12),(13) The graphs of the functions $y = f(x) + 1, y = f(x) + 10$ and $y = f(x) - 10$ are the graph of $y = f(x)$ translated 1 and 10 units upwards and 10 units downwards, respectively (up and down along the y-axis). The graphs of $f(x) = x^2$ and $f(x) = x^3$ along with these translates are given below.

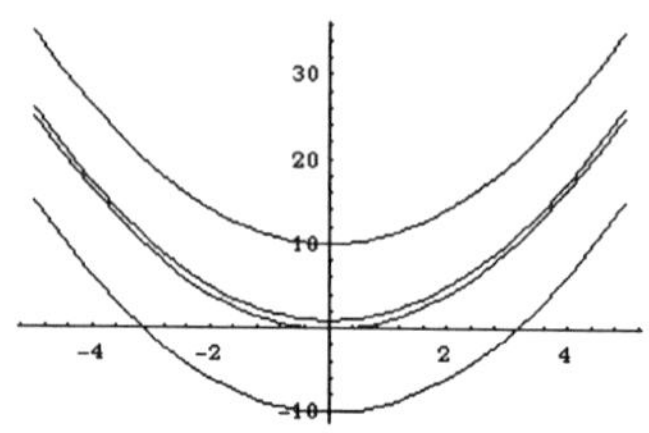

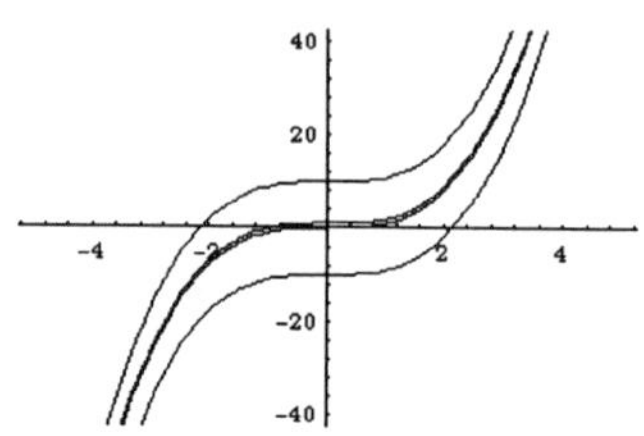

Solutions for §1.4

(1) $\cos\left(\frac{3\pi}{2}\right) = 0$. **(2)** $\sin(6\pi) = 0$. **(3)** $\tan(2\pi) = 0$. **(4)** $\cos\left(\frac{4\pi}{3}\right) = -\frac{1}{2}$. **(5)** $\sin\left(\frac{4\pi}{3}\right) = -\frac{\sqrt{3}}{2}$.
(6) $\tan\left(\frac{5\pi}{3}\right) = -\sqrt{3}$. **(7)** $\sin\left(\frac{5\pi}{3}\right) = -\frac{\sqrt{3}}{2}$. **(8)** $\cos\left(\frac{5\pi}{3}\right) = \frac{1}{2}$. **(9)** $\sin\left(\frac{3\pi}{4}\right) = \frac{1}{\sqrt{2}}$.
(10) $\cos\left(\frac{3\pi}{2}\right) = -\frac{1}{\sqrt{2}}$. **(11)** $\tan\left(\frac{3\pi}{4}\right) = -1$. **(12)** $\cos\left(\frac{5\pi}{4}\right) = -\frac{1}{\sqrt{2}}$.
(13) $\sin\left(\frac{5\pi}{4}\right) = -\frac{1}{\sqrt{2}}$. **(14)** $\tan\left(\frac{7\pi}{4}\right) = -1$. **(15)** $\cos\left(\frac{7\pi}{4}\right) = \frac{1}{\sqrt{2}}$.

(16) $10^{\frac{3}{4}} = 5.623413252, 15^{\frac{11}{13}} = 9.889040198$ and $100^{\frac{6}{7}} = 51.79474679$.

(17) $(10^{\frac{3}{4}})^4 = 1000, (15^{11/13})^{13} = 8.649755859 \times 10^{12}$ and $(100^{\frac{6}{7}})^7 = 10^{12}$.

(18) Assuming that α and β are rational, i.e., $\alpha = \frac{m}{n}, \beta = \frac{p}{q}$ it then follows that $(x^\alpha)^\beta = (x^{\frac{m}{n}})^{\frac{p}{q}} = \sqrt[q]{(x^{\frac{m}{n}})^p} = \sqrt[q]{x^{\frac{mp}{n}}} = \sqrt[nq]{x^{mp}} = x^{\alpha\beta}$.

(19)

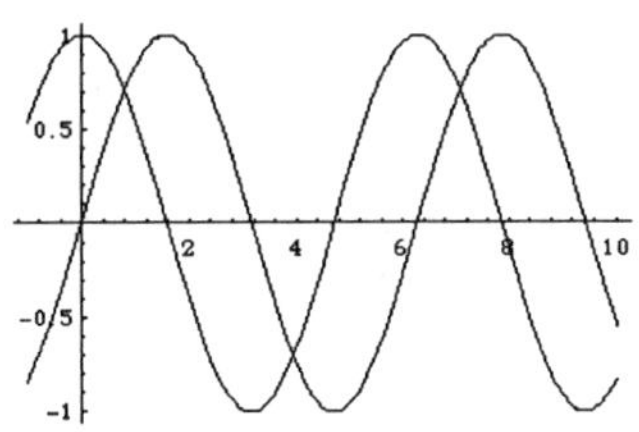

$\cos(x)$ and $\sin(x)$

Both $\cos(x)$ and $\sin(x)$ have period 2π, and this can be seen in the graphs of these functions – the pattern repeats itself after an interval of 2π. In fact $y = \cos(x)$ is the same graph as $y = \sin(x)$ but translated $\frac{\pi}{2}$ to the left. Other symmetries of the graphs that can be noticed are central symmetries (reflections) through any point where the graph cuts the x-axis and reflections through the vertical lines cutting the highest and lowest points of the graphs.

(20)

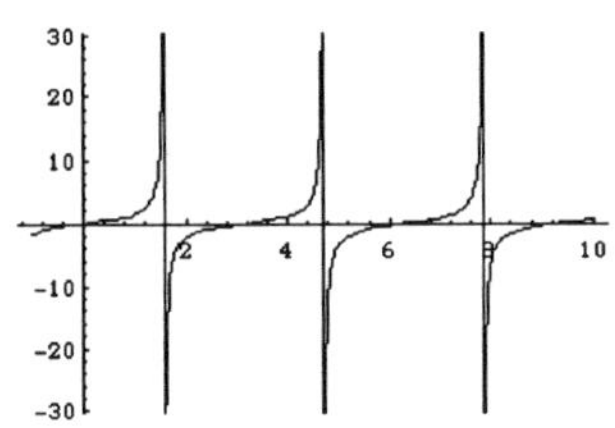

$\tan(x)$

This graph also repeats at intervals of 2π and can be reflected through the points $(n\pi, 0)$ on the x-axis, $(n = \ldots - 2, -1, 0, 1, 2, \ldots)$.

(21) This can be proven using a right angled triangle with hypotenuse H and sides A and B. If the angle between A and H is θ and therefore $\cos(\theta) = \frac{A}{H}$ then we can complete the rectangle $ABA'B'$ and see that the angle between H and B' is $\frac{\pi}{2} - \theta$ with $\sin\left(\frac{\pi}{2} - \theta\right) = \frac{A'}{H} = \frac{A}{H}$.

(22) To show $\cos(\theta + \pi) = -\cos(\theta)$ we note that rotating a point by angle π transforms the x co-ordinate to $-x$. Similarly, $\cos(\theta + 2\pi) = \cos(\theta)$ since rotation by 2π is the identity transformation.

Solutions for Additional Exercises for Chapter 1

(1) $A = f(x) = x^2$.

(2) $V = f(x) = x^3$.

(3)$A = f(r) = \pi r^2$. Let $C = 2\pi r$ be the circumference. Then $A = \frac{c^2}{4\pi}$.

(4) Let $B(n)$ denote the outstanding balance after n months. Then $B(0) = 1025$. After one month we see that: $B(1) = 1025 \cdot 1.01 - 50$. After two months:

$$B(2) = B(1) \cdot 1.01 - 50 = 1025 \cdot 1.01^2 - 50 \cdot 1.01 - 50 = 1025 \cdot 1.01^2 - 50 \cdot \frac{1.01^2 - 1}{.01},$$

and, after n months:

$$B(n) = B(n-1) \cdot 1.01 - 50 = 1025 \cdot 1.01^n - 50\frac{1.01^n - 1}{0.1}.$$

Then $B(12) = 520.87,\ B(20) = 149.75,\ B(21) = 101.24,\ B(22) = 52.25,\ B(23) = 2.78$. So it will take two years to pay off the balance.

(5) Domain $(f) = \{x \mid x \neq 0\}$; Range $(f) = \{\pm 1\}$.

(6) Domain $(f) = \{x \mid x < -2 \text{ or } x \geqslant 1\}$; Range $(f) = \{y \mid 0 \leqslant y < \infty, y \neq 1\}$.

(7) Domain $(f) = \{x \mid -16 < x \leqslant -4 \text{ or } x \geqslant 4\}$; Range $(f) = \{y \mid 0 \leqslant y < \infty\}$.

(8) Domain $(f) = \{x \mid \sin(x) \neq \frac{1}{3}\}$; Range $(f) = \{y \mid y \leqslant -\frac{1}{2} \text{ or } y \geqslant \frac{1}{4}\}$.

(9a) Domain $(f) = \{x \mid x \in \mathbb{R}\}$ and Range $(f) = \{y \mid y \in \mathbb{R}\}$. Domain $(g) = \{x \mid x \in \mathbb{R}\}$ and Range $(g) = \{y \mid y \geqslant 0\}$

(9b) $g(f(x)) = (x+6)^2$; Domain $= \{x \mid x \in \mathbb{R}\}$; Range $= \{y \mid y \geqslant 0\}$. $f(g(x)) = x^2 + 6$; Domain $= \{x \mid x \in \mathbb{R}\}$; Range $= \{y \mid y \geqslant 6\}$. These composite functions do not commute.

(10) Both: We may write $y = \frac{3x-10}{6}$ which determines y as a function of x or $x = \frac{10+6y}{3}$ which determines x as a function of y.

(11) Neither: If we write $y = (25 - x^2)^{\frac{1}{2}}$ then since we may take either of the two square roots, it is not possible to define y as a uniquely defined function of x.

(12) The equation $x = \frac{2}{y^2}$ defines x as a function of y.

(13) Both: $x = \frac{(1-y)}{2}$ determines x as a function of y and $y = 1 - 2x$ determines y as a function of x.

(14) Neither

(15) Both

(16) Domain $(\cos) = \{\theta \mid \theta \in \mathbb{R}\}$; Range $(\cos) = \{\theta \mid -1 \leqslant \theta \leqslant 1\}$ (sine has the same domain and range as cosine). Domain $(\tan) = \{\theta \mid \theta \neq \frac{\pi}{4} + k\frac{\pi}{2},\ k \in \mathbb{Z}\}$; Range $(\tan) = \{\theta \mid \theta \in \mathbb{R}\}$.

(17) $\cos(\theta + \beta) = \cos(\theta)\cos(\beta) - \sin(\theta)\sin(\beta)$;
$\sin(\theta + \beta) = \sin(\theta)\cos(\beta) + \cos(\theta)\sin(\beta)$.

(18)

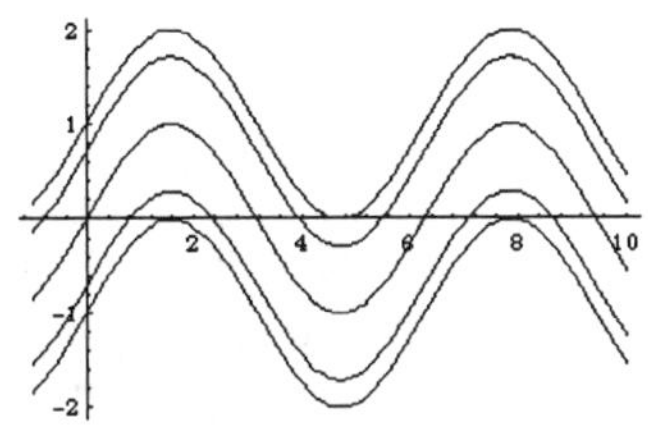

$$f(\theta) = \sin(\beta) + \sin(\theta), \qquad \beta = 0,\ \pm\frac{\pi}{4},\ \pm\frac{\pi}{2}$$

(19)

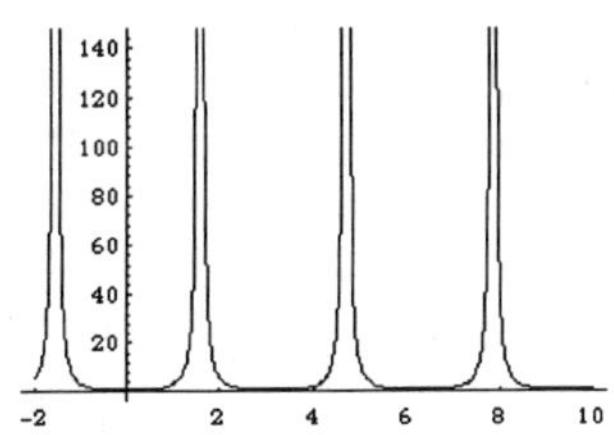

$$f(\theta) = 1 + \tan^2(\theta) = \frac{1}{\cos^2(\theta)}$$

The two graphs are equal, therefore, we can say that $1 + \tan^2(\theta) = \frac{1}{\cos^2(\theta)}$. Further, multiplying both sides of the equation by $\cos^2(\theta)$ yields the identity: $\cos^2(\theta) + \sin^2(\theta) = 1$.

(20) As $h \to 0$, you should notice that $\frac{2^h - 1}{h} \to \ln(2)$

Solutions for §2.2

(1) We get $f + g = x^3 + x^2 + 2x - 2$, $f - g = x^3 - x^2 + 2x + 4$, $f \cdot g = x^5 - x^3 + x^2 - 6x - 3$, $\frac{f}{g} = \frac{x^3+2x+1}{x^2-3}$, $f \circ g = x^6 - 9x^4 + 29x^2 - 32$ and $g \circ f = (x^3 + 2x + 1)^2 - 3$.

(2) We have $f + g = x^4 + 3x^2 - 2x$, $f - g = x^4 - 3x^2 + 2x - 2$, $f \cdot g = 3x^6 - 2x^5 + x^4 - 3x^2 + 2x - 1$, $\frac{f}{g} = \frac{x^4-1}{3x^2-2x+1}$, $f \circ g = 81x^8 - 216x^7 + 324x^6 - 312x^5 + 214x^4 - 104x^3 + 36x^2 - 8x$ and $g \circ f = 3x^8 - 8x^4 + 6$.

(3) We have $f+g = 2+3^x+x^2$, $f-g = 2-3^x+x^2$, $\frac{f}{g} = \frac{2}{3^x}+\frac{x^2}{3^x}$, $f \cdot g = 2\cdot 3^x + 3^x \cdot x^2$, $f \circ g = 2+3^{2x}$ and $g \circ f = 9 \cdot 3^{x^2}$.

(4) We find that $f \circ g = \frac{2}{2^{\frac{2^{4x}}{8}}}$ and $g \circ f = \frac{2^{\frac{8}{2^x}}}{8}$.

(5) $f \circ g = \sqrt{x^2} = |x|$ for $-\infty < x < \infty$ and $g \circ f = (\sqrt{x})^2 = x$.

(6) $f \circ g = \sqrt[3]{x^3} = x$ and $g \circ f = \sqrt[3]{x}^3 = x$ for $-\infty < x < \infty$.

(7) $f \circ g = \cos(x^4+1)$ and $g \circ f = \cos^4(x-1)+2$.

(8) $f \circ f \circ f \circ f = \sqrt{\sqrt{\sqrt{\sqrt{x}}}} = \sqrt[16]{x}$.

(9) By successively applying P_2 we get the sequence 0.2, 0.32, 0.4352, 0.4916, 0.4998... which we see seems to be approaching 0.5. We can prove this as follows: evaluating $P_2\left(\frac{1}{2}+t\right)$ we get $\frac{1}{2}-2t^2$ and since $\frac{|2t^2|}{|t|} < \frac{1}{2}$ for $|t| < \frac{1}{4}$ we see that if x is within a distance $\frac{1}{4}$ of $\frac{1}{2}$ then applying P_2 to x will at least halve this distance. This implies convergence. By applying P_1 to 0.2 we get $0.16,\ 0.1344,\ 0.1163,\ 0.1028,\ 0.0922\ldots$ which can be shown to converge to 0 implying eventual extinction of the population.

(10) Applying $P_{3.2}$ recursively to 0.2 yields 0.512, 0.7995, 0.5128, 0.7994, 0.5130 which appears to be converging to the period two sequence 0.5, 0.8, 0.5, 0.8, 0.5 ... . Taking $r = 3.5$ and working out past 28 iterations of $P_{3.5}$ we notice that it appears to settle down to a period four sequence 0.3828, 0.8269, 0.5009, 0.8750. For $r = 3.57$ the sequence doesn't settle down at all. For $r > 3.57$ the sequences obtained are mostly chaotic with no apparent patterns. Although for some r we eventually get periodic sequences, for example for $r = 3.83$ we get a period three sequence.

Solutions for §2.3

(1) We must take as our domain $\{x \mid 0 \le x\}$. Then the squaring function is one-to-one and $\sqrt{x^2} = (\sqrt{x})^2 = x \Rightarrow$ the squaring and square root functions are inverse.

(2) We have $\sqrt[3]{x^3} = (\sqrt[3]{x})^3 = x$ for $\{x \mid -\infty < x < \infty\}$ which implies the cubing and cube root functions are inverse.

(4) Taking as our domain $\{x \mid 0 \le x\}$ then f is always one-to-one and $(x^N)^{\frac{1}{N}} = (x^{\frac{1}{N}})^N = x$ which implies that $f^{-1}(x) = x^{\frac{1}{N}}$.

(5) If f is one-to-one $\Rightarrow$ for each element $b \in \text{Range}(f)$ there exists a unique $a \in \text{Domain}(f)$ for which $f(a) = b$. Defining $f^{-1}(b) = a$ in this case gives us the inverse of f. If f is not one-to-one then there exist $a, a' \in \text{Domain}(f)$ such that $f(a) = f(a') = b$ (say) and $a \neq a'$. When, however, we define f^{-1} it will never be true that both $f^{-1} \circ f(a) = a$ and $f^{-1} \circ f(a') = a'$ since $f^{-1} \circ f(a) = f^{-1} \circ f(a')$.

(6) The inverse of f exists since f is one-to-one. From the definition we have $f \circ f^{-1} = f^{-1} \circ f = I$ and this also implies that the inverse of f^{-1} is f.

(7) Solving for x we see that (for $m \neq 0$) $x = \frac{1}{m}y - \frac{b}{m}$ which also gives us a line.

(9) The function $f(x) = x$ is clearly its own inverse. Another example is $f(x) = \frac{1}{x}$ since it is clear that $f(f(x)) = \frac{1}{(\frac{1}{x})} = x$.

(10) We have $\arccos(1) = 0$, $\arccos(-1) = \pi$, $\arccos\left(\frac{1}{2}\right) = \frac{\pi}{3}$, $\arccos\left(-\frac{1}{2}\right) = \frac{2\pi}{3}$, $\arccos(\frac{\sqrt{3}}{2}) = \frac{\pi}{6}$, $\arccos\left(-\frac{\sqrt{3}}{2}\right) = \frac{5\pi}{6}$.

(11) We have $\text{Arcsin}(1) = \frac{\pi}{2}$, $\arcsin(-1) = -\frac{\pi}{2}$, $\arcsin\left(\frac{1}{2}\right) = \frac{\pi}{6}$, $\arcsin\left(-\frac{1}{2}\right) = -\frac{\pi}{6}$, $\arcsin\left(\frac{\sqrt{3}}{2}\right) = \frac{\pi}{3}$, $\arcsin\left(-\frac{\sqrt{3}}{2}\right) = -\frac{\pi}{3}$.

(12) To show that $\sin(\arccos(x)) = \sqrt{1-x^2}$ we first let θ equal $\arccos(x)$ then $\cos(\theta) = x = \sqrt{1-\sin^2(\theta)}$. Therefore $x^2 - 1 = \sin^2(\theta)$ and $\sin(\theta) = \sqrt{1-x^2}$.

Solutions for Additional Exercises for Chapter 2

(1) $\tan(x^5) = f(g(x))$ where $g(x) = x^5$; $f(x) = \tan(x)$.

(2) $\frac{2}{6+\cos(x)} = f(g(x))$ where $g(x) = \cos(x)$; $f(x) = \frac{2}{6+x}$.

(3) $\frac{2\sin^2(x)}{6+\cos^2(x)} = f(g(x))$ where $g(x) = \sin^2(x)$; $f(x) = \frac{2x}{7-x}$. Here, we have used the well known identity $\cos^2(x) = 1 - \sin^2(x)$.

(4) The function f does have an inverse. It is given by

$$f^{-1}(x) = \begin{cases} 3x+3 & \text{if } x \le 0 \\ \sqrt{x}+3 & \text{if } x \geqslant 0. \end{cases}$$

(5)

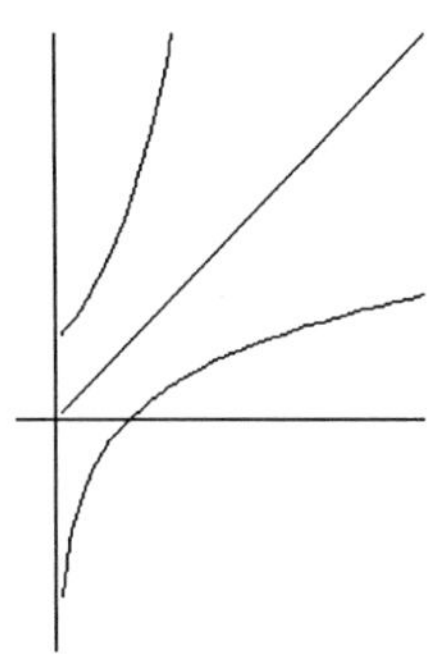

The graph of $y = e^x$ is the reflection (about the line $y = x$) of the graph of $y = \ln(x)$.

(6) $\text{Domain}(f) = \{x \mid x \neq 2, 3\}$; $\text{Range}(f) = \{y \mid y > 0\}$. An inverse for f is not defined on the entire domain. The largest domain for which f^{-1} is defined is $\{x \mid x < 2\}$.

(7) No. Consider $f(x) = x + 1,\ g(x) = x^2$. Then $f(g(x)) = x^2 + 1 \neq g(f(x)) = (x + 1)^2$.

(8) Yes. Consider $f(x) = 2x,\ g(x) = 3x$. Then $f(g(x)) = g(f(x)) = 6x$.

(9) Set $z = 3x + 1$. Then $x = \frac{z-1}{3}$ and $\frac{x}{x^2+1} = \frac{3}{3} \cdot \frac{z-1}{3\left(\frac{z-1}{3}\right)^2+1} = \frac{3z-3}{z^2-2z+4}$. Thus, the original equation $f(3x + 1) = \frac{x}{x^2+1}$ transforms to $f(z) = \frac{3z-3}{z^2-2z+4}$.

(10) $f(x) = (x - 1)^2 - 6(x - 1) + 10 = x^2 - 8x + 17$.

(11) $g(x) = x^2 + 1$ and $f(x) = \sin(\sqrt{x})$.

(12) Since $f(x) = 0$ can only occur for $x = 0$, we see that $f(g(x)) = 0$ implies that $g(x) = 0$. Thus $x = 6\frac{1}{3}$.

(13)

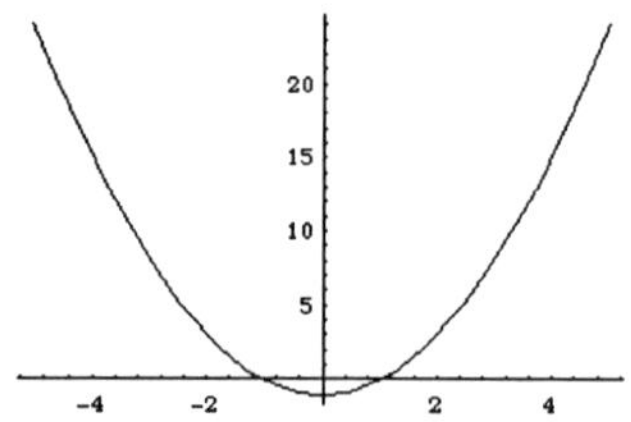

$f(x) = x^2 - 1$

$f(x)$ is even and therefore has symmetry about the $y-$ axis.

(14) f is an even function.

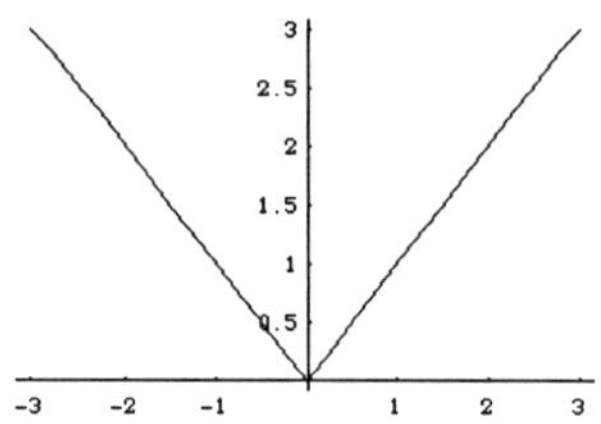

$f(x) = |x|$

(15) f is an odd function.

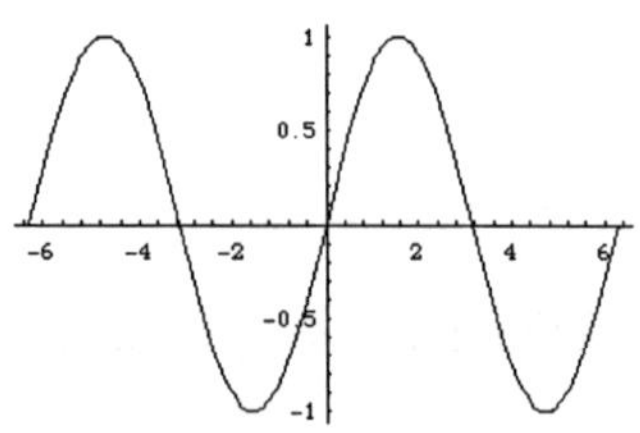

$f(x) = \sin(x)$

(16) f is neither even or odd.

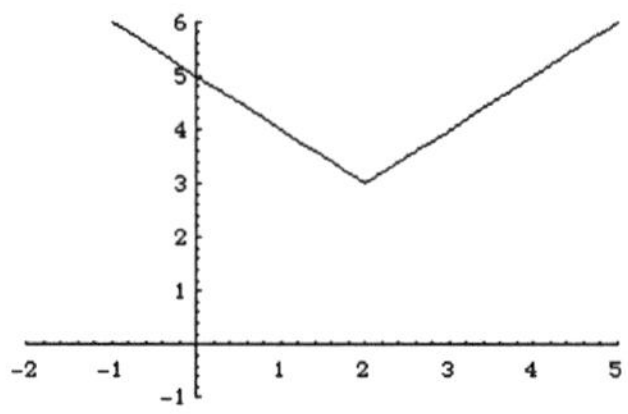

$$f(x) = |x - 2| + 3$$

(17) f is an even function.

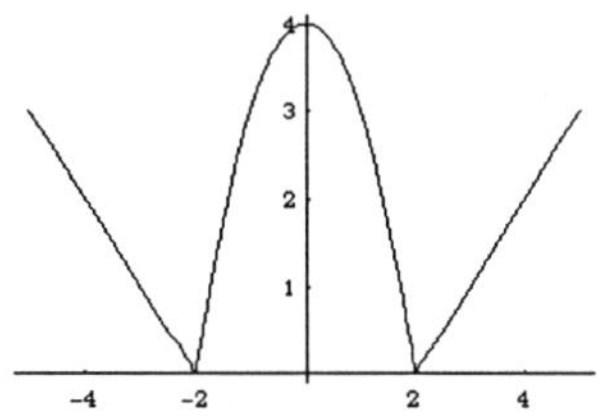

$$f(x) = \begin{cases} -x - 2, & \text{if } x \leq -2 \\ -x^2 + 4, & \text{if } -2 \leq x < 2 \\ x - 2. & \text{if } x \geq 2 \end{cases}$$

(18) If f is odd, then g must be even(odd) so that $g \circ f$ is even(odd).

(19)

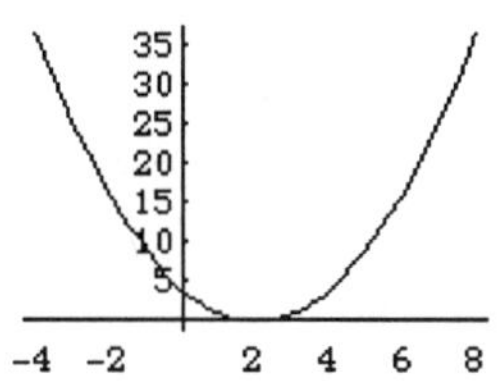

$f(x) = (x - 2)^2$ $\qquad$ $h(x) = e^{f(x)}$

$f(x) = (x - a)^2$ is a parabola. It is symmetric about the line $x = a$. The function $h(x) = e^{f(x)}$ opens up, and gets shifted vertically; to the left for $a > 0$, or to the right for $a < 0$. If $a \neq 0$, then $f(x)$ is neither even nor odd. The composition of f and g has the same basic geometric shape as $f(x)$ but it never equals zero and it rises/falls more rapidly. In general an exponential function grows faster in the "long-run" than power functions.

(20) Assume $f(a - x) = f(a + x)$ for all x. If $a = 0$, then f can be any even function. Otherwise, f must be the constant function, i.e., $f(x) = c$ for some constant c.

Solutions for §3.1

(1) Writing the line as $y = -3x + \frac{11}{2}$ we see that the slope is -3 and the y-intercept is $\frac{11}{2}$.

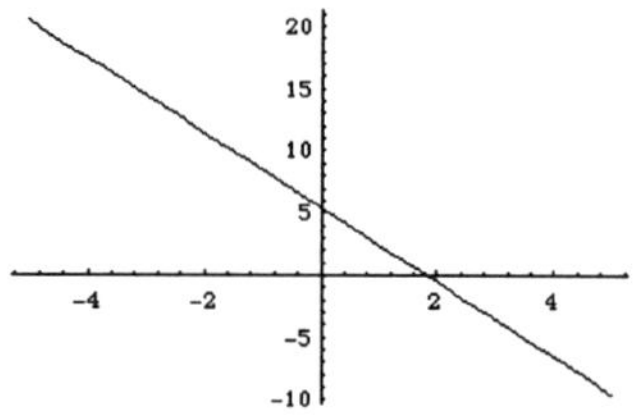

$$6x + 2y = 11$$

(2) This line has slope $\frac{1}{3}$ and y-intercept $-\frac{\pi}{3}$.

(3)

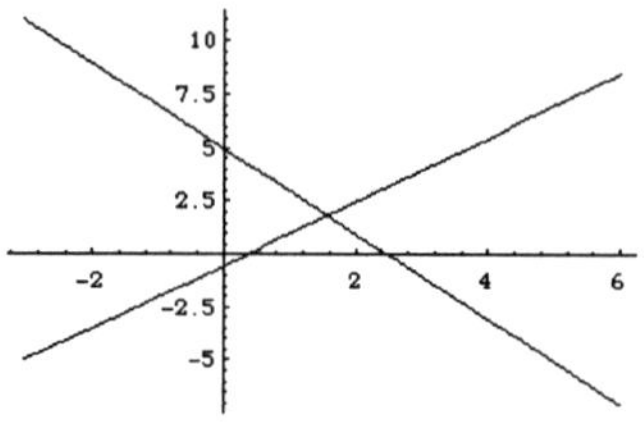

$$3x - 2y = 1 \text{ and } 2x + y = 5 \text{ intersect at } \left(\frac{11}{7}, \frac{13}{7}\right).$$

(4) If $a_1 b_2 - a_2 b_1 = 0 \Rightarrow$ the slopes of L_1 and L_2 are the same and, therefore, the lines are parallel. So either $L_1 = L_2$ or they have no point of intersection. If $a_1 b_2 - a_2 b_1 \neq 0$ then we can solve L_1 and L_2 for x and y:

$$x = \frac{-c_1}{a_1} + \frac{b_1(a_1 c_2 - a_2 c_1)}{a_1(a_1 b_2 - a_2 b_1)}, \quad y = \frac{a_2 c_1 - a_1 c_2}{a_1 b_2 - a_2 b_1}.$$

(5)

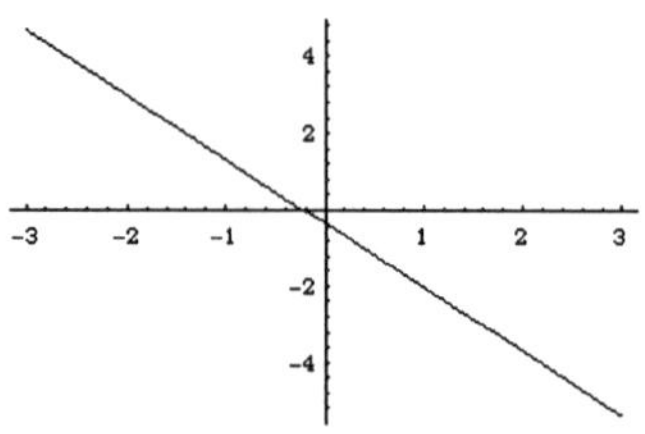

$$y = \frac{(-5x - 1)}{3}.$$

(6)

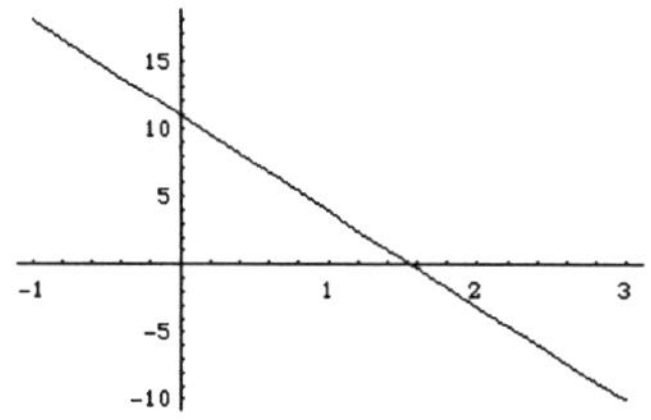

$$y = -7x + 11.$$

(7)

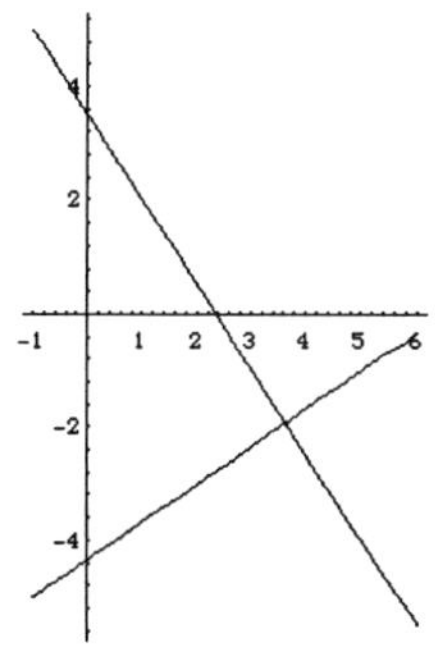

L_1 and L_2.

L_1 and L_2 intersect at the point $\left(\frac{47}{13}, \frac{-25}{13}\right)$. Their slopes are $\frac{2}{3}$ and $-\frac{3}{2}$ respectively $\Rightarrow$ the product of their slopes is -1 and therefore L_1 is perpendicular to L_2, (see question 9).

(8) Two lines are parallel if they make the same angle with the x-axis. This happens if and only if they have the same slope.

(9) Let our two lines be l_1 and l_2 with slope m_1 and m_2. The angles they make with the x-axis are $\tan^{-1}(m_1) = \alpha$ and $\tan^{-1}(m_2) = \beta$. So the angle between the lines is $|\alpha - \beta|$. We evaluate $\tan(\alpha - \beta)$ using the trigonometric identity $\tan(x-y) = \frac{(\tan(x)-\tan(y))}{(1+\tan(x)\tan(y))} \Rightarrow \tan(\alpha-\beta) = \frac{(m_1-m_2)}{(1+m_1 m_2)}$. Now $\alpha-\beta = \pm\frac{\pi}{2}$ if and only if $\tan(\alpha - \beta) = \infty$ and this can only happen when $m_1 m_2 = -1$.

(10) The line $y + x - 4 = 0$ passes through the points (1,3) and (-1,5). But if we set $x = 9, y = -4$ we get $(-4) + (9) - 4 = 1 \neq 0$ so the three points are not collinear.

(11) The slope of the line $2x - 3y = 4$ is $\frac{2}{3}$ so the line perpendicular to that must be of the form $y = -\frac{3x}{2} + c$. Plugging in $y = 2, x = -1$ we find $c = \frac{1}{2}$.

(12) The equation of the line through the given points is $y = -\frac{4x}{3} + \frac{13}{3}$ and a line perpendicular to that must have slope $\frac{3}{4}$. The bisector must pass through the midpoint, $(\frac{5}{2}, 1)$ of the segment and is therefore $y = \frac{3x}{4} - \frac{7}{8}$.

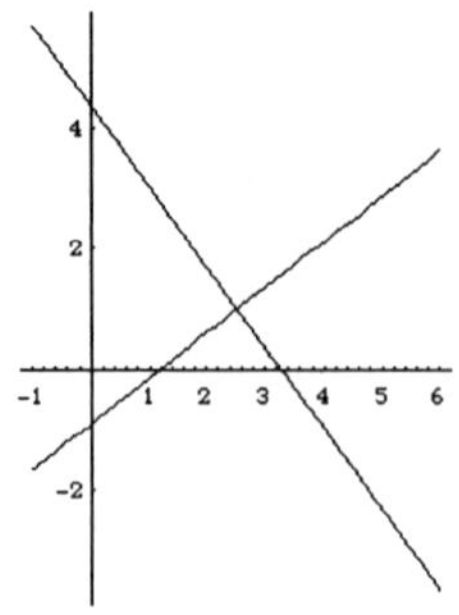

$$y = \frac{1}{3}(-4x+3) \text{ and } y = \frac{3}{4}x - \frac{7}{8}.$$

Solutions for §3.2

(1)

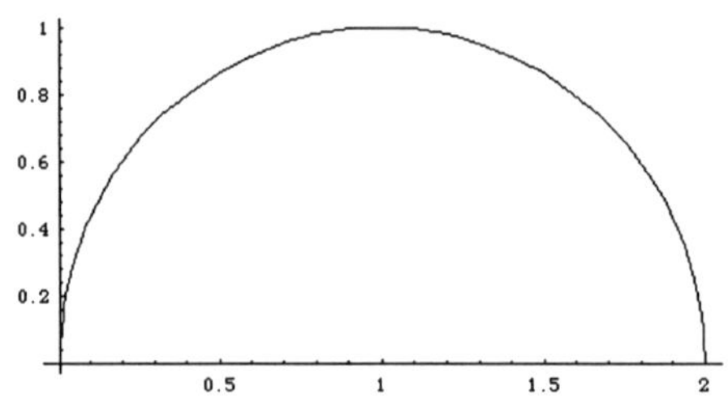

Upper semicircle of radius 1, center $(1,3)$, $y = \sqrt{1-(x-1)^2}$.

The equation for the circle is $(x-1)^2 + (y-3)^2 = 1^2$.

(2) $(x+1)^2 + (y-5)^2 = 3^2$.

(3) $(x+3)^2 + (y+2)^2 = 10^2$.

(4) $(x-13)^2 + (y+2.1)^2 = 5^2$.

(5) $(x-\pi)^2 + (y+1.03)^2 = 2^2$.

(6) Let $d(P_1, P_2)$ be the distance between P_1 and P_2. Then $d\big((1,3),(3,-1)\big) = 2\sqrt{5}$. Furthermore, $d\big((0,1),(-1,-3)\big) = \sqrt{17}$, $d\big((2,3),(-2,11)\big) = 4\sqrt{5}$.

(7) The line perpendicular to $2x-4y=-1$ passing through (1,3) is $y=-2x+5$. This intersects the first line at $(\frac{19}{10}, \frac{6}{5})$ and the distance from this point to $(1,3)$ is $\frac{9}{2\sqrt{5}}$.

(8) The line perpendicular to $y=5x-2$ passing through (-1,2) is $y=-\frac{1}{5}x+\frac{9}{5}$. This intersects the first line at $(\frac{19}{26}, \frac{43}{26})$ and the distance from this point to $(-1,2)$ is $\frac{9}{\sqrt{26}}$.

In problems (9), (10) we use the notation $\sum_{i=1}^{n} f(i) = f(1) + f(2) + \cdots + f(n)$.

(9) Let $d(A, B)$ be the distance function and $f(x) = \sqrt{1 - x^2}$ the function defining the upper semicircle then we wish to evaluate the sum

$$2\sum_{i=1}^{4} d\left(\left(\frac{i-1}{4}, f(\frac{i-1}{4})\right), (\frac{i}{4}, f(\frac{i}{4}))\right)$$

$$= 2\sum_{i=1}^{4} \sqrt{\left(\frac{i-1}{4} - \frac{i}{4}\right)^2 + \left(\sqrt{1 - (\frac{i-1}{4})^2} - \sqrt{1 - (\frac{i}{4})^2}\right)^2}$$

which equals (using our CAS) $3.104496 \approx 3.141592 = \pi$, (to six places).

(10)

$$2\sum_{i=1}^{8} \sqrt{\left(\frac{i-1}{8} - \frac{i}{8}\right)^2 + \left(\sqrt{1 - (\frac{i-1}{8})^2} - \sqrt{1 - (\frac{i}{8})^2}\right)^2}$$

$= 3.128544 \approx 3.141592 = \pi$, (to six places).

Solutions for Additional Exercises for Chapter 3

(1) $y - y_1 = \frac{y_2 - y_1}{x_2 - x_1}(x - x_1)$

(2) $(-1, 3)$

(3) $(-\frac{5}{2}, \frac{13}{2})$

(4) $(\frac{-1+\sqrt{3}}{2}, 1)$

(5) Let $P_1 = (x_1, y_1)$, $P_2 = (x_2, y_2)$. If (x, y) is equidistant from both P_1 and P_2 then, necessarily,

$$\sqrt{(x - x_1)^2 + (y - y_1)^2} = \sqrt{(x - x_2)^2 + (y - y_2)^2}$$

or

$$(x - x_1)^2 + (y - y_1)^2 = (x - x_2)^2 + (y - y_2)^2.$$

A solution is given by $x = \frac{x_1 + x_2}{2}$, $y = \frac{y_1 + y_2}{2}$. Thus, the midpoint is given by $\left(\frac{x_1+x_2}{2}, \frac{y_1+y_2}{2}\right)$.

(6) The distance from $(3, k)$ to $(4, 6)$ is given by $\sqrt{(3-4)^2 + (k-6)^2}$. Similarly, the distance from $(3, k)$ to $(9, 1)$ is $\sqrt{(3-9)^2 + (k-1)^2}$. If these distances are the same, we must have

$$1 + (k - 6)^2 = 36 + (k - 1)^2 \Rightarrow k = 0.$$

(7) For any b, the line $y = 6x + b$ is parallel to $y = 6x - 1$. When $b = 6$, the y–intercept is 6. Thus, $y = 6x + 6$ is the solution to our problem.

(8) The midpoint of the line segment L joining $(2,6)$ and $(3,11)$ is $\left(\frac{5}{2}, \frac{17}{2}\right)$. Furthermore, L has slope $= \frac{11-6}{3-2} = 5$. The perpendicular bisector of L must have slope $-\frac{1}{5}$ and pass through $\left(\frac{5}{2}, \frac{17}{2}\right)$. The equation is $y = -\frac{1}{5}x + 9$.

(9) $y = \frac{1}{9}x + 3$

(10) The graph is symmetric around the line $x = 2$.

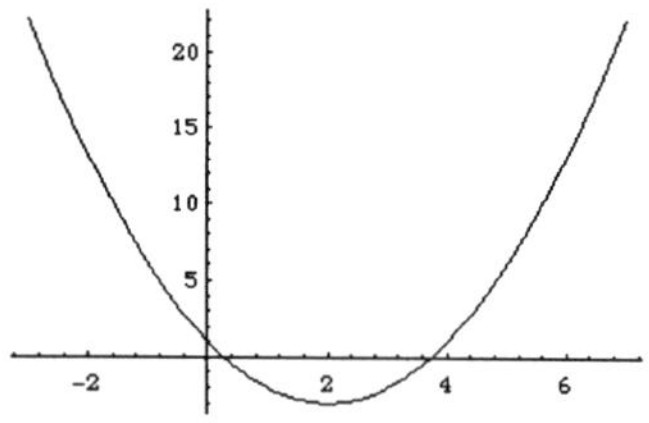

$$y = -(x-2)^2 - 3.$$

(11) If we graph the two parabolas $y = (x-1)^2 + 10$ or $y = -(x-1)^2 + 10$ we obtain the following figures:

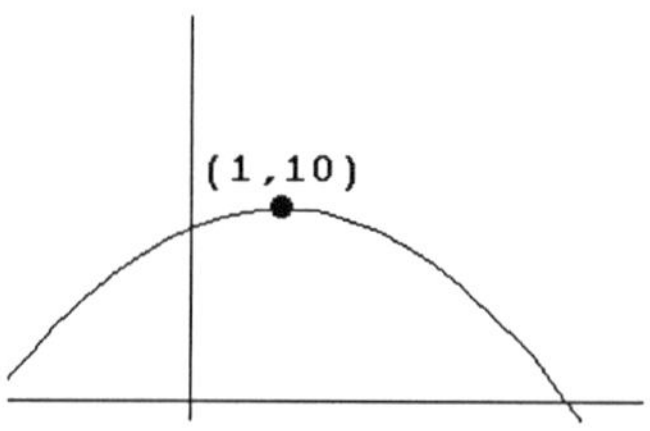

$y = -(x-1)^2 + 10,$

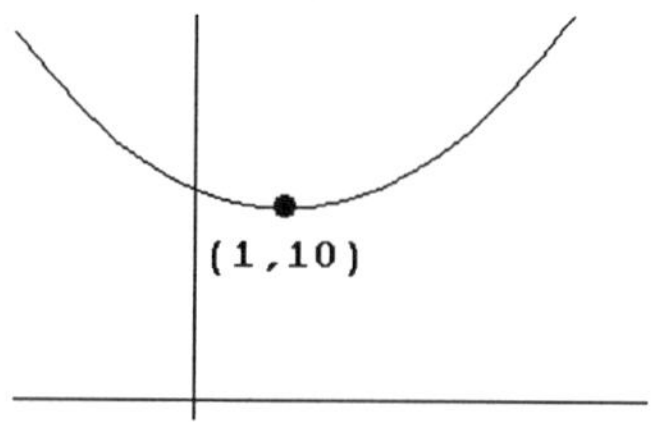

$y = (x-1)^2 + 10$

Thus $y = -(x-1)^2 + 10$ faces down and has maximum value at $y = 10$.

(12) slope $= \tan(\theta)$.

(13) We have $\tan(\beta_1) = m_1$ and $\tan(\beta_2) = m_2$. Also $\theta = \beta_2 - \beta_1$. So

$$\tan(\theta) = \tan(\beta_1 - \beta_2) = \frac{\tan(\beta_2) - \tan(\beta_1)}{1 + \tan(\beta_2)\tan(\beta_1)} = \frac{m_2 - m_1}{1 + m_2 m_1}.$$

(14) The circle with diameter (a, b).

(15a) The equation graphs an ellipse with a horizontal semi-axis of 2. (The horizontal distance has been elongated; it is twice as long).

(15b) Here we also see an ellipse, but this time the vertical semi-axis is 2. (The vertical distance is twice as large.) The center for each graph in question 15 is $(1,1)$.

(16) Both graphs are symmetric about both the x-axis and the y-axis. Their center is $(1, 1)$.

(17) If the right hand side is changed to 0, then the graph degenerates into two lines, given by the equations $y - 1 = \pm\frac{1}{2}(x - 1)$. If the right hand side is replaced by -2, then the left-right hyperbola is changed into an up-down hyperbola, with the axis-crossings moved further away from the origin.

(18) When the right hand side is replaced by zero we obtain two lines:

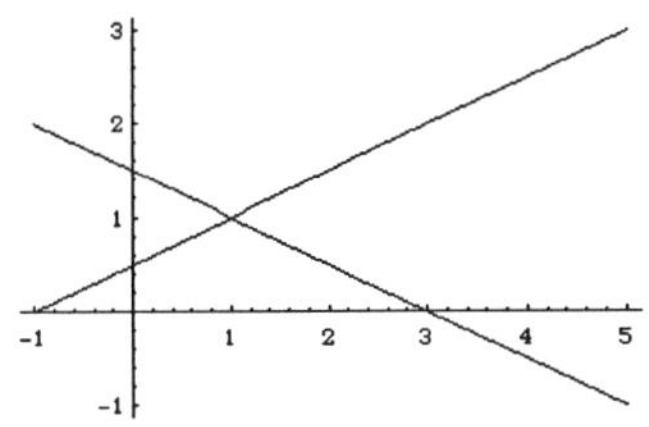

$$y = 1 \pm \frac{x-1}{2}$$

The graph of the original equation is an hyperbola:

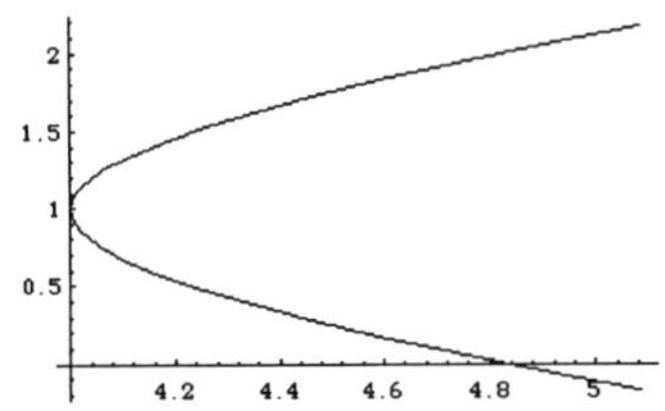

$$\frac{(x-1)^2}{4} - (y-1)^2 = 1$$

(19) Note that $x^2 + 4x = (x+2)^2 - 4$ and $y^2 - 2y = (y-1)^2 - 1$. Thus:

$$x^2 - y^2 + 4x + 2y + 2 = \left((x+2)^2 - 4\right) - \left((y-1)^2 - 1\right) + 2 = 0$$

reduces to: $(x+2)^2 - (y-1)^2 = 1$.

(20) $2(x-1)^2 + (y+3)^2 = 1$.

Solutions for §4.2

(1) We have $\frac{3x}{x^2-2x} = \frac{3}{x-2}$ for $x \neq 0$. Consequently, $\lim\limits_{x\to 0} \frac{3x}{x^2-2x} = \lim\limits_{x\to 0} \frac{3}{x-2} = \frac{3}{-2}$.

(2) $\frac{x^2+1}{2x-7}$ is well defined at $x = 2$ and the limit as $x \to 2$ is $\frac{2^2+1}{2\cdot 2-7} = -\frac{5}{3}$.

(3) $\lim\limits_{x\to 1} \frac{(x-1)^2}{x^3-2x^2+x} = \lim\limits_{x\to 1} \frac{(x-1)^2}{x(x^2-2x+1)} = \lim\limits_{x\to 1} \frac{1}{x} = 1$.

(4) We have $\frac{2x-4}{x^2+x-6} = \frac{2}{x+3}$ for $x \neq 2$. Consequently, $\lim\limits_{x\to 2} \frac{2x-4}{x^2+x-6} = \lim\limits_{x\to 2} \frac{2}{x+3} = \frac{2}{5}$.

In questions 5 to 8 the solutions employ proposition [4.12] on the algebra of limits.

(5) $\lim_{x\to 0} \frac{\sin(5x)}{x} = \lim_{x\to 0} 5 \cdot \frac{\sin(5x)}{5x} = 5 \cdot 1 = 5.$

(6) $\lim_{x\to 0} \frac{\sin(2x)}{x\cos(x)} = \lim_{x\to 0} 2 \cdot \frac{\sin(2x)}{2x} \cdot \frac{1}{\cos(x)} = 2 \cdot 1 \cdot \frac{1}{1} = 2.$

(7) $\lim_{x\to 0} \frac{(\cos(x))^2 \sin(x)}{x+x^2} = \lim_{x\to 0} \cos^2(x) \cdot \frac{\sin(x)}{x} \cdot \frac{1}{1+x} = 1 \cdot 1 \cdot \frac{1}{1} = 1.$

(8) $\lim_{x\to 0} \frac{\sin(\pi x)}{2x+x^2} = \lim_{x\to 0} \pi \cdot \frac{\sin(\pi x)}{\pi x} \cdot \frac{1}{2+x} = \pi \cdot 1 \cdot \frac{1}{2} = \frac{\pi}{2}.$

(9) We compute: $f(0.1) = -1.5789, f(0.01) = -1.5075, f(0.001) = -1.5007, f(-0.1) = -1.4285,$ $f(-0.01) = -1.4925$ and $f(-0.001) = -1.4992$. This agrees with question (1) in that the limits from the left and right as $x \to 0$ are equal and $\lim_{x\to 0} f(x) = \frac{-3}{2}$.

(10) $f(0.1) = 0.8985,\ f(0.01) = 0.9899,\ f(0.001) = 0.999,\ f(-0.1) = 1.0982,\ f(-0.01) = 1.0099$ and $f(-0.001) = 1.001$. This agrees with question (7). The limits from the left and right as $x \to 0$ are equal and $\lim_{x\to 0} f(x) = 1$.

(11) We want to show that $\lim_{x\to a} \sqrt{x} = \sqrt{a}$ for $a > 0$. That is for any sequence $a_1, a_2, a_3 ... \to a$ that $\lim_{n\to\infty} \sqrt{a_n} = \sqrt{a}$. To do this we must show that we can make $|\sqrt{a} - \sqrt{a_n}|$ arbitrarily small. Formally this means that for every $\epsilon > 0$ we can find N such that $|\sqrt{a} - \sqrt{a_n}| < \epsilon$ for all $n > N$. We already know that given any $\delta > 0$ there exists an N such that $|a - a_n| < \delta$ for all $n > N$, so we let $\delta = \epsilon|\sqrt{a}|$. Then we have:

$$|\sqrt{a} - \sqrt{a_n}| = \frac{|a - a_n|}{|\sqrt{a} + \sqrt{a_n}|} < \frac{|a - a_n|}{|\sqrt{a}|} < \frac{\delta}{|\sqrt{a}|} = \epsilon$$

for all $n > N$, and the result is proven.

(12) $g(0.1) = 0.7943,\ g(0.01) = 0.9549,\ g(0.001) = 0.9931$ and $g(0.0001) = 0.999$ so we would guess that the limit of $g(x)$ as x goes to zero from the right is 1.

(13) We find from our CAS that by dividing the quartercircle into 4, 8, 16, 32 pieces we get the following approximations for π: 3.1044, 3.1285, 3.1369 and 3.1399 which are increasing and appear indeed to be approaching $\pi \approx 3.14159$.

Solutions for §4.3

(1)

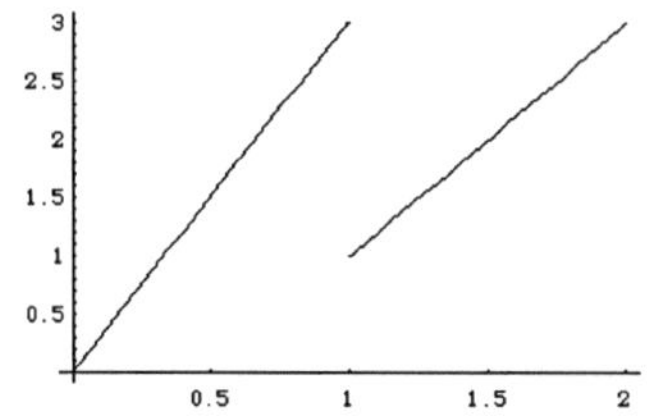

One point of discontinuity at $x = 1$.

$f(x)$ is discontinuous at $x = 1$ because the limit of f from the left (denoted $\lim_{x \to 1^-} f(x)$) is 3, while the limit from the right is $\lim_{x \to 1+} f(x) = 1$.

(2)

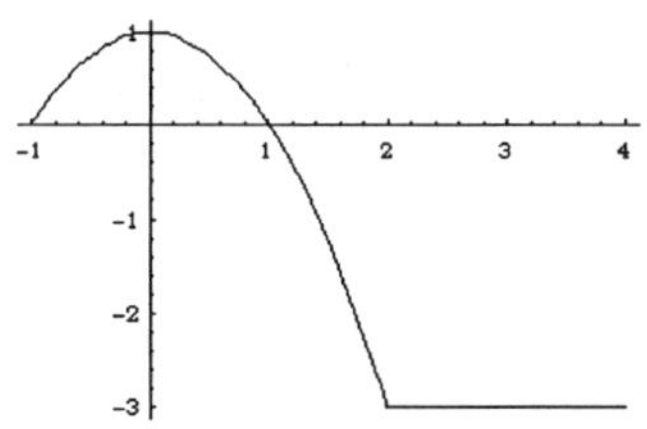

f is continuous: $\lim_{x \to 2^-} f(x) = \lim_{x \to 2+} f(x)$.

(3)

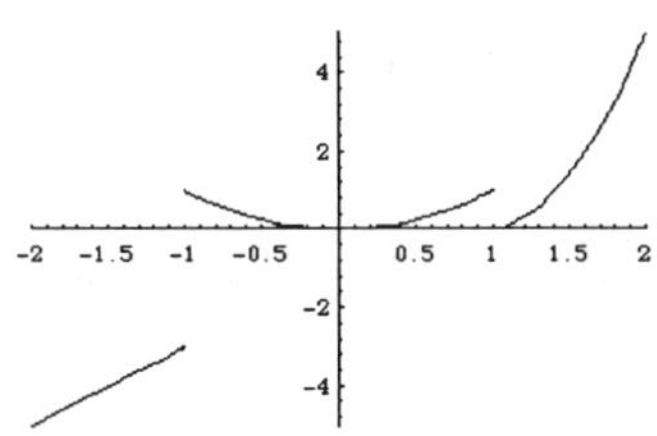

f is discontinuous at -1 and 1.

$f(x)$ is discontinuous at $x = -1$ because $\lim_{x \to -1^-} f(x) = -3 \neq 1 = \lim_{x \to -1^+} f(x)$. Also, $f(x)$ is discontinuous at $x = +1$ because $\lim_{x \to 1^-} f(x) = 1 \neq 0 = \lim_{x \to 1^+} f(x)$.

(4)

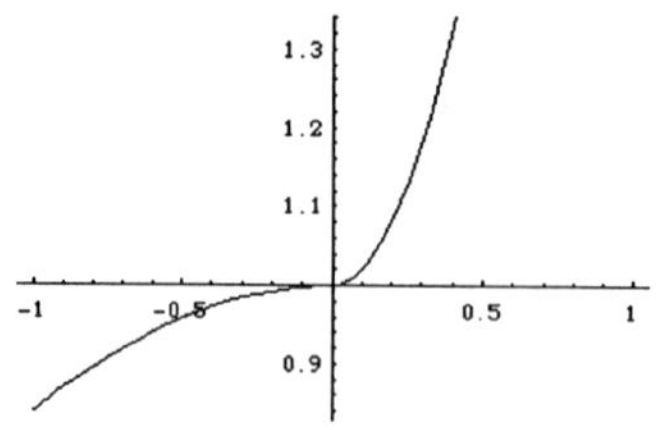

f is continuous: $\lim_{x \to 0^-} f(x) = \lim_{x \to 0^+} f(x)$.

(5)

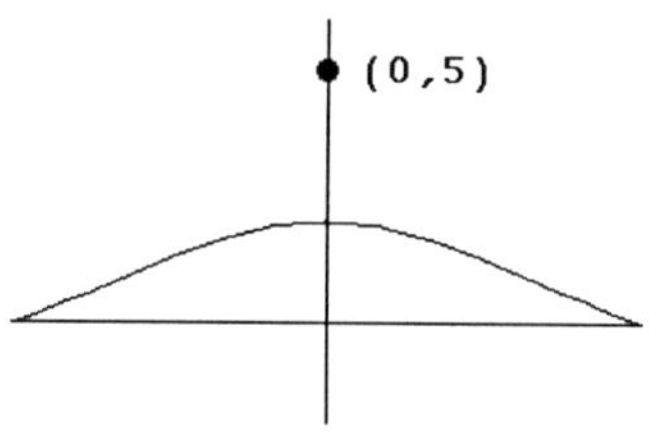

Graph of $\dfrac{\sin(x)}{x}$ for $x \neq 0$.

We have $\lim_{x \to 0^-} f(x) = 1 = \lim_{x \to 0^+} f(x)$. But $f(0) = 5$ so, f is discontinuous at $x = 0$.

(6)

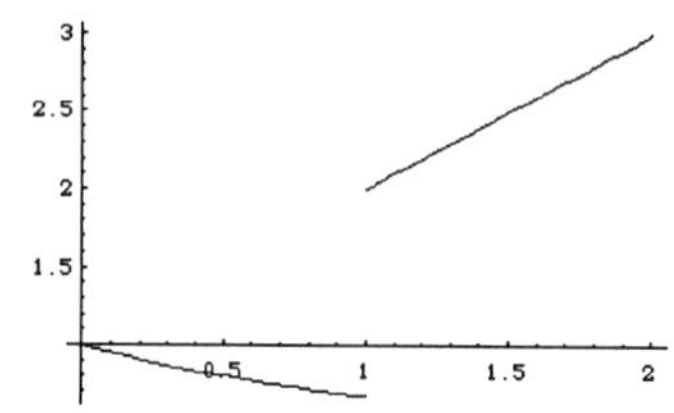

Discontinuity at $x = 1$.

Solutions for Additional Exercises for Chapter 4

(1) We have $f(1.5) \approx .2857$, $f(1.6) \approx .2778$, $f(1.7) \approx .2703$, $f(1.8) \approx .2632$, $f(1.9) \approx .2564$, $f(1.99) \approx .2506$, $f(1.999) \approx .2501$. As x approaches 2, $f(x)$ approaches $\frac{1}{4}$; $f(x)$ is not continuous.

(2) $\lim_{x \to 2} \frac{x-2}{x^2-4} = \lim_{x \to 2} \frac{x-2}{(x-2)(x+2)} = \lim_{x \to 2} \frac{1}{x+2} = \frac{1}{4}$.

(3) $\lim_{x \to 1} \frac{x^4-1}{x-1} = \lim_{x \to 1} \frac{(x-1)(x^3+x^2+x+1)}{x-1} = \lim_{x \to 1} (x^3 + x^2 + x + 1) = 4$.

(4) $\lim\limits_{x\to 9} \frac{x-9}{\sqrt{x}-3} = \lim\limits_{x\to 9} \frac{(\sqrt{x}-3)(\sqrt{x}+3)}{\sqrt{x}-3} = \lim\limits_{x\to 9}(\sqrt{x}+3) = 6$

(5) $\lim\limits_{y\to 4} \frac{4-y}{2-\sqrt{y}} = \lim\limits_{y\to 4} \frac{(2-\sqrt{y})(2+\sqrt{y})}{2-\sqrt{y}} = \lim\limits_{y\to 4}(2+\sqrt{y}) = 4$

(6) $\lim\limits_{x\to -4} \frac{x^2-16}{x+4} = \lim\limits_{x\to -4} \frac{(x-4)(x+4)}{(x+4)} = \lim\limits_{x\to -4}(x-4) = -8$. So if $k = -8$ the function $f(x)$ will be continuous.

(7) The function

$$f(x) = \begin{cases} \frac{1}{4}x - 2, & \text{if } x \le k \\ x^2 - 16x + 64, & \text{if } x > k \end{cases}$$

will be continuous at $x = k$ if $\frac{1}{4}k - 2 = k^2 - 16k + 64$. Thus $k = 8$.

(8) If $\lim\limits_{x\to a} f(x) = b$ then $\lim\limits_{x\to a} \frac{1}{f(x)} = \frac{1}{b}$. Consequently, $\lim\limits_{x\to 0} \frac{\sin(x)}{x} = 1$ implies that $\lim\limits_{x\to 0} \frac{x}{\sin(x)} = 1$

(9) $\lim\limits_{x\to 0} \left(\frac{\sin x}{x} + \frac{x^2}{x(x+1)}\right) = \lim\limits_{x\to 0} \left(\frac{\sin x}{x}\right) + \lim\limits_{x\to 0} \left(\frac{x^2}{x(x+1)}\right) = 1 + 0 = 1.$

(10) $\lim\limits_{x\to 3} f(x) = \frac{1}{42}$; $\lim\limits_{x\to 1} f(x) = -\frac{5}{4}$.

(11) $f(x)$ can be simplified to $\frac{1}{(x+3)(x-2)(x^2+x+1)}$. So $\lim\limits_{x\to 3} f(x) = \frac{1}{78}$ and $\lim\limits_{x\to 1} f(x) = -\frac{1}{12}$.

(12) First consider the statement for addition of functions. If f and g are continuous at b, then $\lim\limits_{x\to b} f(x) = f(b)$ and $\lim\limits_{x\to b} g(x) = g(b)$. It follows that $\lim\limits_{x\to b}(f+g)(x) = \lim\limits_{x\to b}(f(x)+g(x)) = \lim\limits_{x\to b} f(x) + \lim\limits_{x\to b} g(x) = f(b) + g(b) = (f+g)(b)$, so that $f+g$ is in fact continuous at b. This is similarly true for subtraction, multiplication of functions, and for division of functions f and g provided $g(b) \neq 0$.

(13) If $f \circ g$ was not continuous at b, then either g or $f \circ g$ or both would have to be discontinuous at b. Since it is given that both g at b and f at $g(b)$ are continuous, $f \circ g$ must also be continuous at b.

(14) The left and right limits are different as x approaches 1. Further, these limits are different from the value of the function: $f(1) = 3$, at $x = 1$. Thus, the function $f(x)$ cannot possibly be continuous at $x = 1$. A possible sketch is given below.

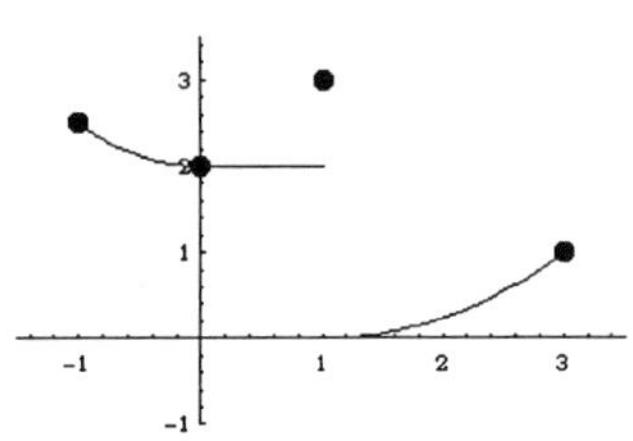

(15) $R(x)$ factors and simplifies to $\frac{x+4}{x-2}$. So $\lim\limits_{x\to -2} R(x) = -\frac{1}{2}$.

(16) $\lim\limits_{x\to\infty} \frac{3x^2+1}{x^2+3} = \lim\limits_{x\to\infty} \frac{3+\frac{1}{x^2}}{1+\frac{3}{x^2}} = \frac{3}{1} = 3.$

(17) $\lim_{x\to\infty} \frac{5x^2+x-11}{6x^2+3} = \lim_{x\to\infty} \frac{5+\frac{1}{x}-\frac{11}{x^2}}{6+\frac{3}{x^2}} = \frac{5}{6}$.

(18) Since $-1 \le \sin(x) \le 1$ it follows that

$$\lim_{x\to\infty} \frac{\sin(x)}{x} \le \lim_{x\to\infty} \frac{1}{x} = 0$$

and

$$\lim_{x\to\infty} \frac{\sin(x)}{x} \ge \lim_{x\to\infty} \frac{-1}{x} = 0.$$

Thus $\lim_{x\to\infty} \frac{\sin(x)}{x} = 0$.

(19) $\lim_{x\to\infty} \frac{x^2+x-2}{3x^3+2x} = \lim_{x\to\infty} \frac{1+\frac{1}{x}-\frac{2}{x^2}}{3x+\frac{2}{x}} = 0$.

(20) $\lim_{x\to+\infty} \frac{x}{|x|} = 1$.

(21) $\lim_{x\to-\infty} \frac{x}{x^2+3} = \lim_{x\to-\infty} \frac{1}{x+\frac{3}{x}} = 0$.

Solutions for §5.3

(1) $\frac{(x+h)^4-x^4}{h} = 4x^3 + 6hx^2 + 4h^2x + h^3 \to 4x^3$ as $h \to 0$.

(2) $\frac{(x+h)^{19}-x^{19}}{h} = 19x^{18} + 171hx^{17} + 969h^2x^{16} + ... + 19h^{17}x + h^{18} = 19x^{18} + hP(x,h)$ where P is a polynomial in x and h. Then $h \cdot P(x,h) \to 0$ as $h \to 0 \Rightarrow f'(x) = 19x^{18}$.

(3) $\frac{(x+h)^{50}-x^{50}}{h} = 50x^{49} + hP(x,h)$ where again P is a polynomial in x and h. Then $h \cdot P(x,h) \to 0$ as $h \to 0 \Rightarrow f'(x) = 50x^{49}$.

(4) $\frac{f(x+h)-f(x)}{h} = 3 + h + 2x \to 3 + 2x = f'(x)$ as $h \to 0$.

(5) $\frac{f(x+h)-f(x)}{h} = 6x + 1800x^{17} + hP(x,h) \to 6x + 1800x^{17} = f'(x)$ as $h \to 0$.

(6) $\frac{f(x+h)-f(x)}{h} = -\frac{2}{1+2h+4x+4hx+4x^2} \to -\frac{2}{1+4x+4x^2}$ as $h \to 0$.

(7) $\frac{f(x+h)-f(x)}{h} = \frac{2}{1+h+2x+hx+x^2} \to -\frac{2}{1+2x+x^2}$ as $h \to 0$.

(8)

$$\left(\frac{\sqrt{x+h}-\sqrt{x}}{h}\right)\frac{\sqrt{x+h}+\sqrt{x}}{\sqrt{x+h}+\sqrt{x}} = \frac{1}{\sqrt{x+h}+\sqrt{x}} \to \frac{1}{2\sqrt{x}} \quad \text{as } h \to 0.$$

(9)

$$\left(\frac{\sqrt{x+11+h}-\sqrt{x+11}}{h}\right)\frac{\sqrt{x+11+h}+\sqrt{x+11}}{\sqrt{x+11+h}+\sqrt{x+11}}$$

$$= \frac{1}{\sqrt{x+11+h}+\sqrt{x+11}} \to \frac{1}{2\sqrt{x+11}} \quad \text{as } h \to 0.$$

(10) We have

$$\begin{aligned}\frac{f(x+h)-f(x)}{h} &= \frac{\sqrt{2x-1}-\sqrt{2x+2h-1}}{h\sqrt{2x+2h-1}\sqrt{2x-1}}\left(\frac{\sqrt{2x-1}+\sqrt{2x+2h-1}}{\sqrt{2x-1}+\sqrt{2x+2h-1}}\right)\\ &= \frac{-2}{\sqrt{2x+2h-1}\sqrt{2x-1}}\left(\frac{1}{\sqrt{2x-1}+\sqrt{2x+2h-1}}\right)\\ &\longrightarrow -(2x-1)^{-\frac{3}{2}} \text{ as } h\to 0.\end{aligned}$$

Solutions for §5.4

(1) For $f(x)=2x^3-x+4$, the equation of the tangent line at $x=0$ is $y=f'(0)(x-0)+f(0)$. Since $f'(0)=-1,\ f(0)=4$, we see that $y=4-x$ is the tangent line.

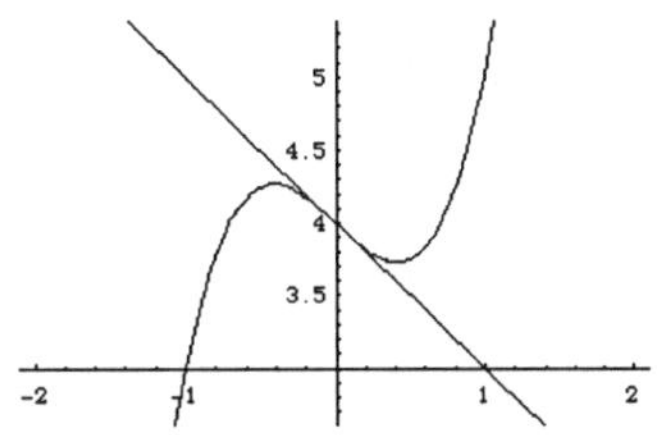

$f(x)=2x^3-x+4$, tangent line at $(0,4)$ is: $y=4-x$

(2) Here $f(x)=x^5-3x^4+2x+9,\ f'(1)=5-12+2=-5,\ f(1)=1-3+2+9=9$. The equation of the tangent line is: $y=f'(1)(x-1)+f(1)$ which simplifies to: $y=14-5x$.

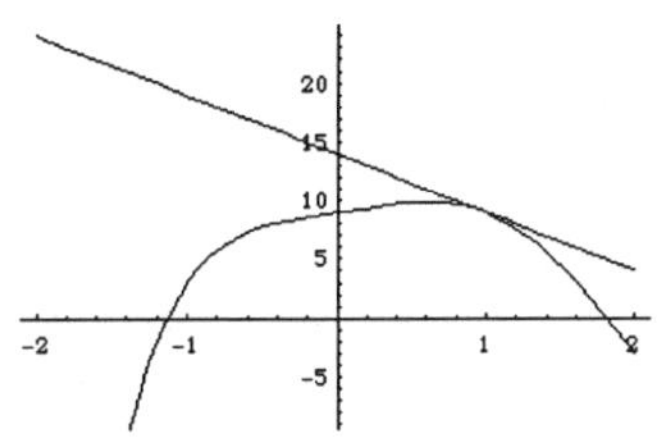

$f(x)=x^5-3x^4+2x+9$, tangent line at $(1,9)$ is: $y=14-5x$

(3) Here $f(x)=\frac{2}{x+1}$ and $f'(x)=\frac{-2}{(x+1)^2}$, so $f'(2)=-\frac{2}{9}$. Also, $f(2)=\frac{2}{3}$. The equation of the tangent line is: $y=f'(2)(x-2)+f(2)$, which simplifies to: $y=\frac{10-2x}{9}$.

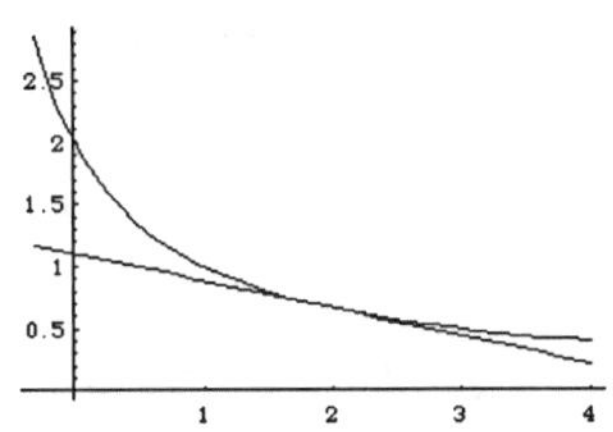

$f(x)=\dfrac{2}{x+1}$, tangent line at $(2,\dfrac{2}{3})$ is: $y=\dfrac{10-2x}{9}$

(4)

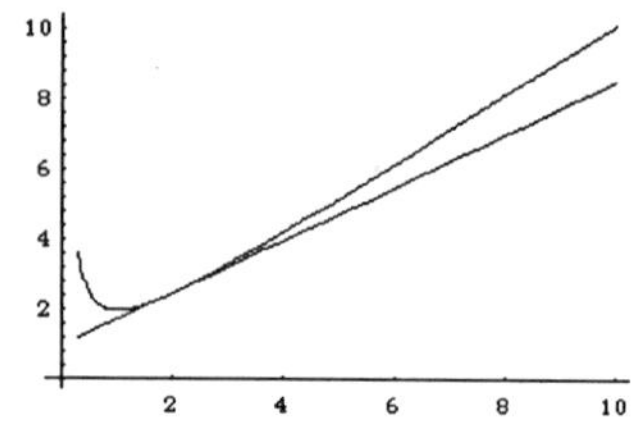

$$f(x) = x + \frac{1}{x}, \text{ tangent line at } \left(2, \frac{5}{2}\right) \text{ is: } y = 1 + \frac{3}{4}x$$

(5)

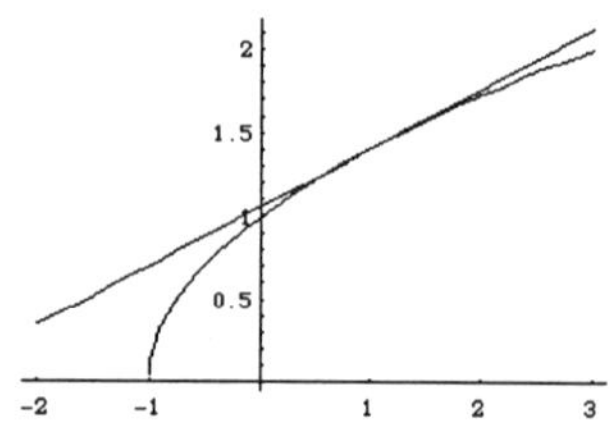

$$f(x) = \sqrt{x+1}, \text{ tangent line at } (1, \sqrt{2}) \text{ is: } y = \frac{x+3}{2\sqrt{2}}$$

(6)

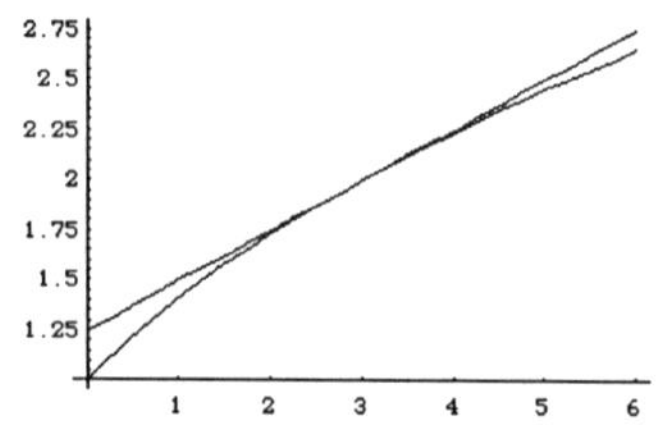

$$f(x) = \sqrt{x+1}, \text{ tangent line at } (3, 2) \text{ is: } y = \frac{5+x}{4}$$

(7) Let $f(x) = x^2 - 2x + 4$. The equation of the tangent line to the graph at the point a is given by the equation: $y = f'(a)x + (f(a) - af'(a)) = (2a-2)x + \left[(a^2 - 2a + 4) - a(2a-2)\right]$. If this line is to go through the origin, then we must have: $(a^2 - 2a + 4) - a(2a - 2) = 0$. We see that $a = \pm 2$. Thus, we obtain two tangent lines: $y = 2x$, $y = -6x$, which pass through the origin.

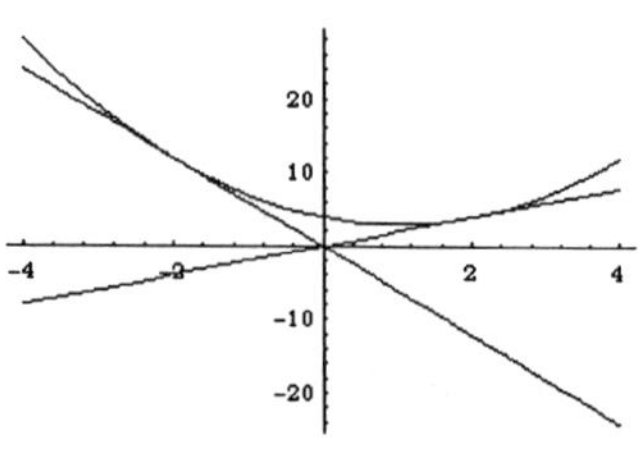

(8) The curves $f(x) = 3x^2$ and $g(x) = 2x^3 + 1$ both meet at the point $(1, 3)$. Since $f'(1) = 6 = g'(1)$, we see that these curves must have the same tangent line: $y = 6x - 3$.

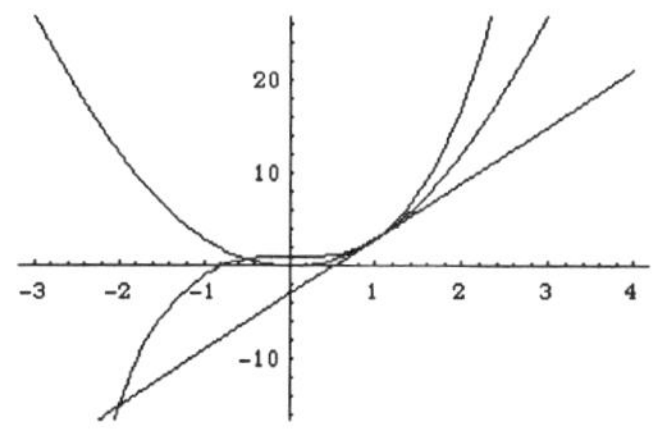

(9) The line $y = 4x + 37$ has slope 4, therefore, (since the slope of the tangent line to the graph of f at a is $f'(a)$) we must solve $f'(a) = 3a^2 + 17 = 4$ but no real value of a satisfies this equation implying no such tangent line exists.

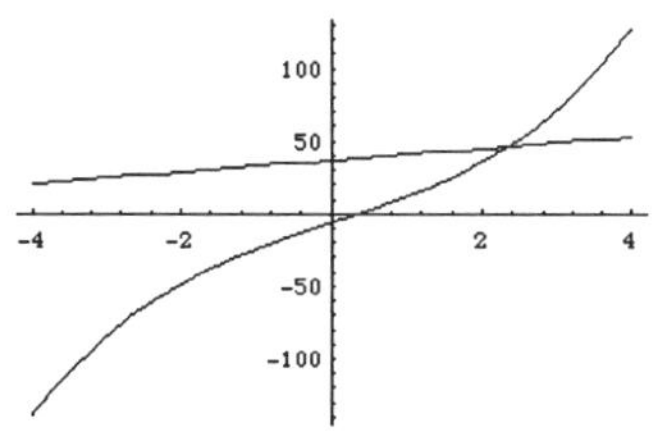

(10) The line $y = 19x + 8$ has slope 19. Let's try to find a point $x = a$ so that the tangent line to the curve $f(x) = x^3 + 7x - 14$ has slope 19. We must solve the equation: $f'(a) = 3a^2 + 7 = 19$. Thus, $a = \pm 2$. The tangent line to the graph at the point $(2, 8)$ has equation $y = 19(x - 2) + 8$.

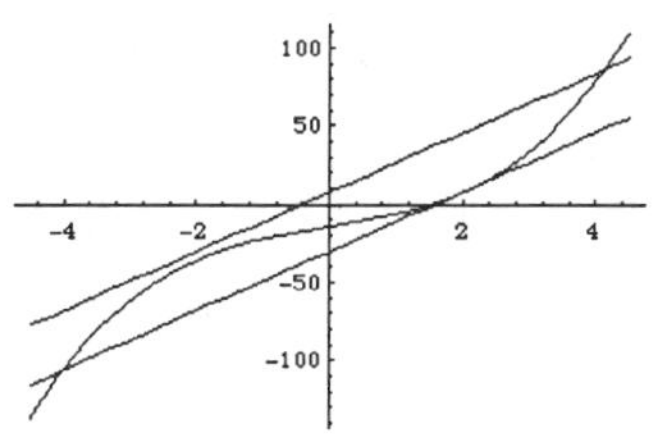

(11) We can derive a formula for the normal line based on the tangent line formula $y = f'(a)(x - a) + f(a)$. The normal line must be perpendicular to the tangent and so has slope $-\frac{1}{f'(a)}$ and the equation is of the form $y = -\frac{x}{f'(a)} + c$. It must also pass through the point $(a, f(a))$ so letting x and y take these values we find $c = f(a) + \frac{a}{f'(a)}$ and so the equation for the normal line to the graph of f at the point $(a, f(a))$ is $y = \frac{a-x}{f'(a)} + f(a)$. This formula can be used in questions 11 to 14.

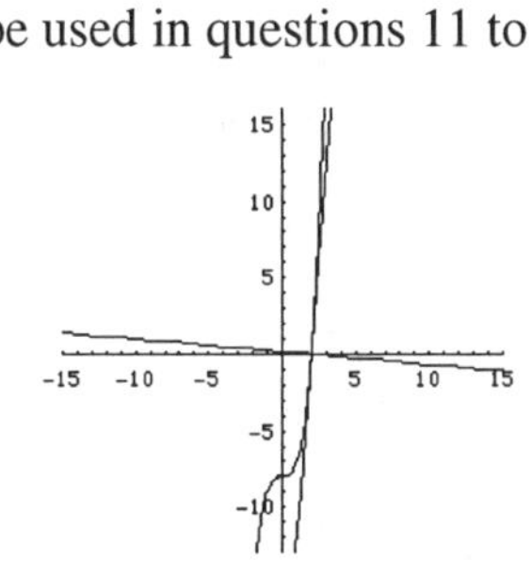

$$f(x) = x^3 - 8, \qquad P = (2, 0)$$

Tangent line: $y = 12x - 24$, Normal line: $y = \frac{1}{12}(2 - x)$

(12)

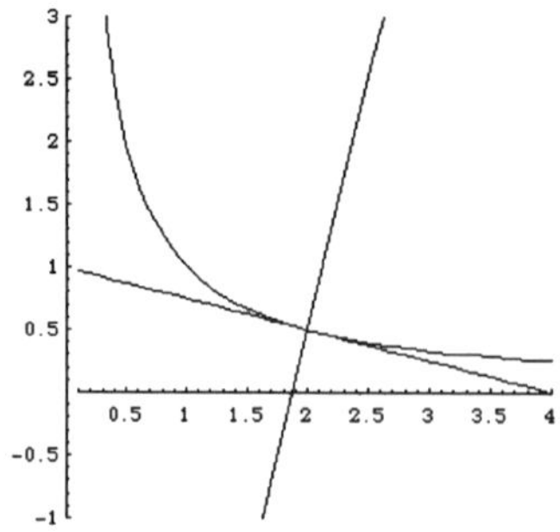

$$f(x) = \frac{1}{x}, \qquad P = \left(2, \frac{1}{2}\right)$$

Tangent line: $y = \frac{1}{4}(4 - x)$, Normal line: $y = \frac{1}{2}(8x - 15)$

(13)

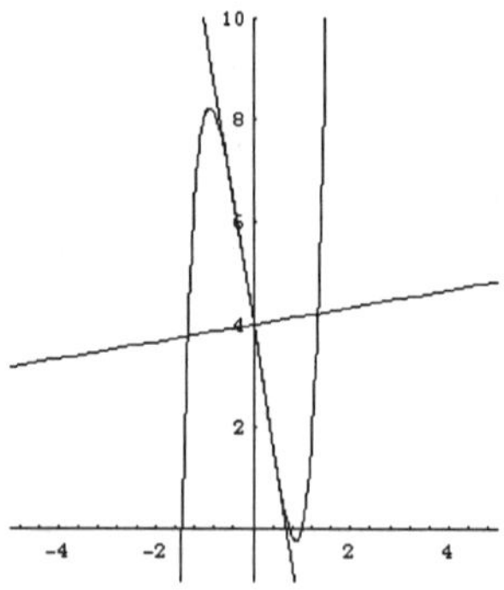

$$f(x) = 2x^5 - 6x + 4, \qquad P = (0, 4)$$

Tangent line: $y = 4 - 6x$, Normal line: $y = 4 + \frac{1}{6}x$

(14)

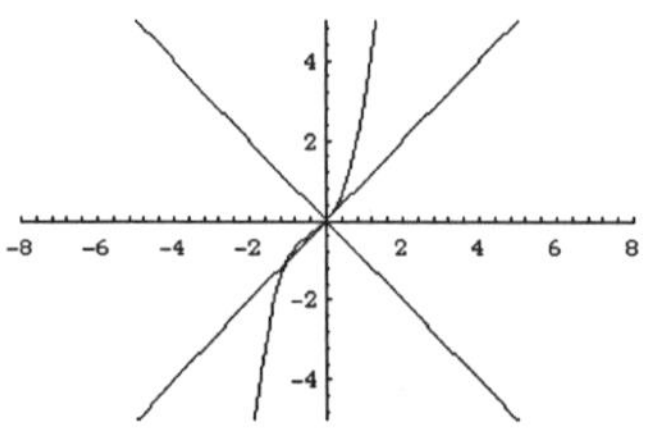

$$f(x) = x + x^2 + x^3, \qquad P = (0, 0)$$

Tangent line: $y = x$, Normal line: $y = -x$

Solutions for §5.5

(1) $f(x) = 2x^2$, $f'(x) = 4x$, $f''(x) = 4$, $f'''(x) = 0$.

(2) $f(x) = 3x^3$, $f'(x) = 9x^2$, $f''(x) = 18x$, $f'''(x) = 18$.

(3) $f(x) = 7x^4$, $f'(x) = 28x^3$, $f''(x) = 84x^2$, $f'''(x) = 168x$.

(4) $f(x) = \frac{2}{x}$, $f'(x) = \frac{-2}{x^2}$, $f''(x) = \frac{4}{x^3}$, $f'''(x) = \frac{-12}{x^4}$.

(5) $f(x) = \sqrt{x}$, $f'(x) = \frac{1}{2}x^{-\frac{1}{2}}$, $f''(x) = -\frac{1}{4}x^{-\frac{3}{2}}$, $f'''(x) = \frac{3}{8}x^{-\frac{5}{2}}$.

(6) $f(x) = 3x^2 - 2x + 1$, $f'(x) = 6x - 2$, $f''(x) = 6$, $f'''(x) = 0$.

(7) $f(x) = 7x^3 - 2x + 3$, $f'(x) = 21x^2 - 2$, $f''(x) = 42x$, $f'''(x) = 42$.

(8) $f(x) = 4x - 3$, $f'(x) = 4$, $f''(x) = 0$, $f'''(x) = 0$.

(9) $f(x) = (2x + 1)^{-1}$, $f'(x) = -2(2x + 1)^{-2}$, $f''(x) = 8(2x + 1)^{-3}$, $f'''(x) = -48(2x + 1)^{-4}$.

(10) $f(x) = 7x^9 - 2x^3 + 3x - 14$, $f'(x) = 63x^8 - 6x^2 + 3$, $f''(x) = 504x^7 - 12x$, $f'''(x) = 3528x^6 - 12$.

(11) If we differentiate $f(x) = 13x^5 - 5x^4 + 2x - 12$ once, we obtain $f'(x) = 65x^4 - 20x^3 + 2$, which is a polynomial of degree 4. After each successive differentiation, the degree goes down by one. It follows that $\frac{d^5 f}{dx^5}$ is a degree 0 polynomial, i.e., the constant function. Thus $\frac{d^n f}{dx^n} = 0$ if $n > 5$.

(12) Let $f(x) = (x - 2)^2$. Then $f'(x) = 2(x - 2)$, $f''(x) = 2$, thus, $f(1) = 1$, $f'(1) = -2$, and $f''(1) = 2$.

Solutions for Additional Exercises for Chapter 5

(1)

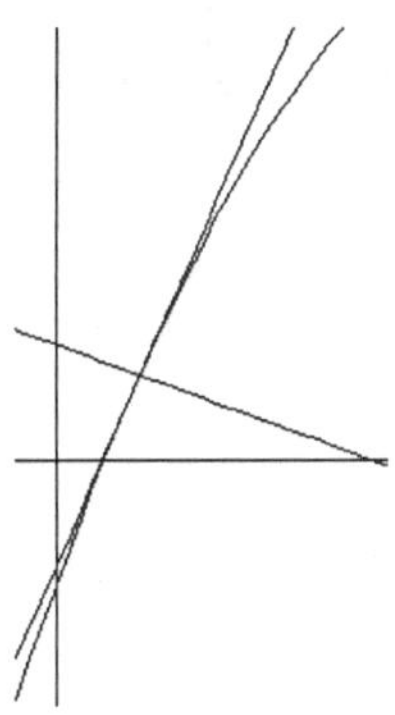

$$f(x) = -x^2 + 12x - 3, \qquad P = (1, 8)$$

$$\text{Tangent line: } y = 10x - 2, \quad \text{Normal line: } y = -\frac{1}{10}x + \frac{81}{10}$$

(2)

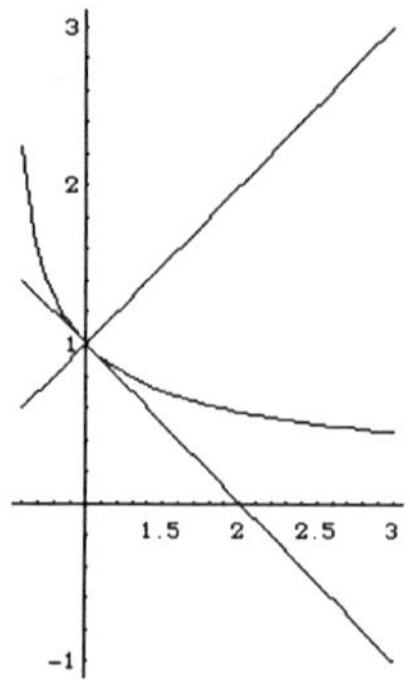

$$f(x) = \frac{1}{\sqrt{2x-1}}, \qquad P = (1,1)$$

Tangent line: $y = 2 - x$, Normal line: $y = x$

(3) The pattern which emerges for the derivative of x^n, n an integer is: nx^{n-1}.
Verification: If $f(x) = x^n$, then, using the limit definition of the derivative, $f'(x) = \lim_{h\to 0} \frac{(x+h)^n - x^n}{h}$. When we expand the numerator we get the expression

$$\frac{x^n + nx^{n-1}h + h^2 \cdot (\text{other terms}) - x^n}{h}.$$

The two x^n terms subtract out, and if we factor out h from all the terms in the numerator and cancel with the h in the denominator (we can do this since h approaches but never actually reaches zero), the only term no longer containing h will be nx^{n-1}. Since all other remaining terms will contain an h after division by the denominator, we can evaluate the limit as $\lim_{h\to 0}(nx^{n-1} + h \cdot (\text{other terms})) = nx^{n-1}$.

(4) Let $f(x) = \frac{ax+b}{cx+d}$ with $ad - bc = 1$. Then

$$f'(x) = \lim_{h\to 0}\left(\frac{a(x+h)+b}{c(x+h)+d} - \frac{ax+b}{cx+d}\right) \cdot \frac{1}{h}$$

$$= \lim_{h\to 0} \frac{(a(x+h)+b)(cx+d) - (ax+b)(c(x+h)+d)}{((c(x+h)+d)(cx+d)} \cdot \frac{1}{h}$$

$$= \lim_{h\to 0} \frac{(ad-bc)h}{(c(x+h)+d)(cx+d)} \cdot \frac{1}{h} = \frac{1}{(cx+d)^2}.$$

(5) The tangent line to the curve $f(x) = x^3 - 2x + 1$ has slope $f'(x) = 3x^2 - 2$. The tangent line is horizontal if $f'(x) = 3x^2 - 2 = 0$. Consequently, the tangent line is horizontal if $x = \pm\sqrt{\frac{2}{3}}$.

(6) The curve $y = f(x) = x^3 + 8x - 11$ has tangent lines whose slope at the point (x, y) is given by $3x^2 + 8$. Since $3x^2 + 8 > 8$. for any value of x this implies that all tangents have slope greater or equal to 8.

(7) f is clearly continuous and differentiable when $x > 0$ and $x < 0$ since f is defined by polynomial equations on each of these intervals. To show that f is continuous everywhere, we, therefore, need only check that the left-handed limit and right-handed limit of the function at 0 are equal. But $\lim_{x \to 0^-} f(x) = \lim_{x \to 0^-} (x^2 + 4x + 4) = 4$, and $\lim_{x \to 0^+} f(x) = \lim_{x \to 0^+} (4x + 4) = 4$. So f is continuous everywhere. Similarly, to show that f is differentiable everywhere, we need only check that the left-handed limit and right-handed limit of the piecewise function derivatives at 0 are equal. But $\lim_{x \to 0^-} f'(x) = \lim_{x \to 0^-} (2x + 4) = 4$, and $\lim_{x \to 0^+} f'(x) = \lim_{x \to 0^+} 4 = 4$. So f is differentiable everywhere, and $f'(0) = 4$.

(8) For continuity, $\lim_{x \to 4^-} f(x) = \lim_{x \to 4^-} (x^3 - 9x^2 + 27x - 27) = 1$, and $\lim_{x \to 4^+} f(x) = \lim_{x \to 4^+} (3x - 11) = 1$. So f is continuous everywhere. For differentiability, $\lim_{x \to 4^-} f'(x) = \lim_{x \to 4^-} (3x^2 - 18x + 27) = 3$, and $\lim_{x \to 4^+} f'(x) = \lim_{x \to 4^+} 3 = 3$. So f is differentiable everywhere. We have $f'(0) = 27$, $f'(4) = 3$, $f'(5) = 3$.

(9) Using the limit definition of the derivative,

$$\frac{d}{dx}\frac{1}{x} = \lim_{h \to 0} \frac{\frac{1}{x+h} - \frac{1}{x}}{h} = \lim_{h \to 0} \frac{\frac{x-(x+h)}{x(x+h)}}{h} = \lim_{h \to 0} -\frac{1}{x(x+h)} = -\frac{1}{x^2}.$$

For $f(x) = \frac{1}{x}$, $f'(x) = -\frac{1}{x^2}$. For $f(x) = \frac{1}{x^2}$, $f'(x) = -\frac{2}{x^3}$. For $f(x) = \frac{1}{x^3}$, $f'(x) = -\frac{3}{x^4}$. In general for $f(x) = \frac{1}{x^n}$, $f'(x) = -\frac{n}{x^{n+1}}$.

(10) We use the limits $\lim_{h \to 0} \frac{\cos(h) - 1}{h} = 0$ and $\lim_{h \to 0} \frac{\sin(h)}{h} = 1$. Now, using the limit definition of the derivative,

$$\begin{aligned} \frac{d}{dx}\sin(x) &= \lim_{h \to 0} \frac{\sin(x+h) - \sin(x)}{h} \\ &= \lim_{h \to 0} \frac{\sin(x)\cos(h) + \cos(x)\sin(h) - \sin(x)}{h} \\ &= \sin(x) \lim_{h \to 0} \frac{\cos(h) - 1}{h} + \cos(x) \lim_{h \to 0} \frac{\sin(h)}{h} = \cos(x). \end{aligned}$$

Thus, the derivative of $\sin(x)$ at $x = 0$ is $\cos(0) = 1$.

(11) The point $x = \frac{5\pi}{7}$ is marked in the graph of $y = \cos(x)$ below. Since the tangent line (at this point) slopes downwards, $f'\left(\frac{5\pi}{7}\right) < 0$.

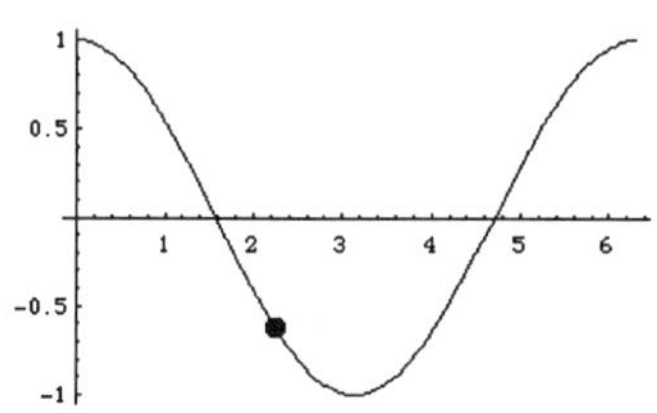

$f(x) = \cos(x)$

(12) Since the tangent line at $(3, 8)$ will be sloping up, $f'(3)$ must be positive.

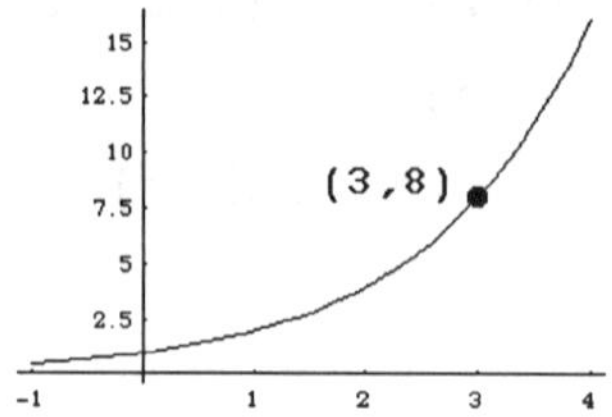

$$f(x) = 2^x$$

(13) The function $f(x) = x^{\frac{1}{3}} + 1$ is not differentiable at $x = 0$ since

$$f'(0) = \lim_{h \to 0} \frac{(h^{\frac{1}{3}} + 1) - 1}{h} = \lim_{h\ to 0} \frac{h^{\frac{1}{3}}}{h} = \lim_{h \to 0} \frac{1}{h^{\frac{2}{3}}} = \infty.$$

In fact, the tangent line is vertical at $x = 0$, so it has infinite slope. On the other hand, $f(x)$ is continuous at $x = 0$ since $\lim_{x \to 0} (x^{\frac{1}{3}} + 1) = 1 = f(0)$.

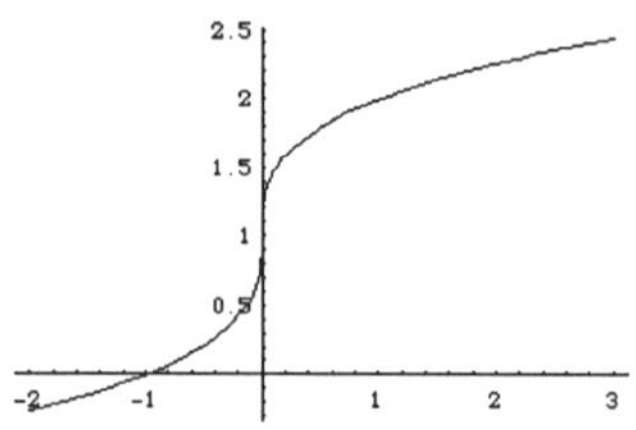

$$f(x) = x^{\frac{1}{3}} + 1$$

(14) $f(2) = 0$ (otherwise $\lim_{h \to 0} \frac{f(2+h)}{h}$ would not exist), and so $f'(2) = \lim_{h \to 0} \frac{f(2+h) - f(2)}{h} = \lim_{h \to 0} \frac{f(2+h)}{h} = 6$.

(15) Suppose that (x_1, y_1) and (x_2, y_2) are two points on the parabola. So $y_1 = ax_1^2$ and $y_2 = ax_2^2$. The tangent line at the point (x_1, y_1) has slope $2ax_1$ and so its equation is $(y - ax_1^2) = 2ax_1(x - x_1)$. Similarly, the tangent line at the point (x_2, y_2) has slope $2ax_2$ and so its equation is $(y - ax_2^2) = 2ax_2(x - x_2)$. We can rewrite these equations as $y = 2ax_1x - ax_1^2$ and $y = 2ax_2x - ax_2^2$. We can solve for the point of intersection of the two tangent lines by solving these two equations simultaneously. We immediately see that x satisfies $2ax_1x - ax_1^2 = 2ax_2x - ax_2^2$. So $x = \frac{ax_1^2 - ax_2^2}{2ax_1 - 2ax_2} = \frac{a(x_1 - x_2)(x_1 + x_2)}{2a(x_1 - x_2)} = \frac{x_1 + x_2}{2}$.

Solutions for §6.1

(1) We have $s(t) = -16t^2 + v_0t + s_0$ where $v_0 = 0$ and $s_0 = 900$. Solving $s(t) = -16t^2 + 900 = 0$ we find the stone hits the ground at time $t = \frac{30}{4}$ seconds (after it is released). Also $v(t) = s'(t) = -32t \Rightarrow$ the stone's velocity on impact is $v\left(\frac{30}{4}\right) = -240$ feet per second.

(2) In this exercise $v_0 = 8, s_0 = 16 \Rightarrow s(t) = -16t^2 + 8t + 16 \Rightarrow s'(t) = -32t + 8$. The ball's vertical speed will be zero when it reaches it's maximum height$\Rightarrow$ $s'(t) = 0$ and $t = \frac{1}{4}$. When this happens the ball is at height $s\left(\frac{1}{4}\right) = 17$ feet.

(3) The equation of motion for the bean bag is $s(t) = -16t^2 + 8t + 16$. Solving $s(t) = 0$ and taking the positive root we find the bean bag hits the ground after $t = \frac{1+\sqrt{5}}{2}$ seconds. At this point it has velocity $s'(\frac{1+\sqrt{5}}{2}) = -8 - 16\sqrt{5}$ feet per second.

(4) $v(t) = s'(t) = -300t^2 + 1000$ and the height is maximized when $v(t) = 0$, i.e., when $t = \sqrt{\frac{10}{3}}$ and this maximum height is $s(\sqrt{\frac{10}{3}})$ which equals $50 + \sqrt{\frac{10}{3}} \cdot \frac{2000}{3} \approx 1267$feet. The velocity after one second is $s'(1) = 700$ feet per second.

(5) The ship lands, i.e., it's height $h(t)$ is zero at time $t = 2$ because $h(t) = 30(t-2)^2$. It's velocity on touching down is $h'(2) = 0$ so the ship makes a perfect landing.

Solutions for §6.2

(1) Let $f(x) = x^3 - 9$ and choose $r_1 = 2$. Then $f'(x) = 3x^2$. We get

$$r_2 = r_1 - \frac{f(r_1)}{f'(r_1)} = 2 - \frac{2^3 - 9}{3 \cdot 2^2} = 2 + \frac{1}{12} = 2.0833\ldots$$

Similarly, $r_3 = r_2 - \frac{f(r_2)}{f'(r_2)} = \frac{25}{12} - \frac{\left(\frac{25}{12}\right)^3 - 9}{3 \cdot \left(\frac{25}{12}\right)^2} = 2.08008\ldots$ and $r_4 = r_3 - \frac{f(r_3)}{f'(r_3)} = 2.080083\ldots$ which is $9^{\frac{1}{3}}$ to six decimal places.

(2) Let $f(x) = x^2 + x - 3$, $f'(x) = 2x + 1$. Choose $r_1 = 1$. Then

$$r_2 = r_1 - \frac{f(r_1)}{f'(r_1)} = 1 - \frac{1 + 1 - 3}{2 + 1} = \frac{4}{3} = 1.3333\ldots$$

Similarly, $r_3 = r_2 - \frac{f(r_2)}{f'(r_2)} = \frac{4}{3} - \frac{\left(\frac{4}{3}\right)^2 + \left(\frac{4}{3}\right) - 3}{2 \cdot \frac{4}{3} + 1} = 1.3030\ldots$ and $r_4 = 1.3027\ldots$, $r_5 = 1.3027\ldots$ which is correct to 5 decimal places, which is one root of the equation $f(x) = 0$. Newton's method breaks down for the starting value $r_1 = -\frac{1}{2}$ because we would then have

$$r_2 = r_1 - \frac{f(r_1)}{f'(r_1)} = -\frac{1}{2} - \frac{\frac{1}{4} - \frac{1}{2} - 3}{0}$$

which is undefined because we have divided by zero.

(3) Starting with the value $r_1 = -1$ we get the sequence -0.8181, -0.7657, -0.7622, 0.7621, 0.7621, an approximation for a real root of the equation.

(4) Starting with the value $r_1 = 0.5$ we get the sequence 1.8089, 1.6180, 1.4780, 1.3876, 1.3439, 1.3330, 1.3324, 1.3324 approximating a real root of the equation.

(5) Newton's method must fail in this case because $f(x)$ has no real roots, i.e., the graph of f does not cross the x-axis.

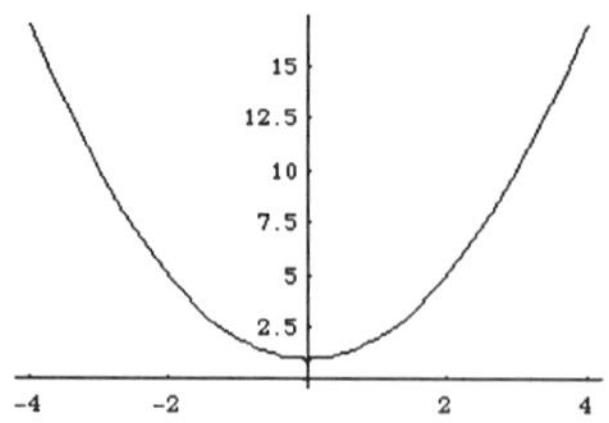

$$f(x) = x^2 + 1$$

Solutions for Additional Exercises for Chapter 6

(1) The fact that the plane is traveling at 500mph at a constant altitude of 2000 feet has no effect on the amount of time it will take a dropped object to reach the ground. We have $s(t) = -16t^2 + v_0 t + s_0$ where $v_0 = 0$, $s_0 = 2000$. The object hits the ground when $s(t) = -16t^2 + 2000 = 0$. Thus $t = \sqrt{\frac{2000}{16}} = \sqrt{125} = 11.18$ seconds. Also, $v(t) = s'(t) = -32t$ so its velocity on impact (when $t = 11.18\ldots$) is 357.77ft per sec, or 251.2mph.

(2) The average velocity of a vehicle traveling for t minutes is given by the ratio

$$\frac{\text{distance traveled}}{\text{time } t} = t^2 + t.$$

Therefore, the instantaneous velocity at $t = 0$ is $\lim_{t \to 0} (t^2 + t) = 0$.

(3) Here $f(x) = x^3 - 3x^2 + 1$, $f'(x) = 3x^2 - 6x$. If $r_1 = 0.5$, then

$$r_2 = r_1 - \frac{f(r_1)}{f'(r_1)} = 0.5 - \frac{(0.5)^3 - 3(0.5)^2 + 1}{3(0.5)^2 - 6(0.5)} = 0.66667$$

Similarly, $r_3 = 0.652778\ldots$, $r_4 = 0.652704\ldots$ and an approximation for the root is given by $x \approx 0.6527\ldots$

(4) $x \approx 0.2586$

(5) The two curves intersect when $x^{-2} = x^3 - 3 + 2x^{-2}$. So we must find a root of $f(x) = x^3 - 3 + x^{-2}$. Note that $f'(x) = 3x^2 - 2x^{-2}$. The following graph suggests an intersection point near $x = \frac{1}{2}$.

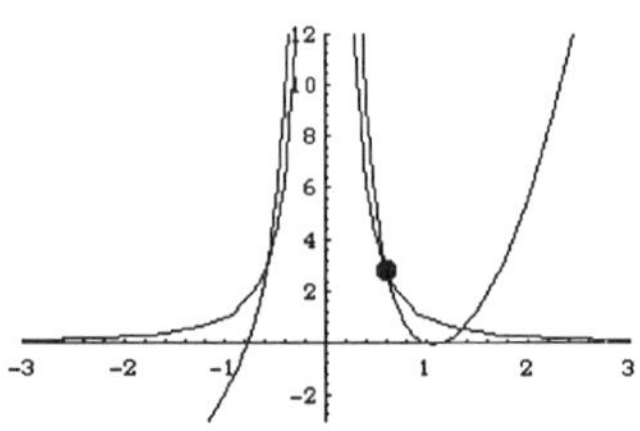

$$y = x^{-2} \quad \text{and} \quad y = x^3 - 3 + 2x^{-2}$$

We now find the root with Newton's method. Let $r_1 = \frac{1}{2}$. Then

$$r_2 = \frac{1}{2} - \frac{f\left(\frac{1}{2}\right)}{f'\left(\frac{1}{2}\right)} = \frac{1}{2} - \frac{\frac{1}{8} - 3 + \frac{1}{4}}{3 \cdot \frac{1}{4} - 2 \cdot 8} = 0.57377\ldots$$

Continuing, we find: $r_3 = 0.597357\ldots$, $r_4 = 0.59923\ldots$, $r_5 = 0.59924\ldots$ When $x = 0.5992$, we have $y = \frac{1}{x^2} \approx 2.7847$. Consequently, a point of intersection is $(x, y) \approx (0.5992, 2.7847)$.

(6) The two curves intersect when $f(x) = x - \cos(x) = 0$. Then $f'(x) = 1 + \sin(x)$. From the graph, we see that there is an intersection point near $x = \frac{1}{2}$.

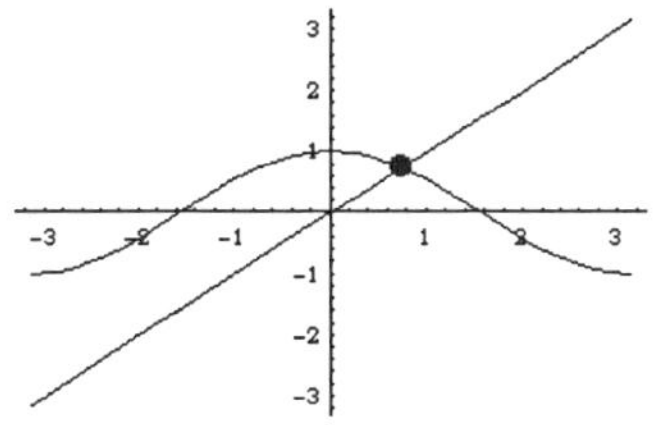

$y = x$ and $y = \cos(x)$

Using Newton's method, we compute: $r_1 = \frac{1}{2}$,

$$r_2 = r_1 - \frac{f(r_1)}{f'(r_1)} = \frac{1}{2} - \frac{\frac{1}{2} - \cos\left(\frac{1}{2}\right)}{1 + \sin\left(\frac{1}{2}\right)} = 0.755222\ldots,$$

$r_3 = 0.739142\ldots$, $r_4 = 0.739085\ldots$, $r_5 = 0.739085\ldots$ We, therefore, obtain the point of intersection $(x, y) \approx (0.7391, 0.7391)$.

(7) Let $f(x) = x^7 + 3x^5 - x^2 - 4$, $f'(x) = 7x^6 + 15x^4 - 2x$. From the graph, we see that a root of $f(x) = 0$ occurs near $x = 1$.

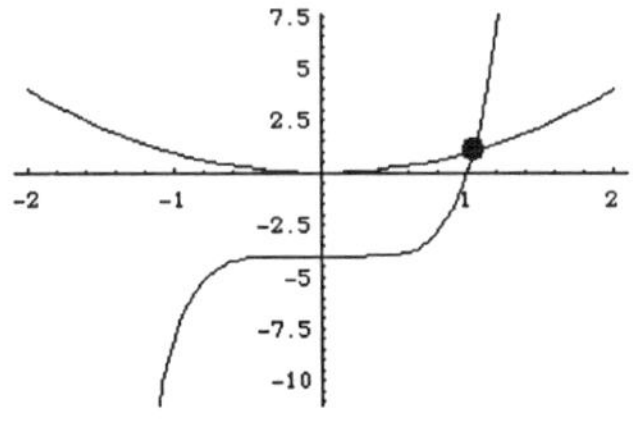

$y = x^2$ and $y = x^7 + 3x^5 - 4$

Applying Newton's method as above, we obtain a point of intersection $(x, y) \approx (1.0447, 1.0914)$

(8) Point of intersection is $(x, y) \approx (-1.5166, -33.0468)$.

(9) Point of intersection is $(x, y) \approx (0.1208, 8.2768)$.

(10) Let $f(x) = x^5 + 9x + 1,\ f'(x) = 5x^4 + 9$. The graph shows that the curve crosses the x–axis near $x = 0$.

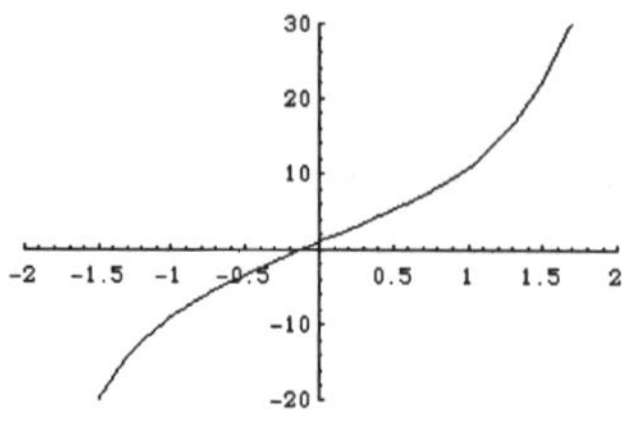

$$f(x) = x^5 + 9x + 1$$

To find a root of $f(x) = 0$, we use Newton's method with $r_1 = 0$. Then

$$r_2 = r_1 - \frac{f(r_1)}{f'(r_1)} = 0 - \frac{f(0)}{f'(0)} = -\frac{1}{9},$$
$$r_3 = \frac{1}{9} - \frac{f\left(\frac{1}{9}\right)}{f'\left(\frac{1}{9}\right)} = -0.111109,$$

$r_4 = -0.111109\ldots$, $r_5 = -0.111109\ldots$ So, we get one real solution at $x = -0.11111\ldots$

(11) One real solution at $x \approx 0.7196$.

(12) One real solution at $x \approx 0.6823$.

(13) One real solution at $x \approx 1.0889$.

(14) Two real solutions at $x = -2$ and $x \approx 2.5943$.

(15) One real solution at $x \approx 1.4840$.

(16) Three real solutions at $x \approx -1.4406$, $x \approx 0.0717$, and $x \approx 1.3848$.

(17) Since $s(t) = 3t^3 - 200t + 1$, we have $v(t) = s'(t) = 9t^2 - 200$. Thus, $v(5) = 9 \cdot 25 - 200 = 25$, so the velocity after 5 minutes is 25. Similarly, $v(10) = 9 \cdot 10^2 - 200 = 700$.

(18) The distance of the projectile above ground is $s(t) = at^2 + bt + c$ with $a < 0$. Then the velocity at time t is $v(t) = s'(t) = 2at + b$. The projectile reaches its maximum height when $v(t) = 0$, i.e., the projectile stops going up (has velocity = 0) and then starts falling. Thus, the projectile reaches maximum height when $t = -\frac{b}{2a}$. At this time, its height is equal to $s\left(-\frac{b}{2a}\right) = c - \frac{b^2}{4a}$.

(19) The drivers's velocity is given by $v(t) = s'(t) = 3at^2 + 2bt + c$. If $a < 0$, the driver reaches his maximum velocity at time $t = -\frac{b}{3a}$, since $v'\left(-\frac{b}{3a}\right) = 0$. At this time his velocity is equal to $c - \frac{b^2}{3a}$. Since $s(t)$ has at most three roots, the driver returns home at most three times.

Solutions for §7.1

(1) $15x^4 - 4x$

(2) $34x$

(3) $100x^{99}$

(4) $15x^{14} - 21x^{-8}$

(5) $-10x^{-6} - 3x^{-2}$

(6) $-100x^{-101}$

(7) $-2.2x^{0.1} + (7.009)5x^{6.009}$

(8) $-7.81(2x^{-8.81}) + 5\pi x^4$

(9) $-0.0031x^{-1.001}$

(10) $0.0012x^{-1.001}$

(11) $-2x^{-3} + 5$

(12) $700x^{-101} - 8x^{-5}$

(13) $\frac{1}{3}x^{\frac{-2}{3}}$

(14) $\frac{-1}{2}x^{\frac{-3}{2}} + \frac{3}{4}x^{\frac{-1}{4}}$

(15) $\frac{5}{2}x^{\frac{-9}{4}} - x^{\frac{-4}{3}}$

Solutions for §7.2

(1) $x^2(18 + 7x^4 + 20x^7)$, $\frac{-6x^2}{(1+2x^3)^2}$, $\frac{x^2(-18+7x^4+8x^7)}{(1+2x^3)^2}$.

(2) $5 - 9x^2 - 90x^5 + 72x^7$, $\frac{1-18x^5}{x^2(-1+3x^5)^2}$, $\frac{-5-3x^2+90x^5-36x^7}{x^2(-1+3x^5)^2}$.

(3) $2 - 8x^{-5} - 12x^{-4} - 22x^{-2} + 33x^2$, $\frac{-2-33x^2}{(1+2x+11x^3)^2}$, $\frac{-(8+20x+154x^3+2x^5+33x^7)}{x^5(1+2x+11x^3)^2}$.

(4) $-2x^{-\frac{9}{10}} + \frac{57}{10}x^{\frac{9}{10}} + 20x$, $\frac{-19x^{\frac{9}{10}}}{10(-2+x^{\frac{19}{10}})^2}$, $\frac{-(20+57x^{\frac{9}{5}}+180x^{\frac{19}{10}}}{10x^{\frac{9}{10}}(-2+x^{\frac{19}{10}})^2}$.

(5) $\frac{-(1+30x^{\frac{9}{2}}+9x^5)}{2\sqrt{x}(-1+x^5)^2}$.

(6) $\frac{-(1+24x^{\frac{11}{3}})}{3x^{4/3}(1+2x^{\frac{11}{3}})^2}$.

(7) $\frac{44}{3(11+x^{\frac{2}{3}})^2x^{\frac{1}{3}}}$.

(8) $\frac{9+6x-108x^5+102x^{\frac{33}{2}}-64x^{\frac{35}{2}}+96x^{\frac{43}{2}}}{2\sqrt{x}(3-2x+4x^5)^2}$.

(9) $\frac{6-x+12x^2-3x^3}{x^3(1+x^2)^2}$.

(10) $\frac{2(0.15-0.1\sqrt{x}+3.8x^{1.6}+2.8x^{2.1}}{(1+\sqrt{x})^2x^{0.7}}$.

(11) $\frac{d(x \cdot x)}{dx} = \frac{dx}{dx} \cdot x + x \cdot \frac{dx}{dx} = 2x$.

(12) $\frac{d(x^2 \cdot x)}{dx} = \frac{dx^2}{dx} \cdot x + x^2 \cdot \frac{dx}{dx} = 2x \cdot x + x^2 = 3x^2$.

(13) $\frac{d(x^N \cdot x)}{dx} = \frac{dx^N}{dx} \cdot x + x^N \cdot \frac{dx}{dx} = Nx^{N-1} \cdot x + x^N = (N+1)x^N$.

(14) $(f \cdot g \cdot h)' = ((f \cdot g) \cdot h)' = (f \cdot g)' \cdot h + (f \cdot g) \cdot h' = f' \cdot g \cdot h + f \cdot g' \cdot h + f \cdot g \cdot h'$.

(15) $\frac{144-126x-252x^5+324x^6-12x^{10}+13x^{11}+40x^{14}-174x^{15}+138x^{16}}{2x^5\sqrt{x-1}}$

(16) $\frac{1+4x+7x^2+6x^3}{(x-1)^4(1+x)^2(1+x+x^2)^2}$

Solutions for §7.3

(1) $10(-4x+38x^{18})(1-2x^2+2x^{19})^9$

(2) $160x^3(-2+8x^4)^4$

(3) $\frac{-60x^2}{(1+2x^3)^{11}}$

(4) $\frac{-3+11x^{10}}{2\sqrt{7-3x+x^{11}}}$

(5) $\frac{-4x}{3(1+x^2)^{\frac{5}{3}}}$

(6) $\frac{\frac{x-1}{(1+x)^2}+\frac{1}{1+x}}{4\left(\frac{x-1}{x+1}\right)^{\frac{3}{4}}}$

(7) $\frac{-18.4x^{39}}{(-3.01+2x^{40})^{1.23}}$

(8) $\pi^2x^{-1+\pi}(1+x^\pi)^{-1+\pi}$

(9) $\frac{42(1+x^2)^{20}}{x^{20}}-\frac{21(1+x^2)^{21}}{x^{22}}$

(10) $\frac{-3x^2}{2(-2+x^3)^{\frac{3}{2}}}$

(11) $\frac{-10(1+5x)(-3+4x^3)}{(1-3x+x^4)^{11}}+\frac{5}{(1-3x+x^4)^{10}}$

(12) $\frac{500(3+x^{\frac{1}{3}})^{499}}{3x^{\frac{2}{3}}}$

(13) $\frac{1}{8x^{\frac{3}{4}}\sqrt{1+x^{\frac{1}{4}}}}$

(14) $\frac{1}{27(1+(1+x^{\frac{1}{3}})^{\frac{1}{3}})^{\frac{2}{3}}(1+\sqrt{x})^{\frac{3}{4}}\sqrt{x}}$

(15) $\frac{1}{16\sqrt{2+(1+\sqrt{x})^{\frac{1}{4}}}(1+\sqrt{x})^{\frac{3}{4}}\sqrt{x}}$

(16) $-3(1+10x)+6(1+10x)(-10+x+5x^2)^2$

(17) $\frac{5(2+\sqrt{x})^4}{2\sqrt{x}}$

(18) $\frac{3x^2(-2+x^3)}{\sqrt{-3+(-2+x^3)^2}}$

(19) $\frac{2(-3+x^{\frac{2}{3}})^2(-222+196x^{\frac{2}{3}}-59x^{\frac{4}{3}}+6x^2)}{3x^{\frac{1}{3}}}$

(20) $150(1+(-2+x)^{\frac{3}{2}})^{99}\sqrt{-2+x}$

(21) $\frac{-330x^{0.1}(3+x^{1.1})^2}{(1+(3+x^{1.1})^{303}}$

Solutions for §7.4

(1) $-\sin(x^4+3x^2-2)(4x^3+6x)$

(2) $\cos(2x^{19}-3x^5-2x+11)(38x^{18}-15x^4-2)$

(3) $\sec^2(x^3+3x^2-6)(3x^2+6x)$

(4) $-2x\csc^2(2+x^2)$

(5) $-\left(\frac{1-x}{(1+x)^2}+\frac{1}{1+x}\right)\sin\left(\frac{x-1}{1+x}\right)$

(6) $\left(\frac{1-x}{(1+x)^2}+\frac{1}{1+x}\right)\cos\left(\frac{x-1}{1+x}\right)$

(7) $\sin(x)\sin(\cos(x))$

(8) $-\sin(x)\sin(\cos(x))\sin(\cos(\cos(x)))$

(9) $2x\cos(2+x^2)\cos(\sin(2+x^2))$

(10) $\frac{3+3\cos(x)\sin^2(x)}{2\sqrt{3x+\sin^3(x)}}$

(11) $\frac{3x^3\cos(2+x^4)}{\sin^{\frac{1}{4}}(2+x^4)}$

(12) $(-6x+10x^9)\sin(2-3x^2+x^{10})\sin(\cos(2-3x^2+x^{10}))$

(13) $\frac{3x^2\sec^2(\sqrt{2+x^3})}{2\sqrt{2+x^3}}$

(14) $\frac{(3-3x^2)\csc^2((2-3x+x^3)^{\frac{1}{3}})}{3(2-3x+x^3)^{\frac{2}{3}}}$

(15) $\frac{\sin(x)(\sin(x)+2\cos(x)\sin(x))}{(\cos(x)-\sin^2(x))^2}+\frac{\cos(x)}{\cos(x)-\sin^2(x)}$

(16) $\frac{-3-\cos(x)}{2(2+\sqrt{x})^2\sqrt{x}}-\frac{\sin(x)}{2+\sqrt{x}}$

(17) $\cos^4(x) - 3\cos^2(x)\sin^2(x)$

(18) $-60(1 + 3\cos(x))^{19}\sin(x)$

(19) $-6x\cos^2(1 + x^2)\sin(1 + x^2)$

(20) $\frac{(1-\cos(x))\sin(x)}{(1+\cos(x))^2} + \frac{\sin(x)}{1+\cos(x)}$

Solutions for Additional Exercises for Chapter 7

(1) $f'(x) = na_n x^{n-1} + (n-1)a_{n-1}x^{n-2} + \ldots + 2a_2 x + a_1$

(2) $f'(x) = 24x^3 - 2x$ and $f'(x) = 0$ for $x = 0, \sqrt{\frac{1}{12}}, -\sqrt{\frac{1}{12}}$.

(3a) The graph of $f(x) = x^4 - 1$ has the general shape of a parabola opening up and is shifted down from the origin by one. $f(x)$ displays even symmetry so $f(x) = f(-x)$.

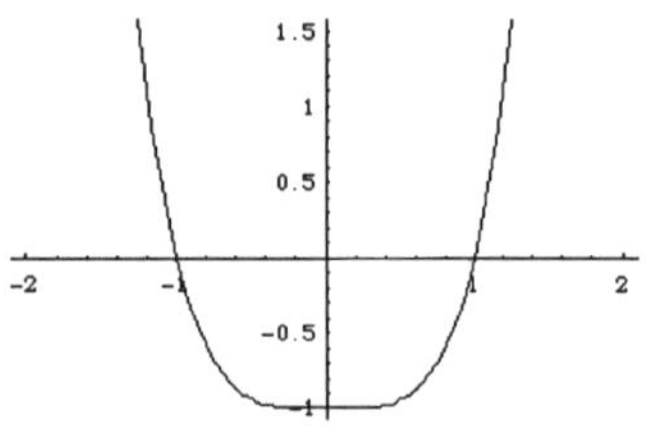

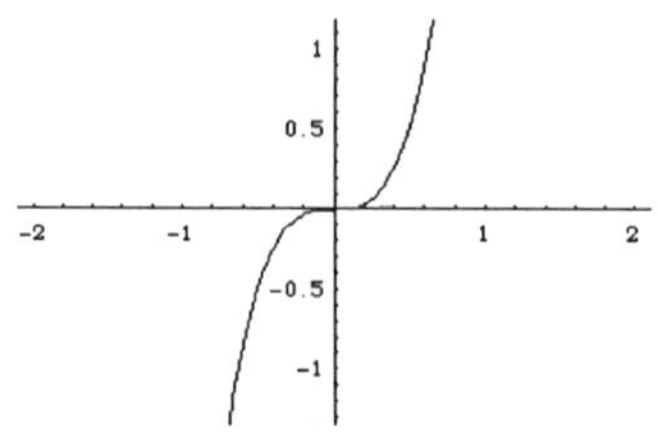

$f(x) = x^4 - 1$, $\quad f'(x) = 4x^3$

(3b) Yes, $f'(x) = -f'(-x)$.

(3c) $f'(x) = 4x^3$ and displays odd symmetry.

(3d) In general for $f(x) = x^n + k; n \in \mathbb{N}, k \in \mathbb{R}$; $f'(x) = \lim_{h \to 0} \frac{(x+h)^n + k - (x^n + k)}{h} = nx^{n-1}$. The k and x^n terms cancel and we are left with $\lim_{h \to 0} \frac{nx^{n-1}h + \ldots + h^n}{h}$. After factoring out an h from the numerator and canceling with the denominator, we get that the limit $= nx^{n-1}$.

(4a) The graph of $f(x) = x^5 + 2x^3 - x$ has the property that $f(-x) = -f(x)$, and it displays odd symmetry.

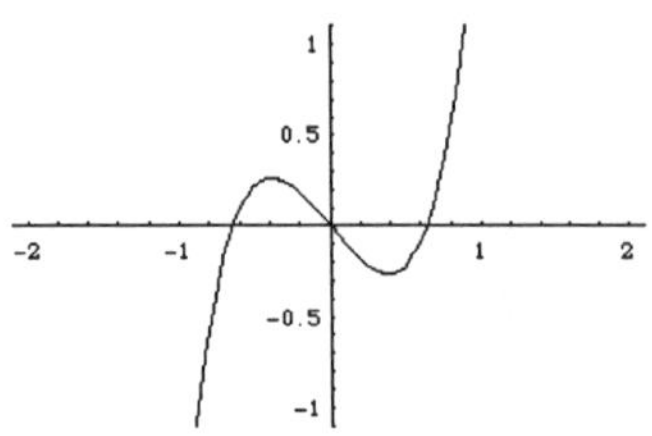

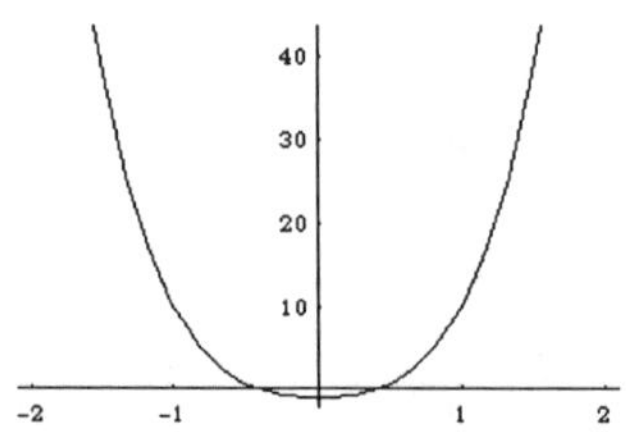

$f(x) = x^5 + 2x^3 - x$, $\quad f'(x) = 5x^4 + 6x^2 - 1$

(4b) Yes, $f'(x) = f'(-x)$.

(4c) $f'(x) = 5x^4 + 6x^2 - 1$ and has even symmetry.

(5) We use the linear approximation $g(x+h) \approx g(x) + hg'(x)$, which is valid for small h. Now, using the limit definition of the derivative,

$$\begin{aligned}\frac{d}{dx}(f \circ g)(x) &= \lim_{h\to 0} \frac{f(g(x+h)) - f(g(x))}{h} \\ &= \lim_{h\to 0} \frac{f(g(x)+hg'(x)) - f(g(x))}{h} \\ &= g'(x) \lim_{h\to 0} \frac{f(g(x)+hg'(x)) - f(g(x))}{hg'(x)} \\ &= g'(x) \lim_{k\to 0} \frac{f(g(x)+k) - f(g(x))}{k} = g'(x)f'(g(x)),\end{aligned}$$

with the substitution $k = hg'(x)$, so that $k \to 0$ as $h \to 0$.

(6) Take for example $f(x) = x^2$ and $g(x) = x^2 + 2$. Here $f'(x) = g'(x)$ but $f(x) \neq g(x)$. In general, any added/subtracted constants to $f(x)$ or $g(x)$ will satisfy; $f'(x) = g'(x)$ and $f(x) \neq g(x)$.

(7) $f^{(1)}(x) = -\frac{2x}{(x^2-a)^2}$
$f^{(2)}(x) = \frac{6x^2+2a}{(x^2-a)^3}$
$f^{(3)}(x) = \frac{-24x^3-24xa}{(x^2-a)^4}$
$f^{(4)}(x) = \frac{120x^4+120x^2a+24a^2}{(x^2-a)^5}$
$f^{(5)}(x) = \frac{-720x^5-2400x^3a-720xa^2}{(x^2-a)^6}$
$f^{(6)}(x) = \frac{5040x^6+25200x^4a+15120x^2a^2+720a^3}{(x^2-a)^7}$
Note that we always have $f^{(n)}(x) = \frac{P_n(x)}{(x^2-a)^{n+1}}$, where $P_n(x)$ is a polynomial of degree n satisfying $P_n(-x) = (-1)^n P_n(x)$. Moreover, the leading coefficient of $P_n(x)$ is $(-1)^n n!$.

(8) We see from the graphs below that $f(1) = 0$ and the tangent line to $f(x)$ at $x = 0$ is horizontal so that $f'(1) = 0$.

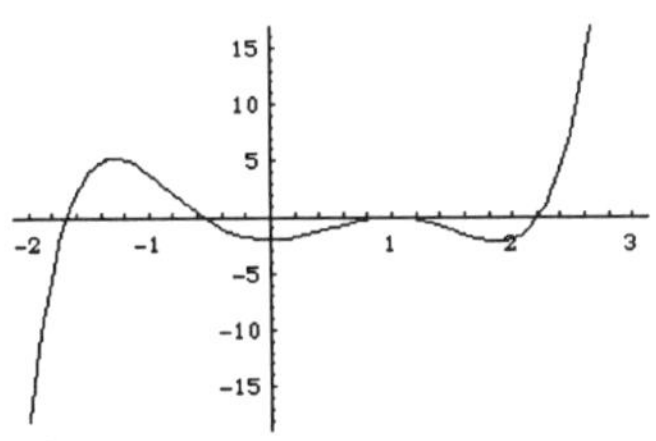

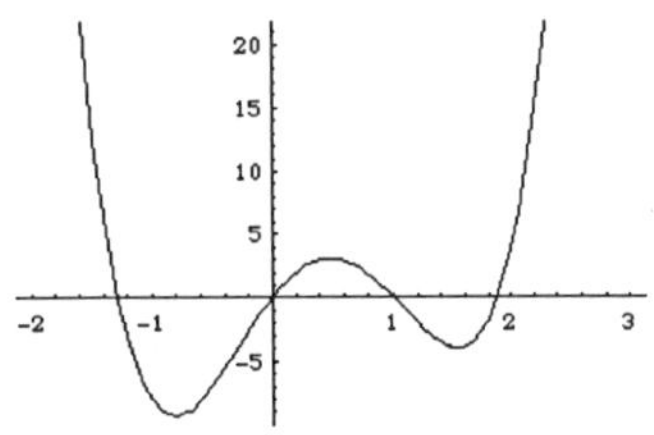

$f(x) = (x-1)^2(x^3 - 4x - 2)$, $\qquad f'(x) = 12\,x - 9\,x^2 - 8\,x^3 + 5\,x^4$

$f(1) = 0$ and $f'(0) = 0$. $f''(x) = 20x^3 - 24x^2 - 18x + 12$, so that $f''(1) = -10$.

(9) $\frac{2\sin((\frac{1}{x-3})^2+3)}{(x-3)^3}$

(10) Observe that $(x-3)^3 = x^3 - 9x^2 + 27x - 27$. Thus $f(x^{\frac{1}{3}} + 3) = x$. Hence $(f \circ g)'(x) = 1$.

(11) Here, as in problem (10), the functions are inverses. Hence, $(f \circ g)(x) = x,\ (f \circ g)'(x) = 1$.

(12) No. Let $f(x) = x^2 + 2$, $g(x) = \sqrt{x}$. Then $f(g(x)) = x + 2$ and $(f \circ g)'(x) = 1$ but f and g are not inverse functions.

(13) No. By the chain rule, $(f \circ g)'(x) = f'(g(x)) \cdot g'(x) = 0$ implies that either $g'(x) = 0$ or $f'(g(x)) = 0$. Thus either g is the constant function or f is constant on the range of g.

(14) Let $f(x) = \cos(\pi(x^2 - 1))$. Then $f'(x) = -2\pi \sin(\pi(x^2 - 2))$. The equation of the tangent line at $\left(\frac{\sqrt{2}}{2}, 0\right)$ is given by

$$y = f'\left(\frac{\sqrt{2}}{2}\right)\left(x - \frac{\sqrt{2}}{2}\right) = -2\pi\sqrt{2}\sin\left(\frac{\sqrt{2}}{2}\right)\left(x - \frac{\sqrt{2}}{2}\right) = 2\pi\sqrt{2}\left(x - \frac{\sqrt{2}}{2}\right).$$

(15) Let $f(x) = \tan(\sqrt{x})$. Then $f'(x) = \frac{1}{2}x^{-\frac{1}{2}} \cdot \frac{1}{\cos^2(\sqrt{x})}$. The equation for the tangent line at $\left(\frac{\pi^2}{9}, \sqrt{3}\right)$ is given by

$$y = f'\left(\frac{\pi^2}{9}\right)\left(x - \frac{\pi^2}{9}\right) + \sqrt{3} = \frac{6}{\pi}\left(x - \frac{\pi^2}{9}\right) + \sqrt{3}.$$

(16) Let $f(x) = \frac{3-x}{2+x}$. Then $f'(x) = \frac{-5}{(x+2)^2}$. The equation of the tangent line at $\left(1, \frac{2}{3}\right)$ is given by

$$y = f'(1)(x - 1) + \frac{2}{3} = \frac{-5}{9}(x - 1) + \frac{2}{3}.$$

(17) Since $f(x) = \sqrt{1 + \sqrt{x}}$, $f'(x) = \frac{1}{2}(1 + \sqrt{x})^{-\frac{1}{2}} \cdot \frac{1}{2}x^{-\frac{1}{2}} = \frac{1}{4}\frac{1}{(1+\sqrt{x})x}$. The equation of the tangent line at $(4, 3)$ is

$$y = f'(4)(x - 4) + \sqrt{3} = \frac{1}{48}(x - 4) + \sqrt{3}.$$

(18) $y = \frac{10\pi}{3}\left(\frac{\sqrt{3}}{2}\right)^9 x + \left(\frac{\sqrt{3}}{2}\right)^{10}$.

Solutions for §8.2

(1) Solving $x + y^2 = 1$ for y, we obtain $y = \pm(1 - x)^{\frac{1}{2}}$ and $y' = \pm\frac{1}{2}(1 - x)^{-\frac{1}{2}}$. Solving by implicit differentiation, we have:

Step 1: $x + f(x)^2 = 1$.

Step 2: $1 + 2f'(x)f(x) = 0$.

Step 3: $1 + 2yy' = 0$

Step 4: $y' = -\frac{1}{2y}$.

(2) Solving $x^2 + y^2 = 1$ for y, we obtain $y = \pm\sqrt{1 - x^2}$, and $y' = \pm x(1 - x^2)^{\frac{1}{2}}$. Solving by implicit differentiation, we have:

Step 1: $x^2 + f(x)^2 = 1$.

Step 2: $2x + 2f(x)f'(x) = 0$.

Step 3: $2x + 2yy' = 0$.

Step 4: $y' = -\frac{x}{y}$.

(3) Solving $xy = 1$ for y, we obtain $y = \frac{1}{x}$, and $y' = -\frac{1}{x^2}$. Solving implicitly:

Step 1: $xf(x) = 1$.

Step 2: $f(x) + xf'(x) = 0$.

Step 3: $y + xy' = 0$.

Step 4: $y' = -\frac{y}{x}$.

(4) $y = \pm x^{-\frac{1}{2}} \Rightarrow y' = \mp\frac{1}{2}x^{-\frac{3}{2}}$ and secondly $2xf(x)f'(x) + f(x)^2 = 0 \Rightarrow y' = -\frac{y}{2x}$.

(5) $y = \pm\sqrt{2}x^{-\frac{1}{2}} \Rightarrow y' = \left(\pm\frac{1}{\sqrt{2}}\right) x^{-\frac{3}{2}}$ and as in question (4), $y' = -\frac{y}{2x}$.

(6) $y = \pm\sqrt[4]{\frac{3}{x}} \Rightarrow y' = \pm\frac{1}{4}\left(\frac{3}{x}\right)^{-\frac{3}{4}} \cdot (-x^{-2})$ or $y' = -\frac{y}{4x}$.

(7) $y = \pm 1 - x \Rightarrow y' = -1 = \frac{-2y-2x}{2y+2x}$.

(8) $y = \left(\frac{x-2}{x^2}\right)^{\frac{1}{3}} \Rightarrow y' = \frac{1}{3}\left(\frac{x-2}{x^2}\right)^{-\frac{2}{3}}\left(\frac{4x-x^2}{x^4}\right) = \frac{1-2xy^3}{3x^2y^2}$.

(9) $y = \frac{3}{2}x^2 \pm \frac{1}{2}\sqrt{8+9x^4} \Rightarrow y' = 3x \pm \frac{9x^3}{\sqrt{8+9x^4}} = \frac{6xy}{(2y-3x^2)}$.

(10) Step 1: $x\cos(x) + f(x)\cos(x) = 13$.

Step 2: $\cos(x) - x\sin(x) + f'(x)\cos(x) - f(x)\sin(x) = 0$.

Step 3: $\cos(x) - x\sin(x) + y'\cos(x) - y\sin(x) = 0$.

Step 4: $y' = -\sec(x)\Big(\cos(x) - x\sin(x) - y\sin(x)\Big)$.

(11) Step 1: $f(x)^7 - 2x^3f(x)^4 + 3xf(x)^2 + x^5 = 3$.

Step 2: $7f(x)^6 \cdot f'(x) - 6x^2f(x)^4 - 8x^3f(x)^3f'(x) + 3f(x)^2 + 6xf(x)f'(x) + 5x^4 = 0$.

Step 3: $7y^6 \cdot y' - 6x^2y^4 - 8x^3y^3 \cdot y' + 3y^2 + 6x \cdot y \cdot y' + 5x^4 = 0$.

Step 4: $(7y^6 - 8x^3y^3 + 6xy)y' - 6x^2y^4 + 3y^2 + 5x^4 = 0$, or

$$y' = \frac{6x^2y^4 - 3y^2 - 5x^4}{7y^6 - 8x^3y^3 + 6xy}.$$

(12) $y' = \frac{2xy + y\sec^2(x)}{3y^2 - \tan(x) - x^2}$

(13) $y' = \frac{6x^2y\sin(x^3y) - y^2e^{xy^2}}{2xye^{xy^2}}$

(14) $y' = \frac{-2y}{-2x-3x^2y^2+3y^4}$

(15) Step 1: $2^{x^2+f(x)^2} = 17$, i.e., $e^{\ln(2)(x^2+f(x)^2)} = 17$.

Step 2: $\ln(2)(2x+2f(x)\cdot f'(x))e^{\ln(2)(x^2+f(x)^2)} = 0$. Since e^A can never be zero for any A, we deduce that $2x + 2f(x) \cdot f'(x) = 0$.

Step 3: $x + y \cdot y' = 0$.

Step 4: $y' = -\frac{x}{y}$.

(16) $y' = \frac{-x^2}{y^2} \Rightarrow y'$ at $(2^{\frac{-1}{3}}, 2^{\frac{-1}{3}}) = -1$.

(17) $y' = \frac{1+2\pi\cos(\pi x)y\sin(\pi x)}{\cos^2(\pi x)+3y^2} \Rightarrow$ the tangent line at $(0,2)$ is $y = \frac{x}{13} + 2$.

(18) $y' = \frac{1-2xy\cos(2x)-y\sin(2x)}{x\sin(2x)+4y^3} \Rightarrow$ the tangent at $(0,1)$ is $y = \frac{x}{4} + 1$.

(19) Take the two circles $x^2 + y^2 - 2x - 4y - 20 = 0$ and $x^2 + y^2 - 20x - 28y + 196 = 0$. Are they tangential at $(4,6)$? Firstly it can easily be checked that both circles pass through this point (i.e., $x = 4$, $y = 6$ satisfies both equations). Then we have $y' = \frac{1-x}{y-2}$ and $y' = \frac{10-x}{y-14}$ for the respective circles and the slopes are the same $(= -\frac{3}{4})$ at $(4,6) \Rightarrow$ the circles are tangential there.

Solutions for §8.3

(1) $3x^2e^{x^3}$

(2) $10xe^{3+5x^2}$

(3) $2x\ln(2)2^{x^2}$

(4) $2(1+\ln(x))x^{2x}$

(5) $(1+2\ln(x))x^{1+x^2}$

(6) $\frac{6x}{2+3x^2}$

(7) $\frac{16x^3\ln(2x^4-1)}{2x^4-1}$

(8) $\ln(x)^{x-1} + \ln(x)^x\ln(\ln(x))$

(9) $(x^2+\cos(x))^{x-1}(2x^2 + x^2\ln(x^2+\cos(x)) + \cos(x)\ln(x^2+\cos(x)) - x\sin(x))$

(10) $\frac{e^{\frac{x}{2}}(2-x\ln(x))}{2xe^x}$

(11) $\frac{-2x\sin(\ln(1+x^2))}{1+x^2}$

(12) $e^x\cot(e^x)$

(13) We have $\cosh^2(x) = \left(\frac{1}{2}\right) + \left(\frac{1}{4}\right)e^{-2x} + \left(\frac{1}{4}\right)e^{2x}$ and $\sinh^2(x) = -\frac{1}{2} + \left(\frac{1}{4}\right)e^{-2x} + \left(\frac{1}{4}\right)e^{2x}$ so their difference is 1.

(14)

$$1 - \tanh^2(x) = 1 - \frac{\sinh^2(x)}{\cosh^2(x)} = \frac{\cosh^2(x) - \sinh^2(x)}{\cosh^2(x)} = \frac{1}{\cosh^2(x)}$$

(15) We have $\sinh(x)\cosh(y) = \left(\frac{1}{4}\right)\left(-e^{-x-y} + e^{x-y} - e^{-x+y} + e^{x+y}\right)$ and

$$\sinh(y)\cosh(x) = \frac{1}{4}\cdot\left(-e^{-x-y} - e^{x-y} + e^{-x+y} + e^{x+y}\right) = \frac{1}{2}\cdot\left(e^{x+y} - e^{-x-y}\right) = \sinh(x+y).$$

The formula for $\cosh(x+y)$ is treated similarly.

(16) $2x\cosh(x^2)$, $\frac{1}{x(\cosh(\ln(x)))^2}$, $-3x^{-4}\sinh(x^{-3})$

(17) Checking that $\lim_{h\to 0}(1+h)^{\frac{1}{h}}$ =e we evaluate $(1+h)^{\frac{1}{h}}$ for $h = \frac{1}{10}, \frac{1}{20}, \frac{1}{50}, \frac{1}{100}$ and get2.593, 2.653, 2.691 and 2.704 ≈ e. And for these values $\frac{e^h-1}{h} = 1.0517, 1.0254, 1.0100, 1.0050 \approx 1$.

Solutions for §8.4

(1) $f^{-1}(x) = \sqrt[4]{x}$ with domain and range $\{x \mid x \geq 0\}$. The derivative of $f^{-1}(x)$ is $\frac{1}{4}x^{\frac{-3}{4}}$.

(2) $f^{-1}(x) = \sqrt[N]{x}$ with domain and range $\{x \mid x \geq 0\}$. The derivative of $f^{-1}(x)$ is $\frac{1}{N}x^{\frac{1-N}{N}}$.

(3) $f^{-1}(x) = \frac{1}{x}$ with domain and range $\{x \mid x \neq 0\}$. $\frac{df^{-1}(x)}{dx} = -\frac{1}{x^2}$.

(4) $f^{-1}(x) = \sqrt{\ln(x)}$ with domain $\{y \mid y \geq 1\}$ and range $\{x \mid x \geq 0\}$. $(f^{-1})'(x) = \frac{1}{2x\sqrt{\ln(x)}}$.

(5) $f^{-1}(x) = \arccos(x)$ with domain $\{x \mid -1 \leq x \leq 1\}$ and range $\{y \mid 0 \leq y \leq \pi\}$. Then $(f^{-1})'(x) = -\frac{1}{\sin(\arccos(x))}$. Also $\sin(x) = \sqrt{1-\cos^2(x)}$ for $0 \leq x \leq \pi$ and this implies $(f^{-1})'(x) = -\frac{1}{\sqrt{1-\cos^2(\arccos(x))}} = -\frac{1}{\sqrt{1-x^2}}$.

(6) $f^{-1}(x) = \arctan(x)$ with domain $\{x \mid -\infty < x < \infty\}$ and range $\{y \mid -\frac{\pi}{2} < y < \frac{\pi}{2}\}$. Then $(f^{-1})'(x) = -\frac{1}{\cos^2(\arctan(x))}$. Using the identity $1 + \tan^2(x) = \sec^2(x)$ we obtain the relation $\cos(x) = (1+\tan^2(x))^{-\frac{1}{2}}$ for $-\frac{\pi}{2} \leq x \leq \frac{\pi}{2} \Rightarrow (f^{-1})'(x) = \frac{1}{(1+x^2)}$.

(7) $f^{-1}(x) = \arcsin(x^2)$ with domain $\{x \mid 0 \leq x \leq 1\}$ and range $\{y \mid 0 \leq y \leq \frac{\pi}{2}\}$. Then

$$(f^{-1})'(x) = \frac{\cos(\arcsin(x^2))}{2\sqrt{\sin(\arcsin(x^2))}} = \frac{\sqrt{1-x^2}}{2x}.$$

(8) $f^{-1}(x) = \frac{\ln(x)}{\ln(a)}$ with domain $\{x \mid 0 < x\}$ and range $\{y \mid -\infty < y < \infty\}$. Then

$$(f^{-1})'(x) = \frac{1}{\ln(a)\cdot a^{\frac{\ln(x)}{\ln(a)}}} = \frac{1}{x\ln(a)}.$$

(9) $f^{-1}(x) = \arctan\left(\frac{1}{x}\right)$ with domain $\{x \mid -\infty < x < \infty\}$ and with range $\{y \mid 0 < y < \pi\}$. So

$$(f^{-1})'(x) = \frac{1}{-\csc^2\left(\arctan\left(\frac{1}{x}\right)\right)} = -\sin^2\left(\arctan\left(\frac{1}{x}\right)\right) = \frac{-1}{x^2+1}.$$

(10) $f^{-1}(x) = \ln(x + \sqrt{x^2+1})$ with no restrictions on its domain or range. We get

$$(f^{-1})'(x) = \frac{1}{\cosh(\ln(x + \sqrt{x^2+1}))} = \frac{2}{x + \sqrt{1+x^2} + \frac{1}{x+\sqrt{1+x^2}}}.$$

(11) $f^{-1}(x) = \ln(x + \sqrt{x^2-1})$ with domain $\{x \mid 1 \leq x < \infty\}$ and with range $\{y \mid -\infty < y < \infty\}$. So

$$(f^{-1})'(x) = \frac{1}{\sinh(\ln(x + \sqrt{x^2-1}))} = \frac{2}{x + \sqrt{-1+x^2} - \frac{1}{x+\sqrt{-1+x^2}}}.$$

(12) $f^{-1}(x) = \ln\left(\sqrt{\frac{x+1}{1-x}}\right)$ with domain $\{x \mid -1 < x < 1\}$ and with range $\{y \mid -\infty < y < \infty\}$. So

$$(f^{-1})'(x) = \cosh^2\left(\ln\left(\sqrt{\frac{x+1}{1-x}}\right)\right) = \frac{1}{1-x^2}.$$

Solutions for Additional Exercises for Chapter 8

We shorten the 4 step method of implicit differentiation to a more compact 3 step method in the following problems

(1) Step 1: $y^3 + 2y - x^4 = 0.$
Step 2: $3y^2 \cdot y' + 2y' - 4x^3 = 0.$
Step 3: $y' = \frac{4x^3}{3y^2+2}.$

(2) Step 1: $6x^2 + 6y - 20 = 0.$
Step 2: $12x + 6y' = 0.$
Step 3: $y' = -2x.$

(3) Step 1: $x^3 + y - xy = 0.$
Step 2: $3x^2 + y' - y - xy' = 0.$
Step 3: $y' = \frac{y-3x^2}{1-x}.$

(4) Step 1: $5xy - (x-y)^2 = 0.$
Step 2: $5y + 5xy' - 2(x-y)(1-y') = 0.$
Step 3: $y' = \frac{7y-2x}{7x-2y}.$

(5) We rewrite the curves as $y = \frac{3}{1+2x}$ and $y = \left(\frac{x+1}{2}\right)^{\frac{1}{3}}$. These curves intersect when $\frac{3}{1+2x} = \left(\frac{x+1}{2}\right)^{\frac{1}{3}}$. We see from the graph and this equation, that the curves intersect when $x = 1$, $y = 1$. For the first curve, $y' = \frac{-6}{(1+2x)^2}$ so the slope of the tangent line at $(1,1)$ is SLOPE $= \frac{-6}{9} = \frac{-2}{3}$. For the second curve: $2y^3 = (x+1)$, $6y^2 \cdot y' - 1 = 0$, $y' = \frac{1}{6y^2}$. So the slope of the tangent line at $(1,1)$ is SLOPE $= \frac{1}{6}$. Thus the angle of intersection is $\arctan\left(\frac{1}{6}\right) - \arctan\left(\frac{-2}{3}\right) \approx 0.75315$.

(6) $\theta = \frac{\pi}{2}$, at intersections $(\sqrt{2}, \sqrt{2})$ and $(-\sqrt{2}, -\sqrt{2})$.

(7) $\theta \approx 1.45$ radians, at intersection $(.80, 1.14)$.

(8) $\theta = \frac{\pi}{4}$ at intersection $(0,0)$.

(9) $\frac{dy}{dx} = \frac{dy}{du}\frac{du}{dx} = \left(-\frac{2u}{(u^2+1)^2}\right)\left(-\frac{1}{(x-1)^2}\right) = -\frac{2x-2}{(x^2-2x+2)^2}$.

(10) $\frac{dy}{dx} = \frac{dy}{du}\frac{du}{dx} = (3u^2-2)\left(\frac{1}{3u^2-2u}\right) = \left(\frac{3u^2-2}{3u^2-2u}\right)$.

(11) Given that $y = xf(x)$, $\frac{dy}{dx} = f(x) + xf'(x)$,
$\frac{d^2y}{dx^2} = 2f'(x) + xf^2(x)$,
$\frac{d^3y}{dx^3} = 3f^2(x) + xf^3(x)$,
$\frac{d^4y}{dx^4} = 4f^3(x) + xf^4(x)$, and so forth; the pattern emerging is:
$\frac{d^ny}{dx^n} = nf^{n-1}(x) + xf^n(x)$.

(12) By implicit differentiation: $2y \cdot y' - 4y' + 2x = 0$. Hence, $y' = \frac{-2x}{2y-4} = \frac{-x}{y-2}$. At the point $(0,1)$, we have $y' = \frac{-0}{1-2} = 0$. So the tangent line has slope 0 and passes through $(0,1)$. It must have equation $y = 1$.

(13) Since $\frac{dy}{dx} = \frac{x}{y}\left(\frac{50-8(x^2+y^2)}{50+8(x^2+y^2)}\right)$, we have that $\left.\frac{dy}{dx}\right]_{(3,1)} = \frac{-9}{13}$, and the equation of the tangent line at $(3,1)$ is $y = \frac{-9}{13}(x-3)+1$.

(14) $\frac{dy}{dx} = -\left(\frac{x}{y}\right)^{\frac{1}{3}}$. At the point $(-1,1)$ on the curve the derivative is therefore $\frac{dy}{dx} = 1$. The equation of the tangent line at $(-1,1)$ is the $y = x+2$.

(15) $\frac{dy}{dx} = \frac{y-8y^2t}{(8yt^2-t-1)\tan t}$.

Solutions for §9.1

(1) We claim that $f'(x) \neq 0$ for any x and hence f has no local maxima or minima. To see that the derivative never vanishes observe that if $f'(x) = 3x^2 - 18x + 30$, then $\frac{f'(x)}{3} = x^2 - 6x + 10 = 0$. But then $x = \frac{6 \pm \sqrt{-4}}{2}$ which is not a real number.

(2) We see that $f'(x) = 4\left(x + \frac{1}{2}\right)(x-2)x = 0$ when $x = -\frac{1}{2}, 0, 2$. From the graph below

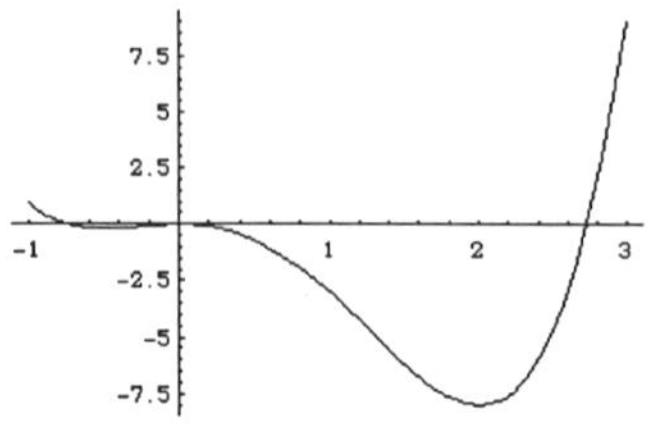

$$f(x) = x^4 - 2x^3 - 2x^2$$

we immediately see that $x = -\frac{1}{2}, 2$ are local minimum and $x = 0$ is a local maximum.

(3) In this example $f'(x) = 5x^4 = 0$ only when $x = 0$. From the graph below

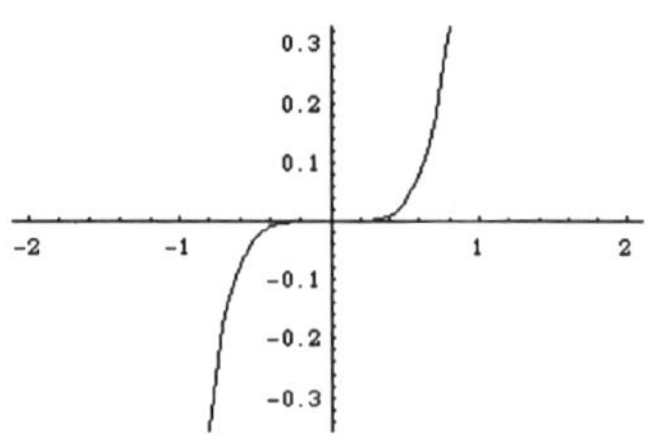

$$f(x) = f(x) = x^5$$

we see that $x = 0$ is neither a local maximum or a local minimum even though the tangent line is horizontal at $x = 0$. The point $x = 0$ is called an inflection point.

(4) Here $f'(x) = 4x^2\left(x + \frac{3}{4}\right)$. From the graph below

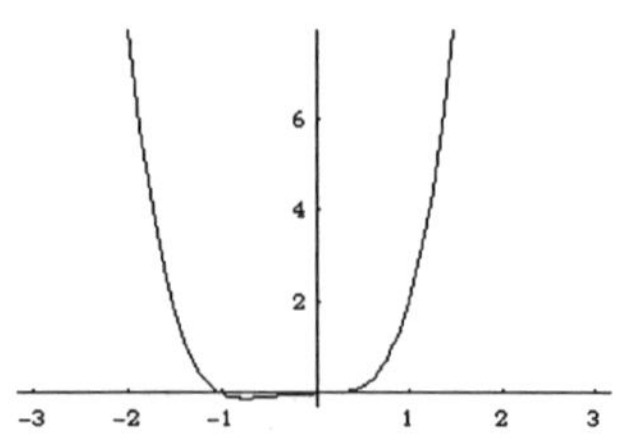

$$f(x) = x^4 + x^3$$

we see that $x = -\frac{3}{4}$ is a minimum and $x = 0$ is an inflection point.

(5) The derivative $f'(x) = -\sin(x) + 1 = 0$ whenever $x = \frac{\pi}{2} + 2\pi n$ for any integer n. In the stated range this gives two points $-3\frac{\pi}{2}$ and $\frac{\pi}{2}$ where the derivative vanishes. From the graph below

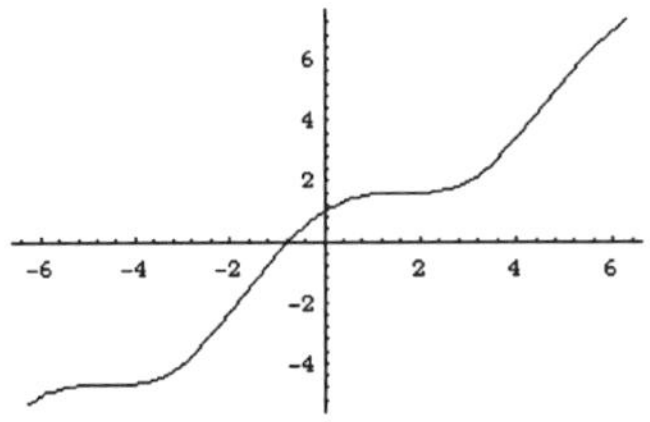

$f(x) = \cos(x) + x$

we see that both of these points are inflection points.

(6) Since the derivative $f'(x) = 2\cos(2x) - 1 = 0$ when $x = \pm\frac{\pi}{6}, \pm\frac{5\pi}{6}, \pm\frac{7\pi}{6}, \pm\frac{11\pi}{6}$ the graph below

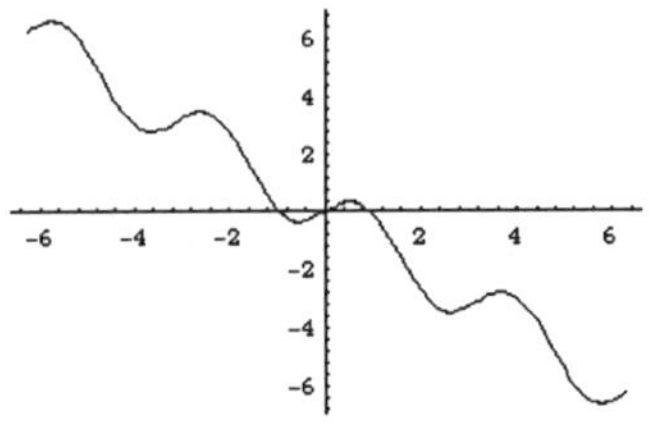

$f(x) = \sin(2x) - x$

indicates that these are alternately maxima and minima.

(7) Here $f'(x) = \frac{1}{x} - 3x^2 = 0$ when $x = 3^{-\frac{1}{3}} \approx 0.6933$. From the graph below

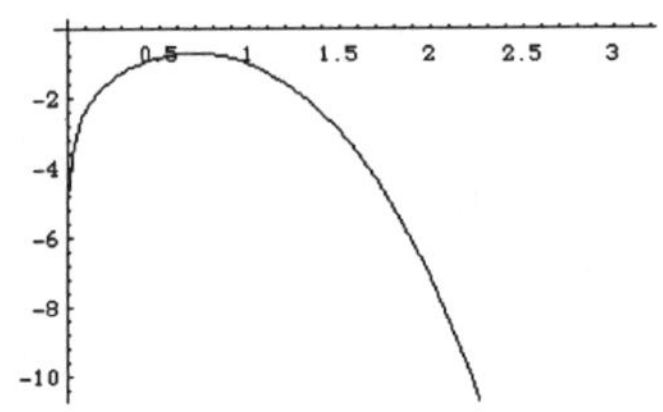

$f(x) = \ln(x) - x^3$

we see that this is a maximum. $f(x)$ has a maximum at $3^{-\frac{1}{3}} \approx 0.6933$.

(8) We see that $f'(x) = \ln(x) \cdot 2^x - 1 = 0$ when $x = \frac{\ln\left(\frac{1}{\ln(2)}\right)}{\ln(2)} \approx 0.5287$. From the graph below

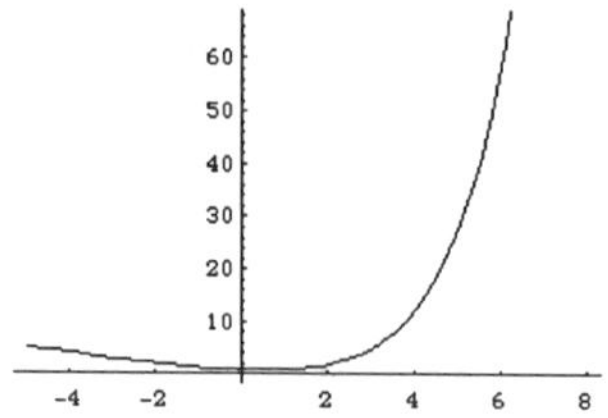

$f(x) = 2^x - x$

we see that this is a maximum.

Solutions for §9.2

(1) f has a critical point at $x = 0$ since $f'(x) = 6x = 0$ when $x = 0$. Further, $f'(x) < 0$ for $x < 0$ and $f'(x) > 0$ for $x > 0$. The first derivative test implies that f has a minimum value at $x = 0$.

(2) f has a critical point at $x = 0$ since $f'(x) = -14x = 0$ when $x = 0$. Further, $f'(x) > 0$ for $x < 0$ and $f'(x) < 0$ for $x > 0$. Thus f has a maximum value at $x = 0$.

(3) Here $f'(x) = 3x^2 - 18x + 30 = 3(x^2 - 6x + 10) = 0$ when $x = \frac{6 \pm \sqrt{-4}}{2}$ which is an imaginary number. Thus $f'(x)$ is never zero and f has no maximum or minimum value in this case.

(4) Here $f'(x) = (x + \frac{1}{2})(x - 2)x = 0$ when $-\frac{1}{2}, 0, 2$. We have: $f'(x) < 0$ for $x < -\frac{1}{2}$, $f'(x) > 0$ for $x > -\frac{1}{2}$ and hence $x = -\frac{1}{2}$ is a minimum; $f'(x) > 0$ for $-\frac{1}{2} < x < 0$, and $f'(x) < 0$ for $0 < x < 2$ hence $x = 0$ is a minimum; and $f'(x) < 0$ for $0 < x < 2$ and $f'(x) > 0$ for $2 < x$ hence $x = 2$ is a minimum.

(5) $f'(x) = 4x^2(x + \frac{3}{4})$ with a minimum at $-\frac{3}{4}$ and an inflection point at $x = 0$.

(6) f has no local maximum or minimum but a point of inflection at $x = 0$.

(7) $f'(x) = 12(x - 1)(x - 2)^2$ with a minimum at 1 and an inflection point at $x = 2$.

(8) f has a maximum at 0 and two minima at -2 and 3.

(9) $f'(x) = -1 - \sin(x)$ which is zero for $x = -\frac{\pi}{2} + 2\pi n$ for all integers n and these are inflection points.

(10) Since $f'(x) = 3\cos(3x) + 1 = 0$ when $x = \pm\frac{1}{3}\arccos(-\frac{1}{3}) + \frac{2\pi n}{3}$ hence the maxima is $x = +\frac{1}{3}\arccos\left(-\frac{1}{3}\right) + \frac{2\pi n}{3}$ and the minima at $x = -\frac{1}{3}\arccos\left(-\frac{1}{3}\right) + \frac{2\pi n}{3}$.

(11) f is decreasing for $x < -\frac{1}{4}$ and increasing for $x > -\frac{1}{4}$.

(12) f increases for negative x and decreases for positive x.

(13) This function is always increasing.

(14) f decreases for $x < -\frac{1}{2}$ and $0 < x < 2$ and increases for $-\frac{1}{2} < x < 0$ and $x > 2$.

(15), (16) $f(x)$ is always increasing since $7x^6 \geq 0$.

(17) This function is increasing because $f'(x) = e^x > 0$ for all x.

(18) f is increasing for $x > 0$ because $f'(x) = \frac{1}{x} > 0$ for $x > 0$.

(19) Sin(x) is increasing for $-\frac{\pi}{2} < x < \frac{\pi}{2}$ and decreasing for $\frac{\pi}{2} < x < \frac{3\pi}{6}$ and sin is periodic with period 2π.

(20) We see that $f'(x) = 1 - 2\sin(x) = 0$ for $x = \frac{\pi}{6}, \frac{5\pi}{6}$ plus $2\pi n$ and so $f(x)$ is decreasing for $\frac{\pi}{6} + 2\pi n < x < \frac{5\pi}{6} + 2\pi n$ and increasing for $\frac{5\pi}{6} + 2\pi n < x < \frac{\pi}{6} + 2\pi(n+1)$.

Solutions for §9.3

(1) Since $f'(x) = 8x$ and $f''(0) = 8 > 0$ we have a minimum at $x = 0$.

(2) Since $f'(x) = 2x + 6$ we have a critical point at $x = -3$. But $f''(-3) = 2 > 0$ so f has a minimum at $x = -3$.

(3) Since $f'(x) = 10x - 3$ we have a critical point at $x = \frac{3}{10}$. But $f'\left(\frac{3}{10}\right) = 10 > 0$, and so $x = \frac{3}{10}$ is a minimum.

(4) $f'(x) = 12(x-1)(x+2)(x+3)$ and $f''(1) = 144$, $f''(-2) = -36$, $f''(-3) = 48$ implies minima at $x = -3, 1$ and a maximum at $x = -2$.

(5) This function has no critical points.

(6) Here f has a maximum at $x = -\frac{4}{3}$ and a minimum at $x = 0$.

(7) $f'(x) = 0$ when $x = \frac{2\pi}{3}, \frac{4\pi}{3}$ $(+2\pi n)$ and $f''\left(\frac{2\pi}{3}\right) = -\sqrt{3} < 0$ and $f''\left(\frac{4\pi}{3}\right) = \sqrt{3} > 0$.

(8) This function has no maxima or minima because $f'(x) = \sqrt{3} - \sin(x) > 0$ for all x.

(9) Here $f'(x) = x^{-1} + 2x > 0$ for all $x \Rightarrow$ no local maxima or minima.

(10) $f'(x) = e^x - 5 = 0$ only when $x = \ln(5)$ and this is a minimum.

(11) The derivative is zero at $x = \frac{\ln\left(\frac{3}{\ln(2)}\right)}{\ln(2)}$ and this is a minimum.

(12), (13), (14) These functions have no local maxima or minima.

(15) This function has a local maximum at $x = \frac{271}{90}$ and a minimum at $x = \frac{136}{45}$.

(16) f has two minima at $x = 0, 1.06$ and a maximum at $x = 1.04$.

(17) This function has a local minimum at $x = 2$ and an inflection point at $x = \frac{101}{50}$.

Solutions for Additional Exercises for Chapter 9

(1) Here $f(x) = x^2 + px + q$. If $f(1) = 3$, then $3 = 1 + p + q$ and so $p + q = 2$. If f has a local maximum or minimum on the interval $0 \leq x \leq 3$ then $f'(x) = 0$ on this interval, i.e., $f'(x) = 2x + p = 0$. But then $-6 \leq p \leq 0$. This value will always be a minimum since $f''(x) = 2 > 0$.

(2) The function is periodic with period 12π. Let $\theta_1 = \arccos\left(\frac{-1+\sqrt{17}}{8}\right) \approx 1.1697$. Let $\theta_2 = \arccos\left(\frac{-1-\sqrt{17}}{8}\right) \approx 2.2658$ All maximum values are achieved at points x of the form $12n\pi + 6\theta_1$ or $12n\pi - 6\theta_1$, where n is an arbitrary integer. All minimum values are achieved at points x of the form $12n\pi$, $12n\pi + 6\theta_2$, or $12n\pi - 6\theta_2$, where n is an arbitrary integer.

(3a) From the graph

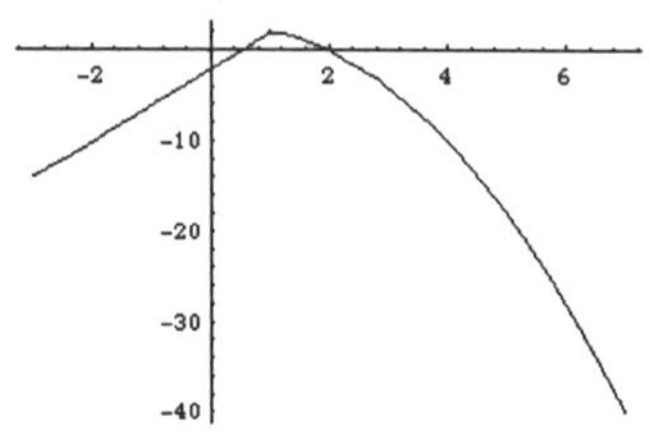

$$f(x) = \begin{cases} 4x - 2, & \text{if } x < 1 \\ -(x-2)(x+1), & \text{otherwise} \end{cases}$$

we see that the maximum value of $f(x)$ for $0 \leq x \leq 6$ is $f(1) = 2$.

(3b) The first and second derivative tests apply only to those functions having continuous second derivatives on all of their domain.

(3c) To find extreme values you must always check endpoints of intervals and break points of piecewise defined functions.

(4) We have

$$f'(x) = \begin{cases} 3, & \text{if } x < 2 \\ 2x - 6, & \text{if } 2 \leq x < 4 \\ -1, & \text{otherwise} \end{cases}$$

The only critical point of f lies at $x = 3$. But $f(3) = 4$ and at the endpoints, $f(2) = f(4) = 5$, $f(0) = -1$. So the maximum of f is 5 (which is achieved at $x = 2$ and $x = 4$). The Minimum is -1 which is achieved at $x = 0$.

(5) $F(x) = 25 \sec x$. Therefore $F'(x) = 25 \sec(x) \tan(x)$. So $F'(0) = 0$. Moreover, on $\left(-\frac{\pi}{2}, \frac{\pi}{2}\right)$ we have $\cos(x) > 0$ so that $\sec(x) + \tan(x) = \frac{1+\sin(x)}{\cos(x)} > 0$. So $F''(x) = 25 \sec^2(x)(\sec(x) + \tan(x)) > 0$. It follows that $x = 0$ is a minimum for F and F has no maximum on $\left(-\frac{\pi}{2}, \frac{\pi}{2}\right)$.

(6) From the graph below

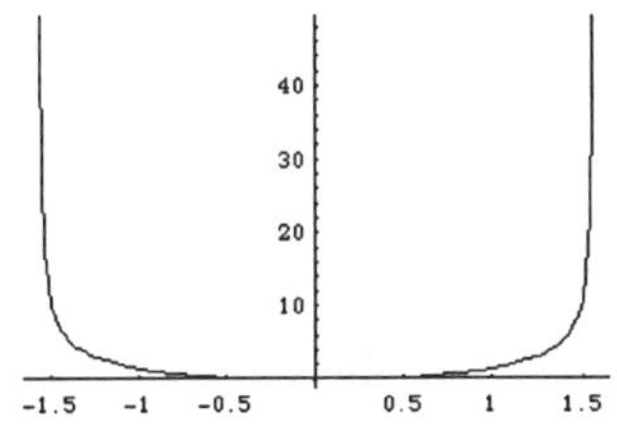

$$f(x) = \sin(x^2) \cdot \sec(x)$$

we see that $f(0) = 0$ is the minimum value and $f(x)$ goes to infinity as $x \to \pm\frac{\pi}{2}$. With our CAS, we compute $f'(0) = 0,\ f''(0) = 2$ which confirms that $x = 0$ is a minimum.

(7) Assume that $f(a) = f(b)$ and $f'(x) \neq 0$ for every x between a and b. For $a \leq x \leq b$, the function $f(x)$ will achieve maximum and minimum values somewhere, because it is continuous on this interval. If its maximum or minimum value is achieved at some c strictly between a and b, then we have that $f'(c) = 0$, which contradicts our assumption. We conclude that the maximum and minimum values must be achieved at the endpoints. But $f(a) = f(b)$ so that we must conclude that f is constant on the interval, and so has zero derivative. But this also is a contradiction of our assumption. Therefore there must be some point c between a and b such that $f'(c) = 0$.

(8) Let $h(x) = f(x) - \frac{f(b)-f(a)}{b-a}(x-a)$. So $h(a) = f(a) - \frac{f(b)-f(a)}{b-a}(a-a) = f(a)$. Consequently, $h(b) = f(b) - \frac{f(b)-f(a)}{b-a}(b-a) = f(a)$. By Rolle's Theorem there must be some c between a and b such that $h'(c) = 0$. But $h'(x) = f'(x) - \frac{f(b)-f(a)}{b-a}$. So we can conclude that $f'(c) = \frac{f(b)-f(a)}{b-a}$, which is what we want to show.

(9) Suppose $f(x)$ is not constant for $a \leq x \leq b$. So there are values c and d with $a < c < d < b$ and $f(c) \neq f(d)$. By the Mean Value Theorem there is some point e between c and d with $f'(e) = \frac{f(d)-f(c)}{d-c}$. So $f'(e) \neq 0$, which is a contradiction.

(10) There are no local maxima or minima.

(11) There is a local maximum at $x = -1$ and a local minimum at $x = 1$.

(12) Local maxima are at points $\frac{\pi}{4} + 2\pi n$, where $n = 0, 1, 2 \ldots$. Local minima are at points $\frac{5\pi}{4} + 2\pi n$, where $n = 0, 1, 2 \ldots$.

(13) Some extreme points are $x = 0$ (maximum), $x \approx 1.8365$ (minimum), and $x \approx 4.8157$ (maximum).

(14) The point $x = 0$ is the only local minimum. There are no local maxima.

(15) There exists no such function. If $f''(x) < 0$ everywhere then in particular f has second derivative everywhere. If $x = 0$ is a minimum then we would have $f''(0) > 0$.

Solutions for §10.1

(1) We have $f(x) = x^4 - 2x^2 + 3 \Rightarrow f'(x) = 4(x-1)x(x+1)$. The zeros of $f'(x)$ lie between -5 and 4 so, computing $f(-5) = 578, f(-1) = 2, f(0) = 3, f(1) = 2$ and $f(4) = 227$ we deduce that 578 and 2 are the maximum and minimum values attained by f.

(2) $f(x) = x^4 - 32x - 148$ and $f'(x) = 4(x-2)(4+2x+x^2) = 0$ when $x = 2$. Thus $x = 2$ is the critical point and $x = 0, 3$ are end points. When we evaluate we see that $f(0) = 148,\ f(2) = 100$, and $f(3) = 133$ and hence, 148 is the maximum, 100 the minimum.

(3) With $f(x) = 2x^3 + 3x^2 - 36x + 11$ we get $f'(x) = 6(x-2)(x+3) = 0 \Rightarrow x = 2$ or -3. But $-3 < -2$ so evaluating $f(-2)$ and $f(2)$ we find a maximum of 79 and a minimum of -33.

(4) Here $f(x) = x^7 - 5x^5 + 13$ and thus, $f'(x) = 7x^6 - 25x^4 = x^4(7x^2 - 25) = 0$ when $x = 0, \pm\frac{5}{\sqrt{7}}$. We compute $f(-3) = -959,\ f(0) = 13,\ f(-\frac{5}{\sqrt{7}}) = 47.4355\ldots,\ \ f(\frac{5}{\sqrt{7}}) = -21.4355\ldots$, and $f(3) = 985$ to obtain a maximum of 985 and a minimum of -959.

(5) $f(x) = \cos^2(x) \Rightarrow f'(x) = -2\cos(x)\sin(x)$. Using the fact that $\cos(x) = 0$ for $x = \frac{\pi}{2} + \pi n$ for all integers n and $\sin(x) = 0$ for $x = \pi n$ we see that when $-\pi \le x \le \pi$, $f'(x) = 0$ for $x = -\pi, -\frac{\pi}{2}, 0, \frac{\pi}{2}, \pi$ giving us maxima and minima of 1 and 0.

(6) Letting $f(x) = \tan(x) \Rightarrow f'(x) = \sec^2(x)$. This never equals zero so it remains to compute $f(-\frac{\pi}{4}) = -1$ and $f(\frac{\pi}{4}) = 1$ giving the minimum and maximum.

(7) When $f(x) = e^{-x} + e^x \Rightarrow f'(x) = -e^{-x} + e^x$ and this is only zero for $x = 0$. So we get a minimum at $f(0) = 2$ and a maximum at $f(6) \approx 403$.

(8) Letting $f(x) = x^2 - \ln(1+x^2) \Rightarrow f'(x) = \frac{2x^3}{1+x^2}$ so $x = 0$ is the only critical point. Computing $f(-1) = f(1) = 1 - \ln(2) \approx 0.3, f(0) = 0$ we have a minimum of 0 and a maximum of $1 - \ln(2)$.

(9) For $f(x) = \cos(2x) + 2\cos(x) \Rightarrow f'(x) = -2\sin(x) - 2\sin(2x)$. Now $\sin(2x) = 2\cos(x)\sin(x)$ so $f'(x) = 0 \Rightarrow \sin(x) = 0$ or $\cos(x) = \frac{-1}{2}$. So we evaluate $f(0), f(\frac{2\pi}{3}), f(\pi), f(\frac{4\pi}{3}), f(2\pi)$ to find a maximum of 3 and a minimum of $-1\frac{1}{2}$.

(10) For $f(x) = \sin(2x) + 2\sin(x) \Rightarrow f'(x) = 2\cos(x) + 2\cos(2x)$. Using $\cos(2x) = 2\cos^2(x) - 1$ to solve $f'(x) = 0$, we see $\cos(x) = -1$ or $\frac{1}{2}$ i.e., in the considered range the zeros are $\{\frac{-5\pi}{3}, -\pi, \frac{-\pi}{3}, \frac{\pi}{3}, \pi, \frac{5\pi}{3}\}$ giving a maximum of $\frac{3\sqrt{3}}{2}$ and a minimum of $\frac{-3\sqrt{3}}{2}$.

(11) Solving for y as a function of x, we find $y = x - \ln(\sin(x))$. Thus $y' = 1 - \cot(x)$ which is zero only when $x = \frac{\pi}{4}$. Computing $f(\frac{\pi}{4}) = 1.13197\ldots,\ \ f(\frac{3\pi}{4}) = 2.70277\ldots$, we find maxima and minima of ≈ 2.7 and 1.13.

Solutions for §10.2

(1) Call the two quantities x and y. Then $x + y = 1000$ and for both terms to remain positive requires that $0 \le x, y \le 1000$. Their product $xy = x(1000 - x) = P(x)$ and $P'(x) = 0 \Leftrightarrow x = 500$. Comparing $P(0) = 0, P(500) = 500^2$ and $P(1000) = 0$ we see that the product is maximized when $x = 500, (y = 500)$ with maximum value 500^2 and minimized when x or y are zero.

(2) Our unknowns are x, y and $y = \frac{1000}{x}$. Then $S(x,y) = x+y = x+\frac{1000}{x} = S(x)$ and $S'(x) = 1-\frac{1000}{x^2} = 0 \Rightarrow x = \sqrt{1000}$ because $x > 0$. Since in this case we are not dealing with x inside a closed interval $a \leq x \leq b$ but instead $x > 0$ it remains to test whether we've found a maximum or a minimum. We see that $S''(\sqrt{1000}) > 0$ so we have found a minimum and it occurs when $S(x,y) = 2\sqrt{1000}$.

(3) Let the dimensions of the base of the closet be $x \times y$, width by depth. Only 3 sides must be built since the wall, ceiling, and floor are already in place. The surface area of the closet is $700 = 14x + 7y$ (with $0 \leq x \leq 50$) $\implies y = 100 - 2x$. The area of the base is $B(x,y) = xy = 100x - 2x^2 = B(x)$. Then $B'(x) = 100 - 4x = 0$ for $x = 25$. When we compute $B(0) = 0, B(50) = 0$, and $B(25) = 1250$, we see that the base has maximum area when it has dimensions 25 by 25.

(4) The distance between (x,y) and $(0,18)$ is $\sqrt{x^2+(y-18)^2} = D(x,y)$. But $D(x,y) = D(x, 4x+1) = \sqrt{x^2+(4x-17)^2}$ and we'll call this just $D(x)$. Here x is free to take all real values.

$$D'(x) = \frac{2x + 2(4x-17)4}{2\sqrt{x^2+(4x-17)^2}} = 0 \Leftrightarrow 2x + 2(4x-17)4 = 0 \Leftrightarrow x = 4.$$

We see (for example from the first derivative test) that this is a minimum and the closest the line comes to the point (0,18) is at the point (4,17).

(5) Letting $D(x) = \sqrt{(u-x)^2+(v-mx-b)^2}$ we compute

$$D'(x) = \frac{-2(u-x) - 2m(v-mx-b)}{2D(x)}.$$

This is zero only when $u - x + m(v - mx - b) = 0$ i.e., when

$$(*) \qquad x = \frac{u+m(v-b)}{m^2+1} \quad \text{and so} \quad y = \frac{b+m(u+mv)}{m^2+1}$$

The line l' has equation $y - v = -\frac{1}{m}(x-u)$ and the coordinates of the intersection of l and l' will be the solution of the system $y - mx = b$ and $my + x = mv + u$ which is indeed $(*)$.

(6) As in question (5) we let $D(x,y) = \sqrt{(x-3)^2+(y-1)^2} = D(x, 1-x^2) = \sqrt{(x-3)^2+(-x^2)^2} = D(x)$. $D'(x) = \frac{4x^3+2x-6}{2D(x)} = 0 \Leftrightarrow x = 1$ (x being real). Computing $D''(1)$ we see that we've found a minimum. The closest point being (1,0).

(7) Since the line l passes through (a,b) we have $y = m(x-a)+b$ so that l is defined in terms of the slope m only. Setting $y = 0$ we see that l intercepts the x axis at $x = a - \frac{b}{m}$ and similarly the y axis at $y = b - ma$. The area of the triangle $A = A(m)$ is $\frac{1}{2}(a-\frac{b}{m})(b-ma)$. $A'(m) = \frac{1}{2}(\frac{b^2}{m^2} - a^2) = \frac{1}{2}(\frac{b}{m}+a)(\frac{b}{m}-a)$ and this is zero for $m = -\frac{b}{a}$ or $m = \frac{b}{a}$. The only condition that we've imposed on m is that $m < 0$, (to ensure l meets the positive x and y axes). The equation of the line l with $m = -\frac{b}{a}$ is $y = -\frac{b}{a}x + 2b$ and an application of the second derivative test gives us that this is indeed the line minimizing A.

(8) Assume the triangle has side length L. Let x denote the length of the side of the rectangle contained in one of these sides and y the other then its area $A(x,y) = xy$. To find y in terms of x we first note that the height of this triangle (using Pythagoras) is $\frac{\sqrt{3}L}{2}$. By noticing that the left half of the equilateral and the triangular section to the left of our rectangle are similar we have, by comparing corresponding lengths that $y = \frac{\sqrt{3}}{2}(L-x)$. This implies $A(x) = \frac{\sqrt{3}}{2}x(L-x)$ and that $A'(x) = 0$ for $x = \frac{L}{2}$. Because $0 \le x \le L$ we must compute $A(0)$, $A(\frac{L}{2})$ and $A(L)$ and we conclude that the area of the largest rectangle must be $\frac{\sqrt{3}L^2}{8}$.

(9) We wish to maximize $A(h)$ with the condition that $7000 \le h \le 30000$. Solving

$$A'(h) = e^{\frac{-h}{2000}}\left(h\left(2 - \frac{-h}{2000}\right)\right) = 0$$

we see that h can only be 0 or 4000 since $e^{\frac{-h}{2000}}$ is never zero. Both of these numbers lie outside the range we're considering so it remains to compute $A(7000) \approx 10^6$ and $A(30000) \approx 275$. Therefore the most data is obtained at 7000 feet.

(10) Let the isosceles triangle have base of length x and height h. Then $\frac{1}{2}xh = 1000$ or $h = \frac{2000}{x}$. The roofs surface area $R(x) = 40 \times 2y$ where y is the length of each side of the triangle. By Pythagoras $y = \sqrt{h^2 + x^2}$ so

$$R(x) = 80\sqrt{\frac{2000^2}{x^2} + x^2} \text{ and } R'(x) = \frac{80}{R(x)}\left(x - \frac{2000^2}{x^3}\right).$$

Then $R'(x) = 0$ implies $x = \pm 20\sqrt{5}$. Because $x > 0$ we conclude that the triangular side of the roof must have base $20\sqrt{5}$ and side $20\sqrt{10}$.

(11) Denote the distance from Q to R by x. The distance from P to R is $\sqrt{x^2 + 3^2}$ and R to C measures $5 - x$. We know that the time taken is the distance traveled divided by your speed so the time taken by the girl is

$$T(x) = \frac{\sqrt{x^2+9}}{1.5} + \frac{5-x}{4} \text{ and } T'(x) = \frac{2x}{3\sqrt{x^2+9}} - \frac{1}{4}.$$

$T'(x) = 0 \Rightarrow x = \pm\sqrt{\frac{81}{53}}$. Clearly $0 \le x \le 5$ so calculating $T(0), T\left(\sqrt{\frac{81}{53}}\right)$ and $T(5)$ gives that the minimum time taken is when she swims to a point R, $\frac{9}{\sqrt{53}}$ from Q.

(12a) Let x be the distance from the fence to where the wire meets the ground and suppose h is the height above the ground the wire is attached to the warehouse.

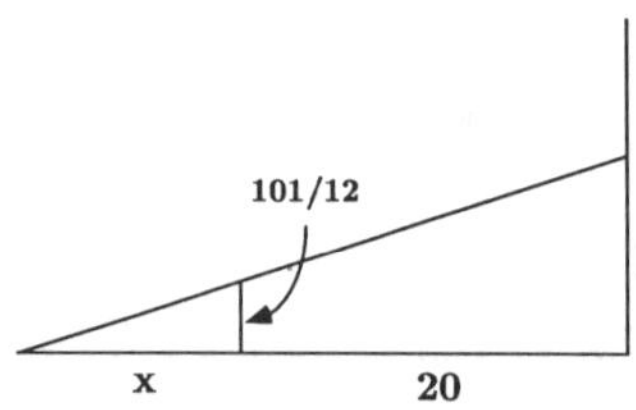

We see (using the fact that similar triangles have proportional corresponding sides) that $\frac{8\frac{5}{12}}{x} = \frac{h}{x+20}$ (*). Then

$$l(x) = \sqrt{\left(\frac{8\frac{5}{12}(x+20)}{x}\right)^2 + (x+20)^2}$$

where $l(x)$ is the length of the FBI wire. Therefore

$$l'(x) = \frac{6(40 - \frac{510050}{9x^3} - \frac{51005}{18x^2} + 2x)}{\sqrt{\frac{(20+x)^2(10201+144x^2)}{x^2}}}.$$

Using our CAS we see that the only $x > 0$ for which $l'(x) = 0$ is $x \approx 11.2$ and this is a minimum.

(12b) The condition $h \leq 20 \Rightarrow$ by (*) $x \geq \frac{400}{20-8\frac{5}{12}} = k \approx 36$. We see from (12a) that $l(x)$ is increasing for $x \geq 11.2$. Therefore $x = k$ minimizes $l(x)$ for $x \geq k$.

(13) If the function $f(x) = 50 - x^3 - 3x^2 + 2x$ describes the height of a mountain then $f'(x)$ describes the slope at the point $(x, f(x))$. We're looking for the point where the slope is greatest and this will be where $f''(x) = 0$, i.e., $-6x - 6 = 0$. So the slope is steepest at $x = -1$, (this is a maximum because $f'''(1) = -6 < 0$).

(14) We want to minimize the posynomial $p(x) = a_0 + a_1x^{r_1} + a_2x^{r_2}$ where the $a_i > 0$. If both $r_1, r_2 > 0$, then the minimal value of $p(x)$ occurs at the endpoint $x = 0$. If both $r_1, r_2 < 0$, the the minimum value occurs at $x = \infty$. Otherwise: $p'(x) = r_1a_1x^{r_1-1} + r_2a_2x^{r_2-1}$. To find the zeros of $p'(x)$ we next solve $r_1a_1x^{r_1-1} = -r_2a_2x^{r_2-1}$. If $r_2a_2 \neq 0$ we obtain:

$$-\frac{r_1a_1}{r_2a_2} = x^{r_2-r_1} \quad \text{or equivalently} \quad x = \left(-\frac{r_1a_1}{r_2a_2}\right)^{\frac{1}{r_2-r_1}}.$$

and this is the x for which $p(x)$ attains it's minimum value.

(15) Let the length of the side of the rectangle be equal to the diameter of the semicircle be x, the other side y. The perimeter is then $\frac{2\pi\frac{x}{2}}{2} + 2y + x = 4$. The surface area

$$A(x,y) = A(x) = x^2(-\frac{\pi}{8} - \frac{1}{2}) + 2x \quad \text{and} \quad A'(x) = 2x\left(-\frac{\pi}{8} - \frac{1}{2}\right) + 2.$$

$A'(x) = 0$ when $x = \frac{8}{\pi+4}$. also $x, y \geq 0 \Rightarrow 0 \leq x \leq \frac{8}{\pi+2}$. On evaluating $A(0), A(\frac{8}{\pi+4})$ and $A(\frac{8}{\pi+2})$ we find that the pendant with maximum area has $x = \frac{8}{\pi+4}$ and $y = \frac{4}{\pi+4}$.

(16) The two pieces of the thread are of length x and $10 - x$ with $0 \leq x \leq 10$. If R is the radius of the circle, we have $2\pi R = 10 - x \implies R = \frac{10-x}{2\pi}$. The areas of the square and circle are $(\frac{x}{4})^2$ and $\frac{(10-x)^2}{4\pi}$ respectively.

$$A(x) = \frac{x^2}{16} + \frac{(10-x)^2}{4\pi} \quad \text{and} \quad A'(x) = \frac{x}{8} - \frac{(10-x)}{2\pi}.$$

$x = \frac{40}{\pi+4}$ is the only solution of $A'(x) = 0$. Clearly $A(\frac{40}{\pi+4}) = \text{Max}\Big\{A(0),\ A(\frac{40}{\pi+4}),\ A(10)\Big\}$ implying that to maximize the area of the two shapes the square should have side $\frac{10}{\pi+4}$ and the circle a diameter of $\frac{10}{\pi+4}$.

(17) Suppose A and B are distances m and n from the x–axis and the horizontal distance between A and B is d. To make things clearer we may take A to be the point $(0, m)$ and B to be (d, n). Let Q be a point on the x-axis, $Q = (x, 0)$. Let $D(x)$ be the sum of the distances AQ plus BQ, then $D(x) = \sqrt{m^2 + x^2} + \sqrt{n^2 + (b-x)^2}$. Solving $D'(x) = 0$ we find that $x = \frac{dm}{m-n}$ or $\frac{dm}{m+n}$. We put these two values into D and get:

$$D\left(\frac{dm}{m-n}\right) = \sqrt{m^2 + \frac{dm^2}{(m-n)^2}} + \sqrt{n^2 + \frac{d^2m^2}{(m-n)^2}}, \tag{1}$$

(2) $$D\left(\frac{dm}{m+n}\right)=\sqrt{m^2+\frac{dm^2}{(m+n)^2}}+\sqrt{n^2+\frac{d^2m^2}{(m+n)^2}}.$$

And we notice that (2) is smaller than (1) because $m,n>0\Rightarrow \frac{1}{(m+n)^2}\le\frac{1}{(m-n)^2}$. Therefore $Q=P=(\frac{dm}{m+n},0)$ is the point we want minimizing $AQ+BQ$. To prove that the angles are equal we see that,

$$\tan\alpha=\frac{m}{\frac{dm}{m+n}}=\frac{m+n}{d}=\frac{n}{d-\frac{dm}{m+n}}=\tan\beta$$

which implies $\alpha=\beta$.

(18) Let $S(x)=(x-a_1)^2+....+(x-a_n)^2$. Then $S'(x)=2(x-a_1)+....+2(x-a_n)$ and

$$S'(x)=0\Leftrightarrow x=\frac{a_1+a_2+\cdots+a_n}{n}\text{ the average of the }a_i.$$

This is the x that minimizes the sum because $S''(x)=2n>0$.

Solutions for §10.3

(1) Let $x=$ length of the side of the square base. Let $y=$height of the box. The volume of the box $V=x\cdot x\cdot y$ and the surface area is $32=x^2+4xy\Rightarrow y=\frac{32-x^2}{4x}$. So $V(x)=\frac{1}{4}x(32-x^2)$ and $V'(x)=0\Rightarrow x=\sqrt{\frac{32}{3}}$. Also $x,y\ge 0\Rightarrow 0\le x\le\sqrt{32}$. $V(0)=V(\sqrt{32})=0\Rightarrow V(\sqrt{\frac{32}{3}})=\frac{64}{3}\sqrt{\frac{2}{3}}$ is the maximum volume.

(2) The can has radius r and height $h\Rightarrow h=\frac{V}{\pi r^2}$ where V is the (fixed) volume. The metal needed is proportional to the surface area of the can $A(r)=\frac{2V}{r}+2\pi r^2$. Then $A'(r)=0\Rightarrow r=\sqrt[3]{\frac{V}{2\pi}}$ which yields the minimum surface area by the second derivative test. The dimensions are $r=\sqrt[3]{\frac{V}{2\pi}}$, $h=\sqrt[3]{\frac{V}{\pi}}$.

(3) We assume that the cylinder C, (with radius x and height $2h$) is carved symmetrically out of the bead $\Rightarrow r^2=h^2+x^2, 0\le x\le r$. Then C has surface area $A(x)=4\pi x\sqrt{r^2-x^2}$ and $A'(x)=0\Rightarrow x=\frac{r}{\sqrt{2}}$ giving a (maximum) area of $2\pi r^2$.

(4) If the cone has height h and radius r then $h^2+r^2=R^2, 0\le h\le R$. Therefore the volume $V(h)=\frac{1}{3}\pi h(R^2-h^2)$ and $V'(h)=0\Rightarrow h=\frac{R}{\sqrt{3}}$. We have $V(0)=V(R)=0\Rightarrow V(\frac{R}{\sqrt{3}})=\frac{2\pi R^3}{9\sqrt{3}}$ is the largest possible volume.

(5) If the dimensions of the starter are $x\times x\times y$ then $y=\frac{50}{x^2}$. The area of plastic needed, $A(x)=x^2+\frac{8\cdot 50}{x}$ and $A'(x)=0\Rightarrow x=\sqrt[3]{200}$ and this minimizes $A(x)$. Here $y=\frac{50}{(200)^{\frac{2}{3}}}$.

(6) Assume the cylindrical interior of the silo has radius $r>0$ and height h, (the exterior has radius $r+4$ and height $h+8$). Then the cost $C(r,h)=20(h\pi r^2+2\pi r^2)+35((h+8)(\pi(r+4)^2)+2\pi(r+4)^2)$ and $h=\frac{120}{\pi r^2}$.

$$\text{So } C'(r)=\frac{20(56+11r)(\pi r^3-60)}{r^3}=0\Rightarrow r=\sqrt[3]{\frac{60}{\pi}}\approx 2.7,$$

giving $h\approx 14$ feet.

(7) The volume of the duct $V(\theta) = 36L(\sin(\theta)(\cos(\theta) + 1)$ and $V'(\theta) = 6L(3\cos(\theta) - 2)(4\cos(\theta) + 3) = 0 \Rightarrow \cos(\theta) = \frac{2}{3}$ or $\frac{-3}{4}$. We assume $0 \leq \theta \leq \frac{\pi}{2}$ so let $\alpha = \cos^{-1}(\frac{2}{3})$. Then calculating $V(0) = 0, V(\alpha) = 27\sqrt{3}L \approx 47L$ and $V(\frac{\pi}{2}) = 36L$ we see that angle α maximizes V.

(8) The volume of the bin, (radius r, height h) is $64 = \frac{2}{3}\pi r^3 + h\pi r^2$ and the cost of producing it $C(r,h)$ is proportional to $3.2\pi r^2 + 2\pi rh = \frac{128}{r} + \frac{14\pi r^2}{3} = C(r)$ and $C'(r) = 0$ for r real$\Rightarrow r = 2\sqrt[3]{\frac{12}{7\pi}} \approx 1.6$. This is the minimum cost since $C''(2\sqrt[3]{\frac{12}{7\pi}}) = 28\pi > 0$.

Solutions for §10.4

(1a) Solving $6000a + b = 350$ and $8000a + b = 300$ we obtain $a = \frac{-1}{40}, b = 500$. Thus the demand function is $p(x) = 500 - \frac{x}{40}$.

(1b) The profit function $P(x) = x \cdot p(x) - 50x$ and $P'(x) = 450 - \frac{x}{20} = 0 \Rightarrow x = 9000$. Therefore to maximize profits the company should charge $p(9000) = \$275$ per card.

(2) The demand function $p(x) = 40,000 - 15x$. The profit function $P(x) = 5x(5600 - 3x)$ and $P'(x) = 0 \Leftrightarrow x = 933\frac{1}{3}$. The optimum price is then $p(933\frac{1}{3}) = \$26,000$.

(3a) $P(x) = x(5 - \frac{x}{40}) - 110 - x$ and $P'(x) = 4 - \frac{x}{20} = 0 \Leftrightarrow x = 80$, i.e., they should produce 80 gallons a day and sell it at $p(80) = \$3$ per gallon. The daily profit will be $P(80) = \$50$ per day.

(3b) The optimum price is still \$3 per gallon and production should remain at 80 gallons a day. The daily profit will now be \$45 per day.

(4) $P(x) = x \cdot p(x) - \frac{1}{4}x - 50$. $P'(x) = 0 \Leftrightarrow x = 300$. So they should charge $p(300) = \$1.20$ per muffin and make a profit of $P(300) = \$235$ a day.

(5a) $P(x) = x \cdot p(x) - 5x - 10,000$ and $P'(x) = 75 - \frac{2x}{25}$. So $P'(\frac{25.75}{2}) = 0$ but $\frac{25.75}{2} \approx 937 < 1000$. $P(1000) = 25,000$ and $P(3000) = -145,000$ implying that the profit is maximized (for $1000 \leq x \leq 3000$) when $x = 1000$ and the company should charge \$40 per unit.

(5b) Four months since $P(1000) = \$25,000$.

(6a) If x telephones are produced the average cost per telephone $= \frac{2000\sqrt{x}}{x} = \frac{2000}{\sqrt{x}}$ which is a decreasing function.

(6b) $P(x) = x \cdot p(x) - 2000\sqrt{x} \Rightarrow P'(x) = \frac{-x}{10} + 200 - \frac{1000}{\sqrt{x}}$ and profit is maximized when production, x, satisfies $P'(x) = 0$.

(6c) Plotting the function $P'(x)$ we see the graph crossing the x-axis near $x = 26$.

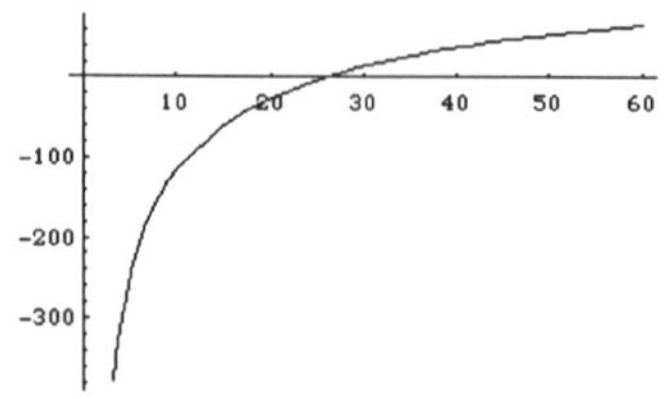

$$p'(x) = \frac{-x}{10} + 200 - \frac{1000}{\sqrt{x}}$$

In fact $P'(25) \approx -2.5 < 0$ and $P'(26) \approx 1.2 > 0$ so this function does have a zero between 25 and 26 and it's closer to 26.

Solutions for Additional Exercises for Chapter 10

(1) Step 1. Critical points: $f'(x) = \frac{x+1}{(2x+1)^{\frac{3}{2}}} = 0$ when $x = -1$.

Step 2. $f(0) = 0$; $f(10) = \frac{10}{\sqrt{21}}$.

Step 3. $f(0) =$ minimum; $f(10) =$ maximum.

(2) Step 1. Critical points: $f'(x) = \frac{1+2\cos(x)}{(2+\cos(x))^2} = 0$ when $x = \frac{2\pi}{3}, \frac{4\pi}{3}$.

Step 2. $f(0) = 0$; $f\left(\frac{2\pi}{3}\right) = \frac{\sqrt{3}}{3}$; $f\left(\frac{4\pi}{3}\right) = -\frac{\sqrt{3}}{3}$; $f(2\pi) = 0$.

Step 3. $f\left(\frac{2\pi}{3}\right) =$ minimum; $f\left(\frac{4\pi}{3}\right) =$ maximum.

(3) Step 1. Critical points: $f'(x) = (\sec(x))^2 \cdot (2\sin(x) - 1) = 0$ when $x = \frac{\pi}{6}$.

Step 2. $f(0) = 2$; $f\left(\frac{\pi}{6}\right) = \sqrt{3}$; $f\left(\frac{\pi}{4}\right) = 2\sqrt{2} - 1$.

Step 3. $f(0) =$ minimum; $f\left(\frac{\pi}{6}\right) =$ maximum.

(4) Step 1. Critical points: Since

$$f'(x) = \begin{cases} x, & \text{if } x < 1 \\ 2x - 5, & \text{if } x \geq 1 \end{cases}$$

we have that $f'(x) = 0$ when $x = \frac{5}{2}$.

Step 2. $f\left(\frac{1}{2}\right) = 0$; $f(1) = 2$; $f\left(\frac{5}{2}\right) = -\frac{1}{4}$; $f(4) = 2$.

Step 3. $f\left(\frac{5}{2}\right) =$ minimum; $f(1)$ and $f(4) =$ maxima.

(5) Step 1. Critical points: Here

$$f'(x) = \begin{cases} -1, & \text{if } x \leq -2 \\ -2x, & \text{if } -2 < x \leq 2 \\ 1, & \text{if } x \geq 2 \end{cases}$$

and we have that $f'(x) = 0$ when $x = 0$.

Step 2. $f(-4) = 2$; $f(-2) = 0$; $f(0) = 4$; $f(2) = 0$. $f(4) = 2$.

Step 3. $f(0) =$ maximum; $f(-2)$ and $f(2) =$ minima.

(6) In step two, we must also compute the function values of break points of piece-wise defined functions.

(7a) We must minimize the function $f(x) = x + \frac{1}{x}$ on the interval $\frac{1}{4} \leq x \leq \frac{3}{4}$. Now $f'(x) = 1 - \frac{1}{x^2} = 0$ when $x = 1$ (which is outside our interval), and since $f\left(\frac{1}{4}\right) = 4\frac{1}{4}$, and $f\left(\frac{1}{4}\right) = \frac{4}{3} + \frac{3}{4} = \frac{25}{12}$ we see that $x = \frac{3}{4}$ is the required solution.

(7b) $\frac{1}{4}$

(8) Side lengths of inscribed triangle with maximum area are $10\sqrt{2}, 10\sqrt{2}, 20$.

(9) Let the right circular cone have fixed base radius R and fixed height H. Let the inscribed right circular cylinder have variable base radius r and variable height h. Then $\text{Vol}_{\text{cone}} = \frac{1}{3}\pi R^2 H$ and $\text{Vol}_{\text{cylinder}} = \pi r^2 h$. By comparing similar triangles, we have the equation $\frac{H-h}{r} = \frac{H}{R}$. So in terms of the variable height h, the volume of the cylinder is $\text{Vol}_{\text{cylinder}} = V(h) = \pi\left(\frac{R}{H}(H-h)\right)^2 h = \frac{\pi R^2}{H^2}(H-h)^2 h^2, 0 \leqslant h \leqslant H$. Then $V'(h) = 0$ if and only if $H - h = h$. So the inscribed cylinder of greatest volume has dimensions $h = \frac{H}{2}$ and $r = \frac{R}{H}\left(H - \frac{H}{2}\right) = \frac{R}{2}$ Therefore $\text{Vol}_{\text{cylinder}} = \pi\left(\frac{R}{2}\right)^2\left(\frac{H}{2}\right) = \frac{3}{8} \cdot \frac{1}{3}\pi R^2 H = \frac{3}{8} \cdot \text{Vol}_{\text{cone}}$.

(10) Radius of semicircle to maximize window area $= \frac{p}{\pi+4}$.

(11) Length of sides of squares to cut to maximize volume $= \frac{L}{6}$.

(12) Time to sell crop to maximize profit $= 10$ weeks.

(13) $C'(x)$ is called the marginal cost function because it tells how production costs change with small changes in product output.

(14) $C'(300) = 39.6$

(15) Cost is maximized at $x = 5250$.

(16) Maximum length of pipe $= 4\sqrt[3]{2}$ feet.

(17) Greatest value of abscissa $= 4$. Least value of abscissa $= -4$.

(18) Dimensions of strongest rectangular beam are $6\sqrt{3}$ inches (base) and $6\sqrt{6}$ inches (depth).

(19) Dimensions of largest inscribed rectangle are $6\sqrt{3}$ (width) and $50\sqrt{5}$ (height).

Solutions for §11.1

(1) $y' = 2x - 4$ so it has a local minimum at $x = 2$.

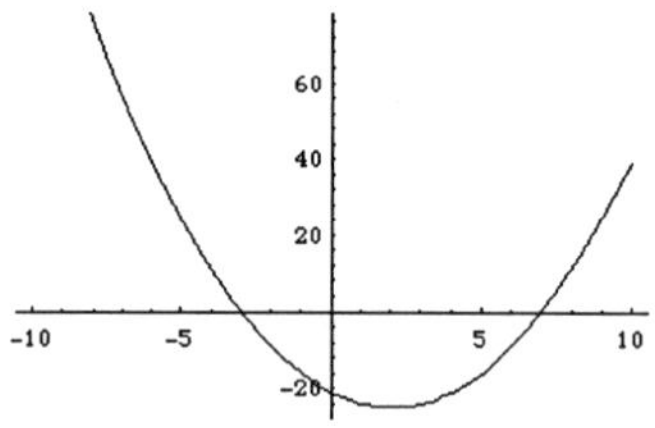

$y = x^2 - 4x - 21$

(2) $y' = -5x^4$ giving an inflection point at $x = 0$.

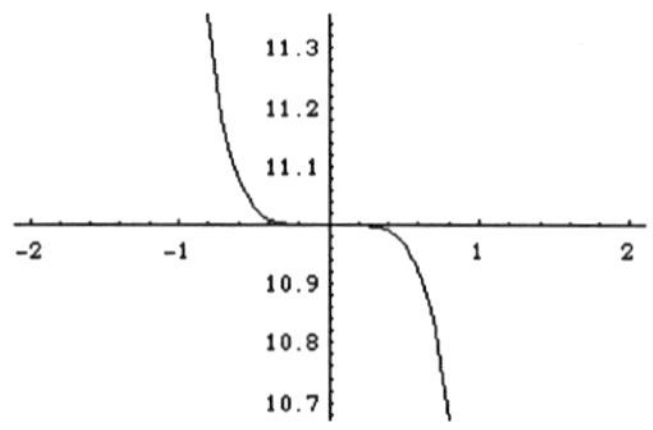

$y = 11 - x^5$

(3) $y' = \frac{1}{x} - 3x^2 \Rightarrow$ local maximum at $x = 3^{-\frac{1}{3}}$.

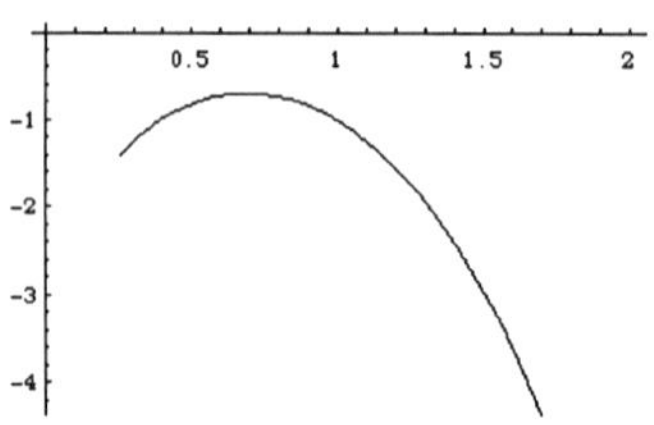

$y = y = \ln(x) - x^3$

(4) $y' = -2x \cdot \sin(x^2) = 0$ for $x = \sqrt{n\pi}$ with $n = 0, 1, 2, \ldots, 31$. These critical points will alternately be local max's and min's.

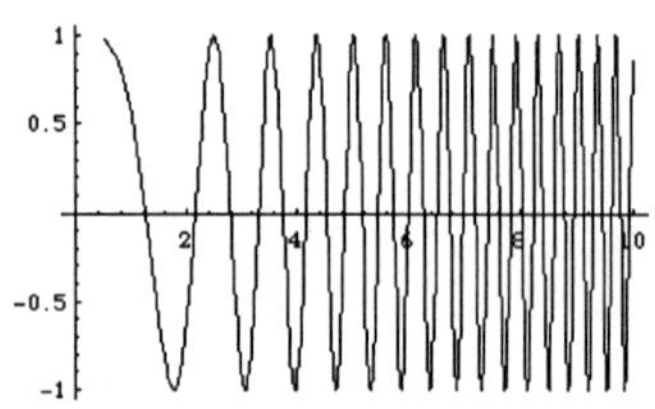

$y = \cos\left(x^2\right)$

(5) $y' = (1 + x)e^x \Rightarrow$ local minimum at $x = -1$.

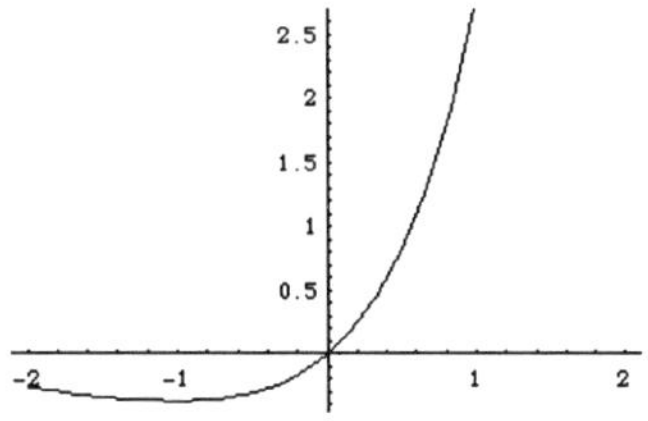

$y = x \cdot e^x$

(6) $y' = 2xe^{x^2} = 0$ for $x = 0$. This function has a minimum at $x = 0$.

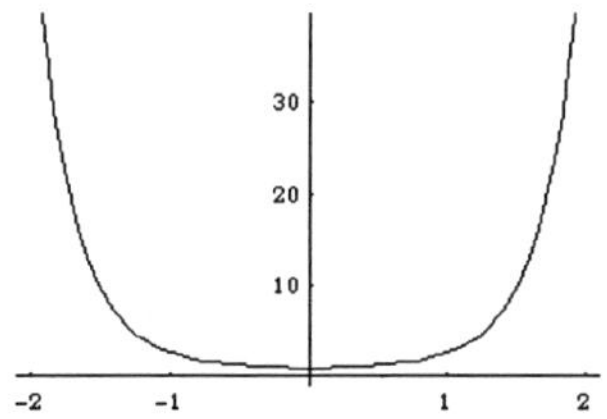

$y = e^{x^2}$

(7) This function has a local maximum at $\frac{9-\sqrt{11}}{7}$, a local minimum at $\frac{9+\sqrt{11}}{7}$ and a point of inflection at 0.

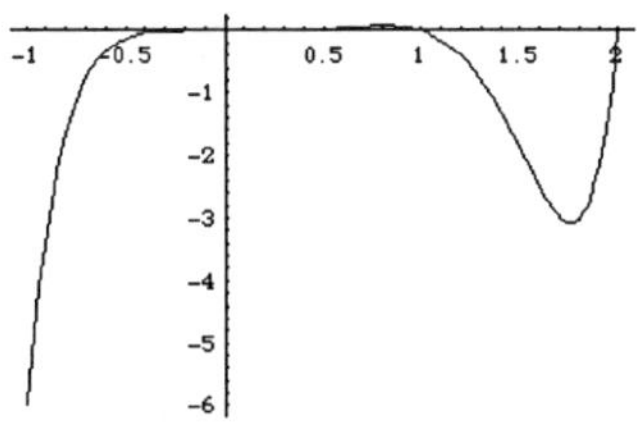

$y = x^7 - 3x^6 + 2x^5$

(8) Using the trigonometric identity $\sin(A + B) = \cos(A)\sin(B) + \cos(B)\sin(A)$ we see that

$$\sqrt{2}\sin\left(x + \frac{\pi}{4}\right) = \cos(x) + \sin(x).$$

Therefore, we have a local maximum at $\frac{\pi}{4}$ and a local minimum at $\frac{5\pi}{4}$.

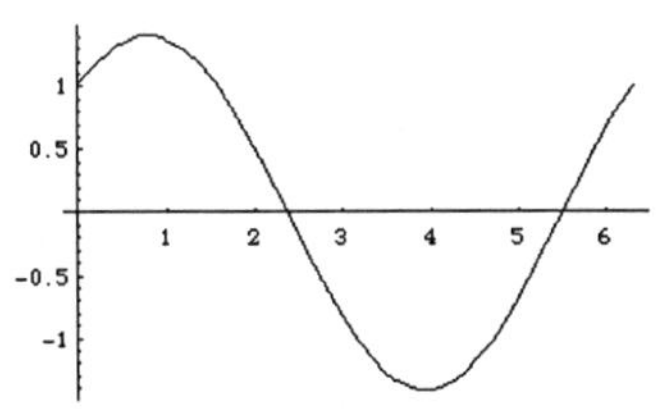

$y = \sin(x) + \cos(x)$

(9) Local maximum at $x = -1$, local minimum at $x = 1$.

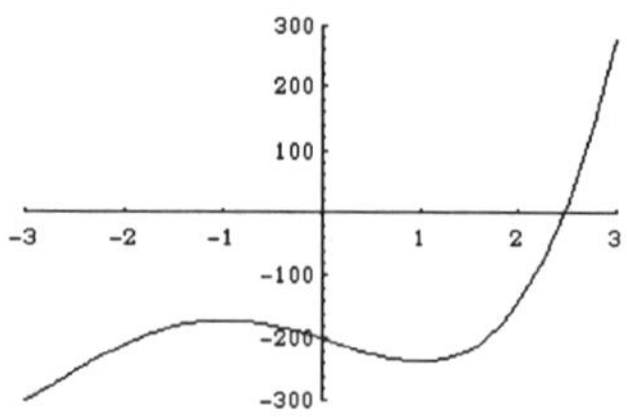

$$y = 3x^4 + 16x^3 - 6x^2 - 48x - 200$$

(10) This function has a local maximum at $x = -1$.

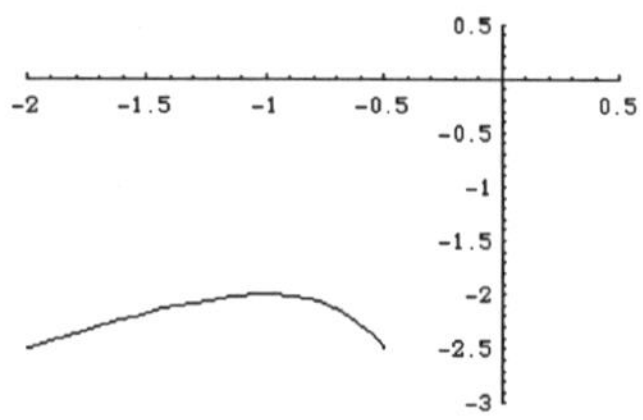

$$y = x + \frac{1}{x}$$

(11) In this case we have local minimums at $x = -3, 0, 3$ and local maximums at $x = -2, 2$.

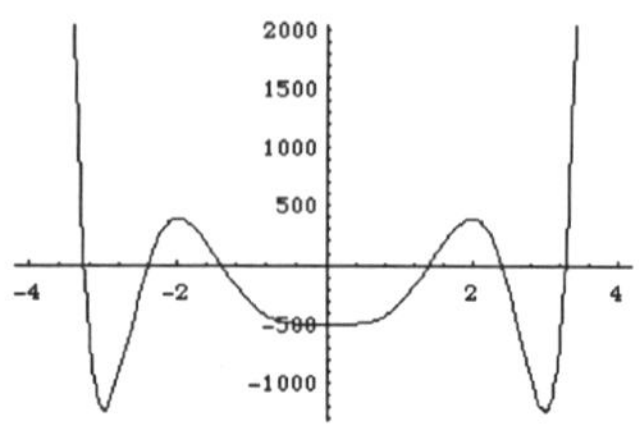

$$y = 3x^8 - 52x^6 + 216x^4 - 500$$

(12) $y' = x^x(1 + \ln(x))$ giving a local minimum at $x = \frac{1}{e}$.

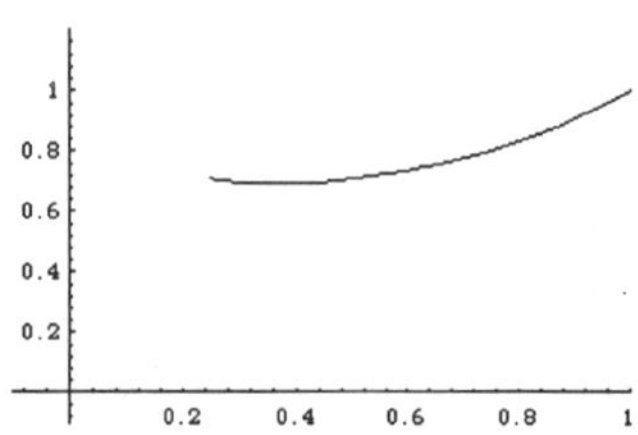

$$y = x^x$$

Solutions for §11.2

(1) $y' = \frac{4}{5\sqrt[5]{x}} \Rightarrow y'(0) = \infty$ and we have a cusp at $x = 0$. Here, y is decreasing for $x < 0$ and increasing for $x > 0$.

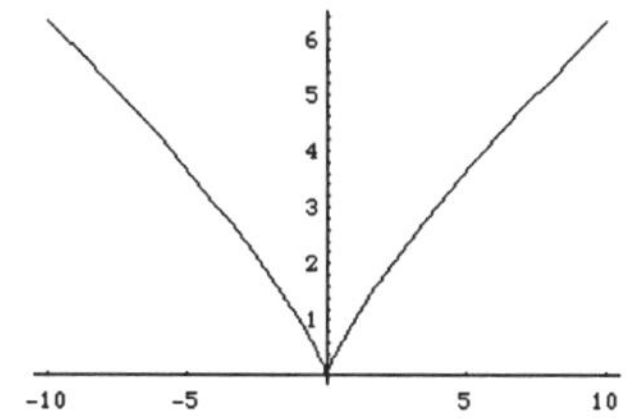

$$y = x^{\frac{4}{5}}$$

(2) In this case we have vertical tangents at $x = \pm 2$, since $y' = \frac{-x}{\sqrt{(4-x^2)}}$ becomes infinite when $x = \pm 2$. In fact, this is just the graph of a semicircle about the origin of radius 2.

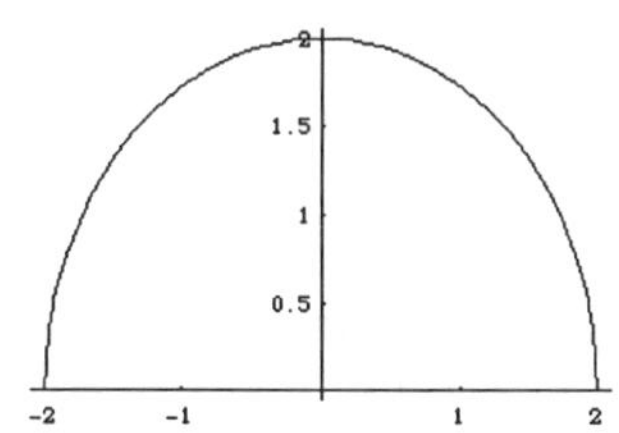

$$y = \sqrt{4 - x^2}$$

(3) $y = x^2 - x + 2$ for $x \geq 2$ and $y = x^2 + x - 2$ for $x < 2 \Rightarrow y' = 2x - 1$ for $x > 2$ and $y' = 2x + 1$ for $x < 2$. The derivative does not exist at $x = 2$ since it has different limits from the left and right (5 and 3) and the function, therefore, has a corner there. Also, $y' = 0$ only at $x = -\frac{1}{2}$, and this point is a local minimum.

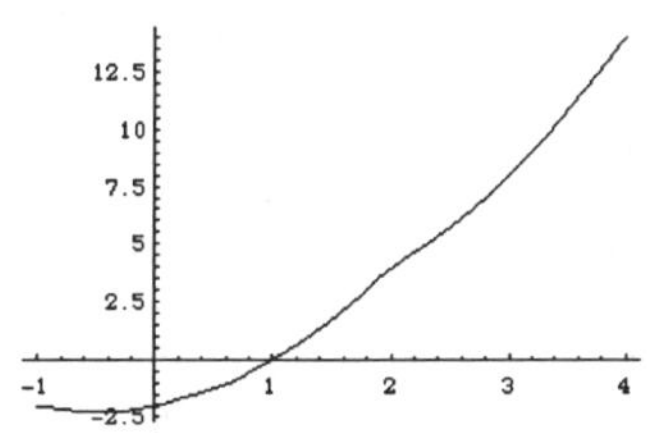

$$y = x^2 - |2 - x|$$

(4) $y' = \frac{2(x-5)^{-\frac{5}{7}}}{7} = \infty$ at $x = 5$ so that's a cusp. Further, y is decreasing for $x < 5$ and increasing for $x > 5$.

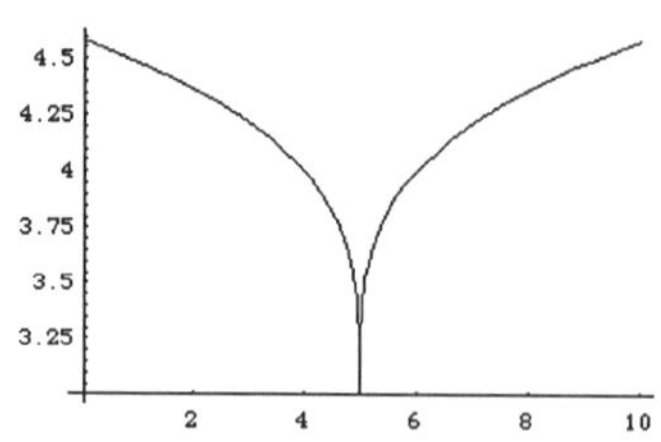

$$y = (x-5)^{\frac{2}{7}} + 3$$

(5) $y' = (\frac{2}{7x} - \frac{9}{7})x^{\frac{2}{7}}$ giving a cusp at $x = 0$ and a local maximum at $x = \frac{2}{9}$.

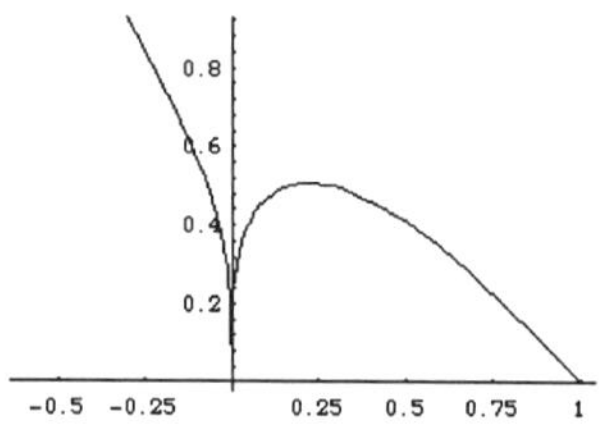

$$y = x^{\frac{2}{7}}(1-x)$$

(6) $|2-x|$ adds a corner at $x = 2$ to the previous figure.

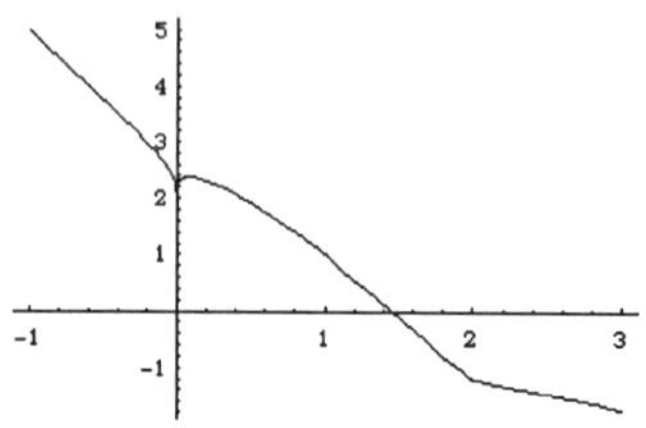

$$y = x^{\frac{2}{7}}(1-x) + |2-x|$$

(7) $|\cos(x)| = \cos(x)$ for $0 \le x \le \frac{\pi}{2}$ and $|\cos(x)| = -\cos(x)$ for $\frac{\pi}{2} \le x \le \pi$ giving a corner at $x = \frac{\pi}{2}$.

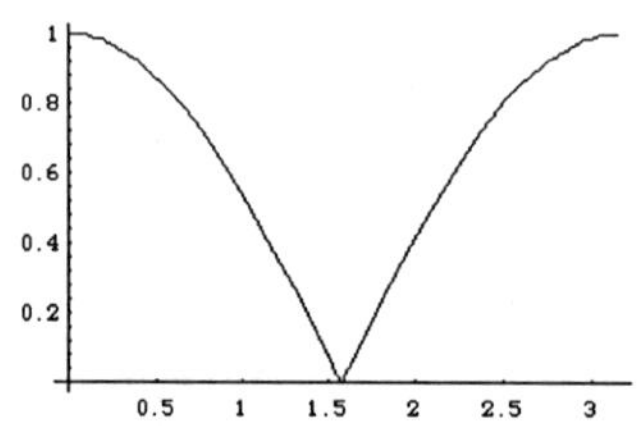

$$y = |\cos(x)|$$

(8) $y = x^x + x - \frac{1}{2}$ for $x \geq \frac{1}{2}$ and $y = x^x - x + \frac{1}{2}$ for $x < \frac{1}{2} \Rightarrow y' = x^x(1 + \ln(x)) + 1$ for $x > \frac{1}{2}$ and $y' = x^x(1 + \ln(x)) - 1$ for $x < \frac{1}{2} \Rightarrow$ we have a corner here at $x = \frac{1}{2}$.

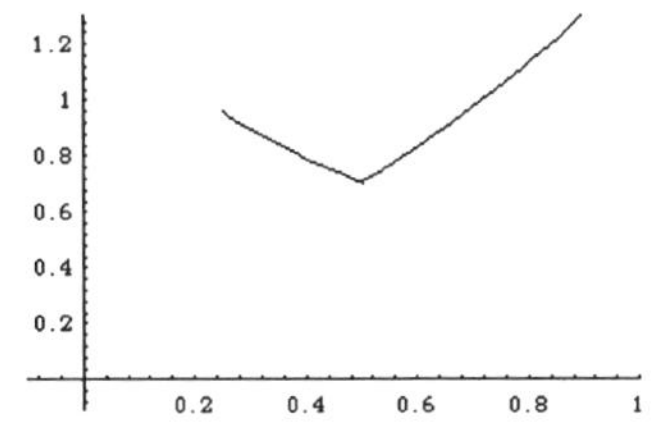

$$y = x^x + |x - \frac{1}{2}|$$

Solutions for §11.3

(1) $y'' = 20(x-2)^3$. Since $y''(2) = 0$, we see that $x = 2$ is a possible inflection point. To check that it is an actual inflection point, we must verify that y'' changes sign around $x = 2$. This is the case since $y'' < 0$ for $x < 2$ and $y'' > 0$ for $x > 2$. So $x = 2$ must be an inflection point and the graph is concave down (second derivative $y'' < 0$) for $x < 2$ and concave up for $x > 2$.

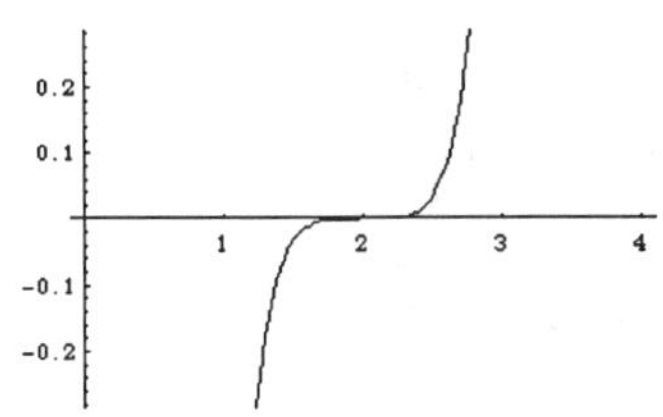

$$y = (x-2)^5$$

(2) $y'' = 24x - 6 = 0$ when $x < \frac{1}{4}$, and y'' changes sign at $x = \frac{1}{4}$, so $x = \frac{1}{4}$ is an inflection point. The graph is concave down for $x < \frac{1}{4}$ since $y'' < 0$ for $x < \frac{1}{4}$. Similarly, the graph is concave up for $x > \frac{1}{4}$.

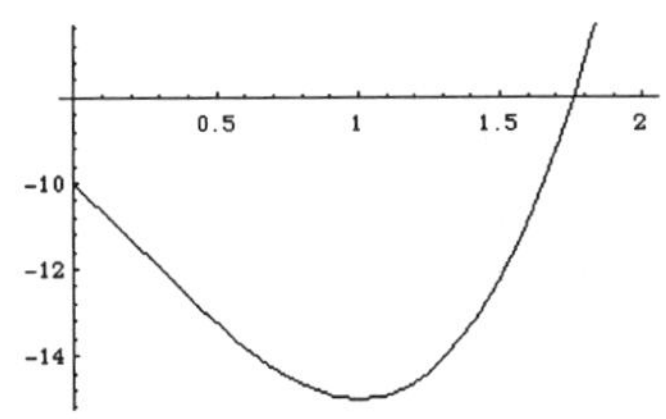

$$y = 4x^3 - 3x^2 - 6x - 10$$

(3) $y'' = 2$, so there are no inflection points and the graph is always concave up.

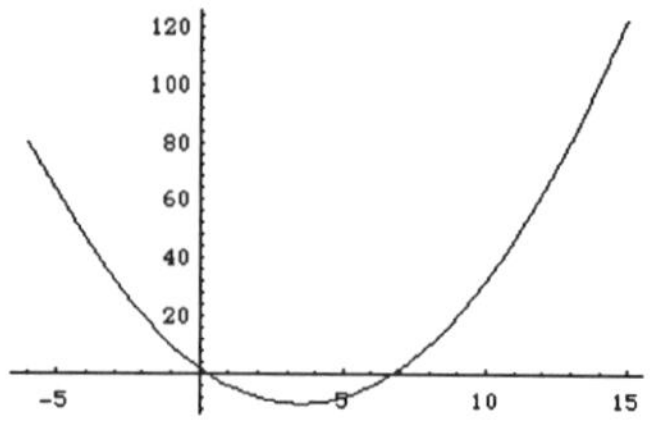

$$y = x^2 - 7x + 2$$

(4) $y'' = -\frac{1}{x^2}$ so there are no inflection points (since y'' is never 0). The graph is concave down for $\frac{1}{2} \leq x \leq 10$ since $y'' < 0$ on this region.

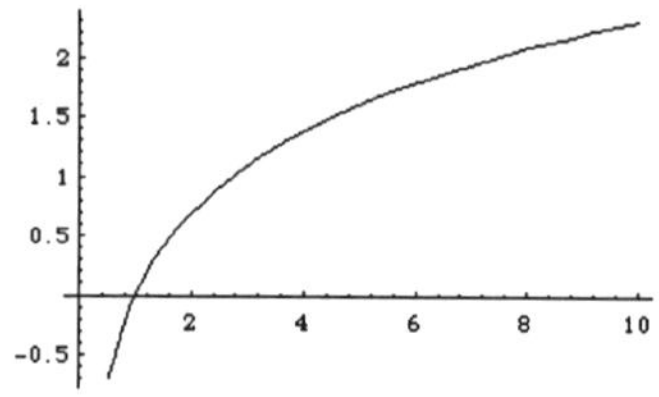

$$y = \ln(x)$$

(5) $y'' = 30(2-x)^4$. The point $x = 2$ is not an inflection point since y'' doesn't change sign around $x = 2$. Since $y'' > 0$ for $x \neq 2$, we see that the graph is always concave up.

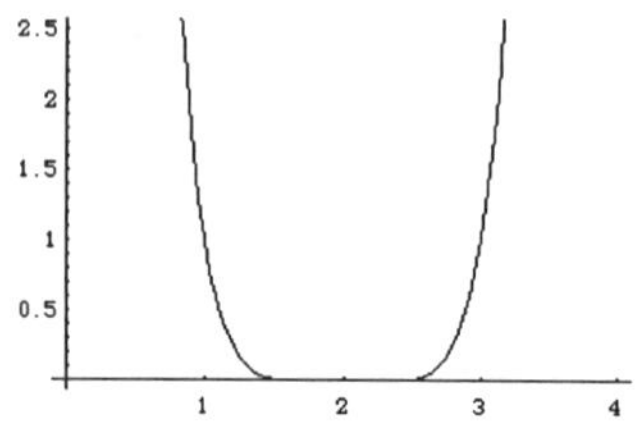

$$y = (2-x)^6$$

(6) We find $y'' = -\sin(x)$ giving inflection points for $x = n\pi$, n any integer. $\Rightarrow$ the graph is concave down for $\{x \mid 2n\pi < x < (2n+1)\pi\}$ and concave up for $\{x \mid (2n+1)\pi < x < (2n+2)\pi\}$.

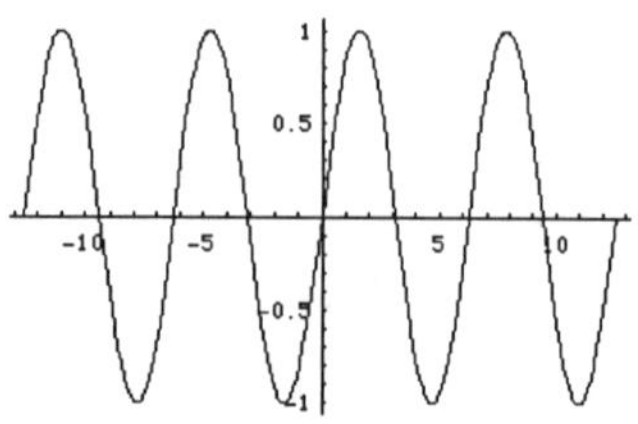

$$y = \sin(x)$$

(7) $y = e^x = y' = y'' > 0$ so the graph is always concave up.

(8) Referring back to question 8 in §11.1, we see $y'' = -\sqrt{2}\sin\left(x + \frac{\pi}{4}\right)$ giving inflection points for $x = \left(n - \frac{1}{4}\right)\pi$, n any integer. $\Rightarrow$ the graph is concave up for $\{x \mid \left(2n - \frac{1}{4}\right)\pi < x < \left(2n + \frac{3}{4}\right)\pi\}$ and concave down for $\{x \mid \left(2n + \frac{3}{4}\right)\pi < x < \left(2n + \frac{7}{4}\right)\pi\}$.

(9) $y = x^{\frac{2}{3}}, y' = \frac{2x^{-\frac{1}{3}}}{3}, y'' = \frac{-2}{9\sqrt[3]{x^4}} < 0$ so the graph is concave down on both sides of the cusp at $x = 0$.

(10) The second derivative is negative in this range implying the graph (a semicircle) is concave down.

(11) The second derivative is positive in this range implying the graph (a semicircle) is concave up.

(12) $y'' = 2(1 + 2x^2)e^{x^2} > 0 \Rightarrow$ the graph is concave up.

(13) $y'' = \frac{36}{x^3}$ giving a vertical asymptote at $x = 0$ and implying that the graph is concave down for negative x, concave up for x positive.

(14) $y'' = 2$ for $x \neq 3 \Rightarrow$ the graph is concave up with a corner at $x = 3$.

(15) $y'' = 6ax + 2b \Rightarrow$ (for $a \neq 0$) that the concavity changes at $x = -\frac{b}{3a}$ the unique point of inflection.

(16) $y' = 2ax + b \Rightarrow$ the general quadratic has a single critical point at $x = -\frac{b}{2a}$. $y'' = 2a \Rightarrow$ the graph has no inflection points and is concave up for positive a, concave down for negative a.

Solutions for Additional Exercises for Chapter 11

(1) If $f(x) = a_0 + a_1x + \ldots + a_nx^n$, then inflection points of f will be zeros of the twice differentiated polynomial $f''(x)$, which has degree $n - 2$. So there are at most $n - 2$ of them.

(2) If $f(x) = ax^3 + bx^2 + cx + d$, then $f(x)$ is always increasing or always decreasing if the quadratic polynomial $f'(x) = 3ax^2 + 2bx + c$ has at most one solution. By the quadratic formula this will happen if and only if $b^2 \leqslant 3ac$.

(3) We have $f'[x] = \frac{x}{\cos^2(x)} + \frac{\sin(x)}{\cos(x)} = 0$ only for $x = 0$. Since $f''(x) = \frac{2}{\cos^3(x)}(\cos(x) + x \cdot \sin(x))$ and $f''(0) = 2$, we see that $x = 0$ cannot be an inflection point. Thus, $x = 0$ is a critical point and corresponds to a local minimum of f.

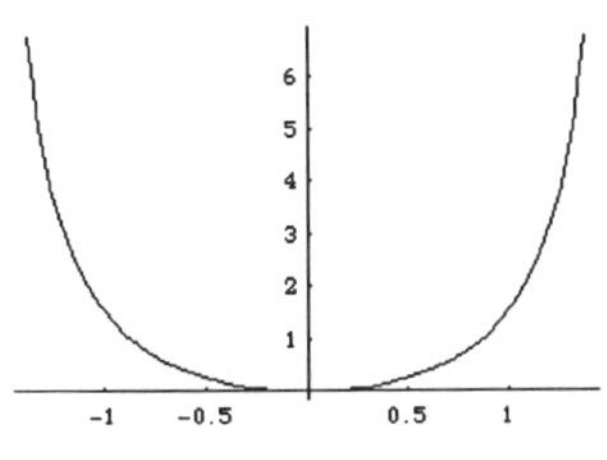

$f(x) = x \cdot \tan(x)$

(4) There are thirteen critical values: $x = 0$, and $x = \sqrt{\frac{\pi}{2}}, \sqrt{\frac{3\pi}{2}}, \sqrt{\frac{5\pi}{2}}, \cdots \sqrt{\frac{25\pi}{2}}$. None of them are inflection points. $x = 0$ is a relative minimum, as well as $x = \sqrt{\frac{3\pi}{2}}, \sqrt{\frac{7\pi}{2}}, \ldots, \sqrt{\frac{23\pi}{2}}$. The other critical points are relative maxima.

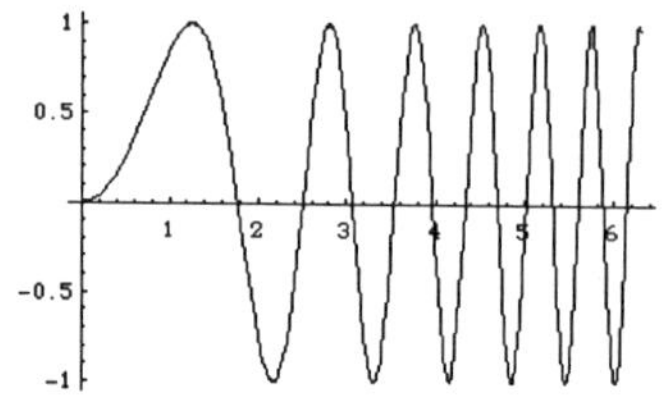

$$f(x) = \sin(x^2)$$

(5) We have $f'(x) = \frac{(4+x)^{\frac{1}{3}}(4+5x)}{3x^{\frac{2}{3}}}$. The critical points are $x = -\frac{4}{5}$, -4 (where f' vanishes) and $x = 0$ (where $f'(0) = \infty$). We see from the graph below that $x = -4$ is a local max and $x = -\frac{4}{5}$ is a local minimum. The point $x = 0$ is a point where the tangent has infinite slope.

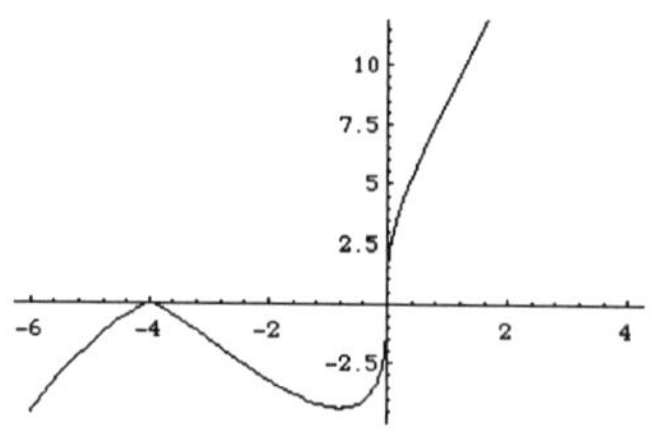

$$f(x) = x^{\frac{1}{3}}(4 + x)^{\frac{4}{3}}$$

(6) We have

$$f'(x) = \begin{cases} -1, & \text{if } x \leq 1 \\ -2(x - \frac{3}{2}), & \text{otherwise.} \end{cases}$$

Critical points are $x = 1$ (point of non-differentiability) and $x = \frac{3}{2}$ (point of vanishing derivative). We see from the graph below that $x = 1$ is a local minimum and $x = \frac{3}{2}$ is a local maximum.

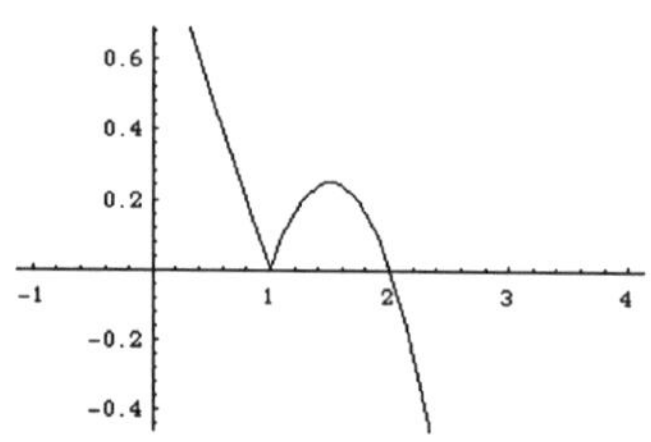

$$f(x) = \begin{cases} -x + 1, & \text{if } x \leq 1 \\ -(x - \frac{3}{2})^2 + \frac{1}{4}, & \text{otherwise.} \end{cases}$$

(7a) Critical values of f are $x = 1$ and $x = 2$.

(7b) f is increasing on $\{x \mid x \leqslant 2\}$. f is decreasing on $\{x \mid x \geqslant 2\}$.

(7c) f is concave up on $\{x \mid 1 \leqslant x \leqslant \frac{3}{2}\}$. f is concave down on $\{x \mid x \leqslant 1 \text{ or } x \geqslant \frac{3}{2}\}$.

(7d)

$$f(x) = \begin{cases} -\frac{1}{2}x^2 + x, & \text{if } x \leqslant 1 \\ \frac{1}{3}\left(x - \frac{3}{2}\right)^3 + \frac{1}{4}x, & \text{otherwise} \end{cases}$$

(8) We have $f'(x) = (4 + 6x - 3x^2)e^{-x}$. There is a relative maximum at $\frac{3+\sqrt{21}}{2}$ and a relative minimum at $\frac{3-\sqrt{21}}{2}$.

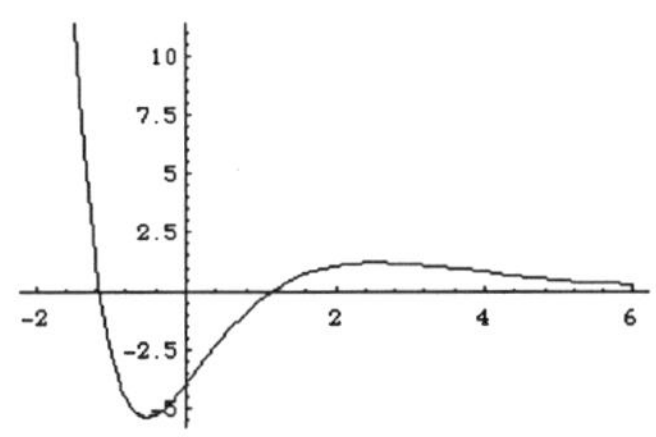

$$f(x) = \frac{3x^2 - 4}{e^x}$$

(9) First, x_0 must be a critical point for $f + g$. (It is certainly a minimum. So if $f + g$ is differentiable at x_0, then $(f + g)'(x_0) = 0$.) However, x_0 need not be a critical point for $f - g$. (For example, if $f = g$, then $f - g \equiv 0$.) Similarly, x_0 must be a critical point of $f \cdot g$, but need not be a critical point of $\frac{f}{g}$. (If $f = g$, then $\frac{f}{g} \equiv 1$.)

(10a)

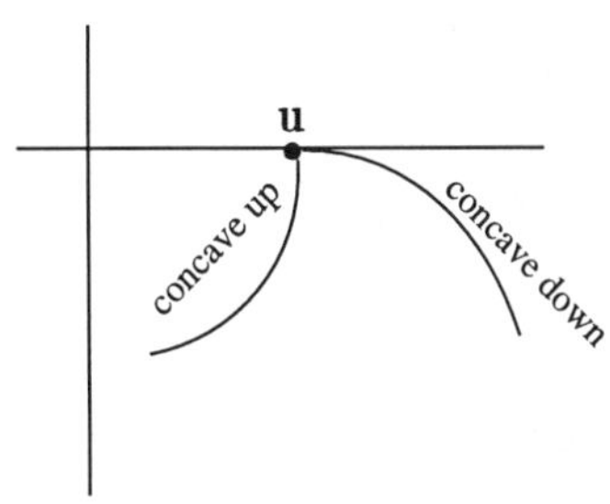

(10b)

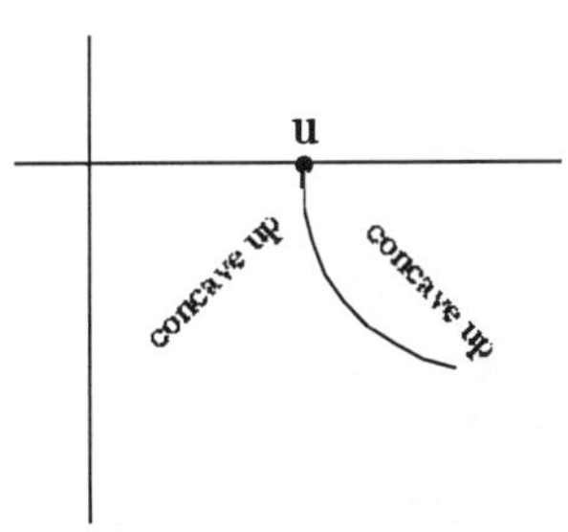

(11) The function $f(x) = -(x-1)^2$ satisfies the required conditions. The graph is displayed below.

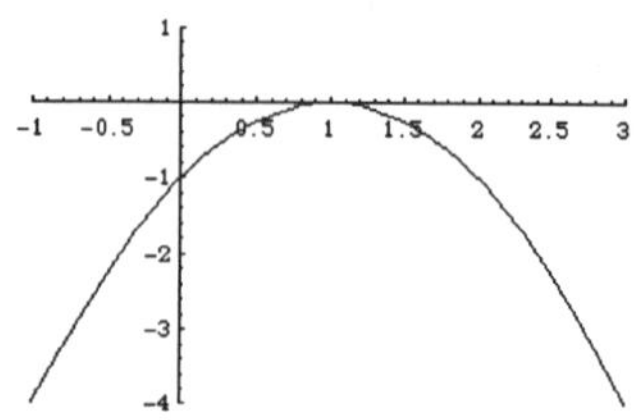

$f(x) = -(x-1)^2$

(12) The function $f(x) = -(x-1)^3 + 2$ satisfies the required conditions. The graph is displayed below.

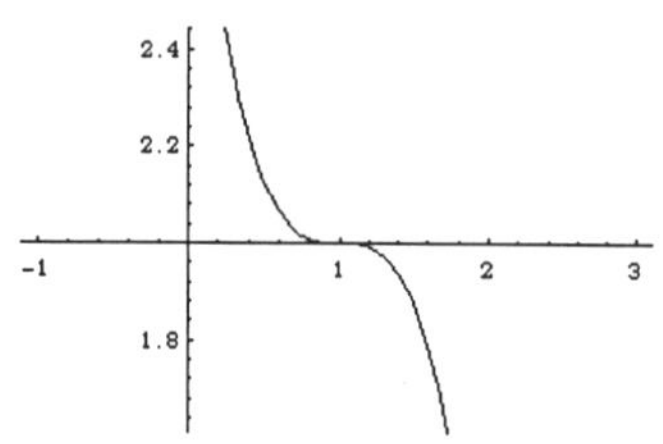

$f(x) = -(x-1)^3 + 2$

(13) $b = -9$

(14) $a = \frac{1}{4}, b = -\frac{3}{2}$

Solutions for §12.2

(1) The function $f(x) = \frac{3x-1}{x^2-3x-4}$ becomes infinite when the denominator $x^2 - 3x - 4 = (x-4)(x+1) = 0$. We, thus, have vertical asymptotes at $x = 4, -1$.

(2) The function $f(x) = \frac{4x^3+11}{2x^2-3x-1}$ becomes infinite when the denominator $2x^2 - 3x - 1 = 0$. We, thus, have vertical asymptotes at $x = (3 \pm \sqrt{17})/4$.

(3) The function $f(x) = \frac{\cos(x)}{x-11}$ becomes infinite when the denominator $x - 11 = 0$. We, thus, have one vertical asymptote at $x = 11$.

(4) The function $f(x) = e^{\frac{1}{x}}$ becomes infinite as $x \to 0$ with $x > 0$. Thus, there is a vertical asymptote at $x = 0$. **Warning:** The function $e^{\frac{1}{x}}$ approaches 0 as $x \to 0$ with $x < 0$.

(5) We compute:

$$\lim_{x\to 0} \frac{\sin(x)}{x^2 + 2x} = \lim_{x\to 0} \frac{1}{x+2} \frac{\sin(x)}{x} = \lim_{x\to 0} \frac{1}{x+2} = \frac{1}{2},$$

so $f(0) \neq \infty$. On the other hand, $f(-2) = \infty$, so f has one vertical asymptote at $x = -2$.

(6) After canceling a factor of $(x-1)$ we see that $f(x) = \frac{x-1}{x^3-x} = \frac{1}{x\cdot(x+1)}$. Thus, $f(x)$ becomes infinite when $x \cdot (x+1) = 0$, and we have vertical asymptotes at $x = 0, -1$.

(7) The function $f(x) = \frac{x^2+2}{\ln(x)}$ becomes infinite when the denominator $\ln(x) = 0$. Since $\ln(1) = 0$, we have a vertical asymptote at $x = 1$.

(8) The denominator of $f(x) = \frac{2x+1}{\cos(x)}$ is zero for $x = \pi(n + \frac{1}{2})$ for all integers n. So we have vertical asymptotes at these values.

(9) The function $f(x) = \frac{x-1}{x^3-1} = \frac{1}{1+x+x^2}$ becomes infinite when $1 + x + x^2 = 0$. Since there are no real solutions to this last equation, there can be no vertical asymptotes.

(10) f has a vertical asymptote at $x = 1$.

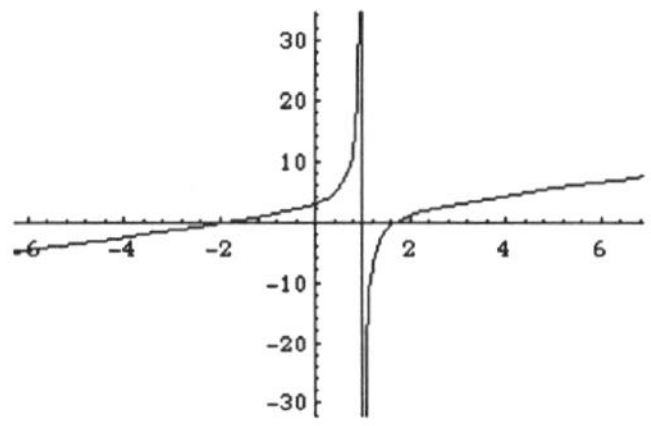

$$f(x) = \frac{x^2-3}{x-1}$$

(11) The function has vertical asymptotes when $x^2 - 3x - 4 = (x+1)(x-4) = 0$, i.e., when $x = -1, 4$.

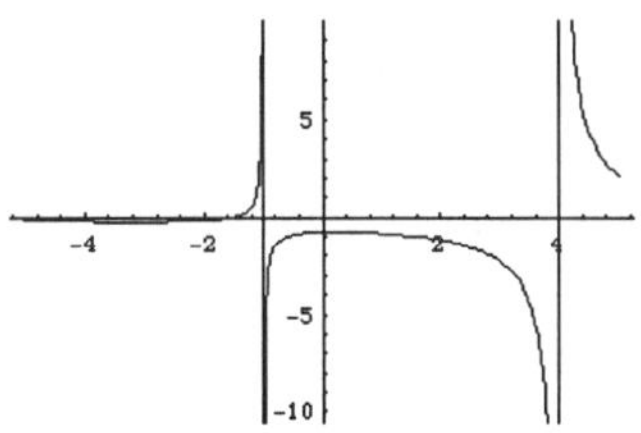

$$f(x) = \frac{2x+3}{x^2-3x-4}$$

(12) The function has vertical asymptotes when $x^2 - 1 = (x-1)(x+1) = 0$, i.e., when $x = \pm 1$.

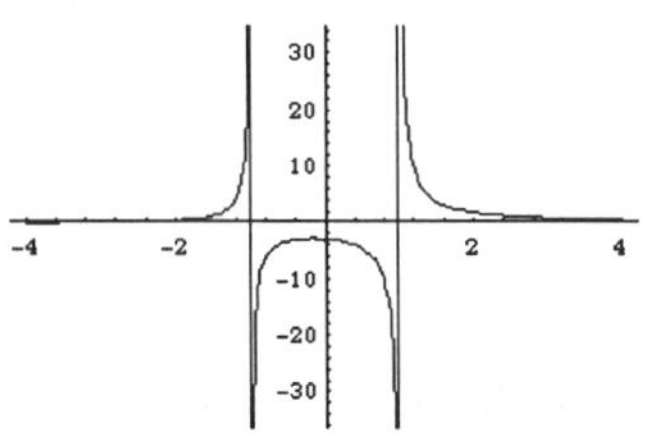

$$f(x) = \frac{x+3}{x^2-1}$$

(13) f has vertical asymptotes when $2x^2 - 3x - 1 = 0$, i.e., at $x = (3 \pm \sqrt{17})/4$.

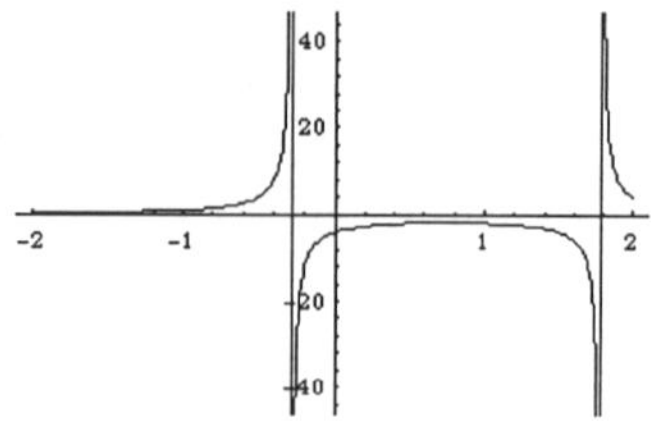

$$f(x) = \frac{4}{2x^2 - 3x - 1}$$

(14) f has vertical asymptotes when $x^3 - 3x^2 = x^2(x-3) = 0$, i.e., at $x = 0, 3$.

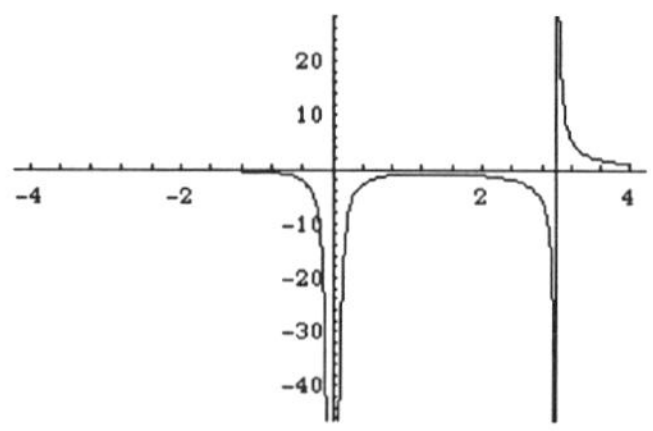

$$f(x) = \frac{x^2 + 1}{x^3 - 3x^2}$$

(15) Vertical asymptote at $x = 2$.

(16) $\tan(x - \pi/2) = \pm\infty$ at $x = n\pi$ for all integers n, so we have vertical asymptotes at these values.

(17) The function $f(x) = e^{\frac{1}{x}}$ has a vertical asymptote at $x = 0$. (The graph is close to zero for $x < 0$).

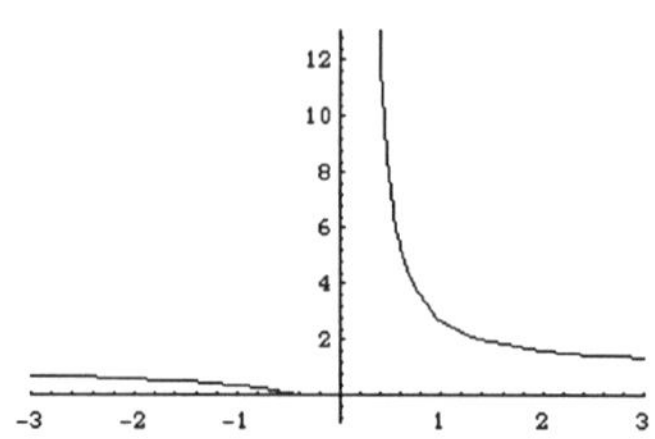

$$f(x) = e^{\frac{1}{x}}$$

(18) This graph has an asymptote at $x = 1$.

(19) The function $f(x) = \frac{1}{\sin(x)}$ becomes infinite when $\sin(x) = 0$. This occurs for $x = n\pi$ (with n an arbitrary integer), so we have the vertical asymptote at $x = 0$ (on the stated interval).

Solutions for §12.3

(1) $f(x) = \frac{3-\frac{2}{x}+\frac{11}{x^2}}{1+\frac{1}{x^2}} \to 3$ as $x \to \pm\infty$.

(2) $f(x) = \frac{-6+\frac{1}{x}+\frac{9}{x^2}}{2+\frac{3}{x^2}} \to -3$ as $x \to \pm\infty$.

(3) $f(x) = \frac{1+\frac{2}{x^2}+\frac{9}{x^3}}{2-\frac{1}{x^3}} \to 1/2$ as $x \to \pm\infty$.

(4) We have $\lim_{x\to-\infty} e^x = 0$. Similarly, $\lim_{x\to\infty} e^x = \infty$. We see that the graph of $y = e^x$ has a horizontal asymptote at $y = 0$ (on the left) as $x \to -\infty$.

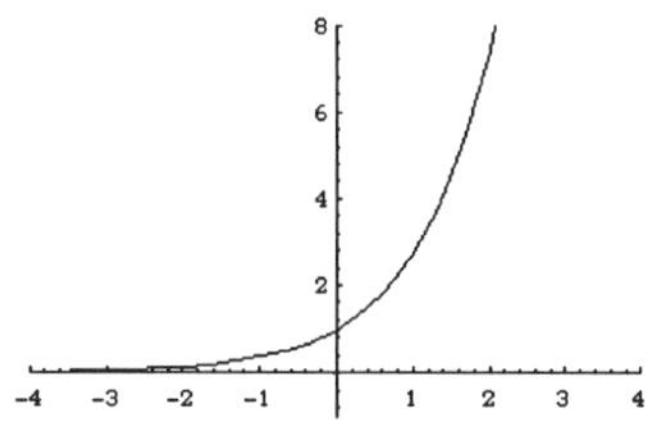

$f(x) = e^x$

(5) We have: $\lim_{x\to\pm\infty} \frac{\ln(x^2)}{x-1} = 0$ since for large x, the function $\ln(x)$ grows more slowly than x. Thus, the graph has horizontal asymptotes as $x \to \pm\infty$.

(6) $f(x) \to 0$ as $x \to \pm\infty$ so there are horizontal asymptotes as $x \to \pm\infty$.

(7) $\lim_{x\to\pm\infty} e^{\frac{1}{x}} = 1$. Thus, $e^{\frac{1}{x}}$ has a horizontal asymptote at $y = 1$ as $x \to \pm\infty$.

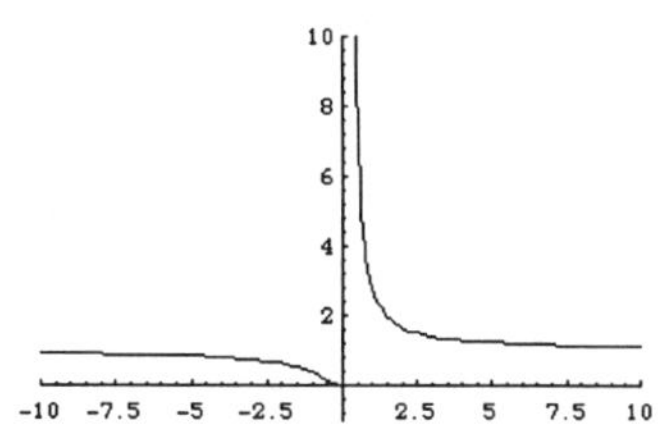

$f(x) = e^{\frac{1}{x}}$

(8) After factoring the x^2, we see that $\frac{\cos(x)}{3-\frac{2}{x}+\frac{3}{x^2}}$ is asymptotic to $\cos(x)/3$ as $x \to \pm\infty$. But since this does not have a limit as $x \to \pm\infty$, the function f doesn't have a horizontal asymptote.

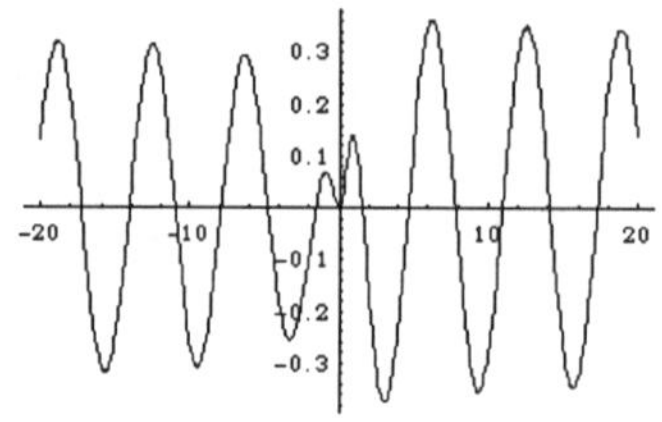

$$f(x) = \frac{x^2 \cdot \cos(x)}{3x^2 - 2x + 3}$$

(9) Here there is a vertical asymptote at $x = -4$ and a horizontal one at $y = 2$.

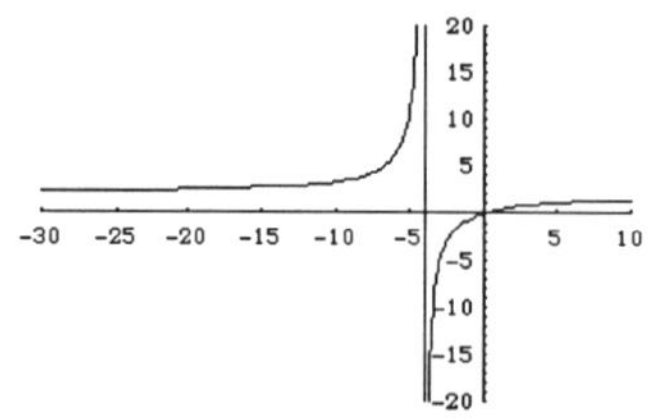

$$f(x) = \frac{2x}{x + 4}$$

(10) f has a vertical asymptote at $x = 0$ and no horizontal asymptote.

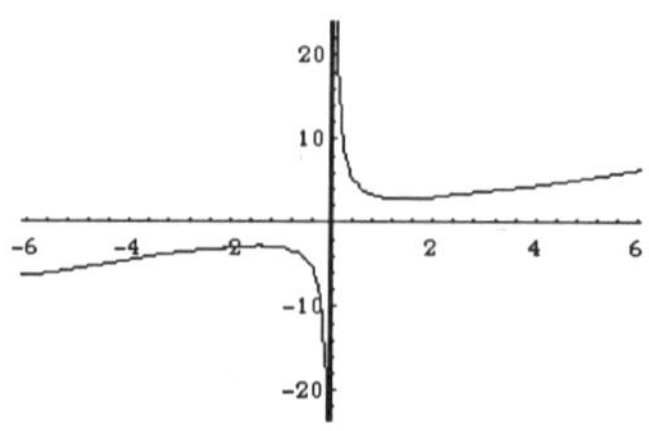

$$f(x) = x + \frac{2}{x}$$

(11) f has a vertical asymptote at $x = -3$ and a horizontal asymptote at $y = 1$.

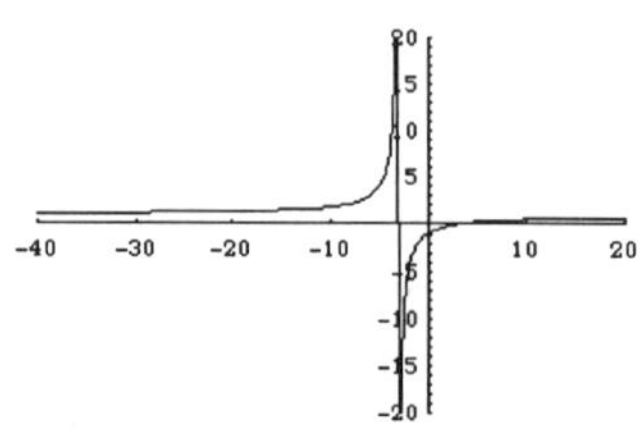

$$f(x) = \frac{x - 3}{x + 3}$$

(12) f has a vertical asymptote at $x = 5/2$ and a horizontal asymptote at $y = 1/2$.

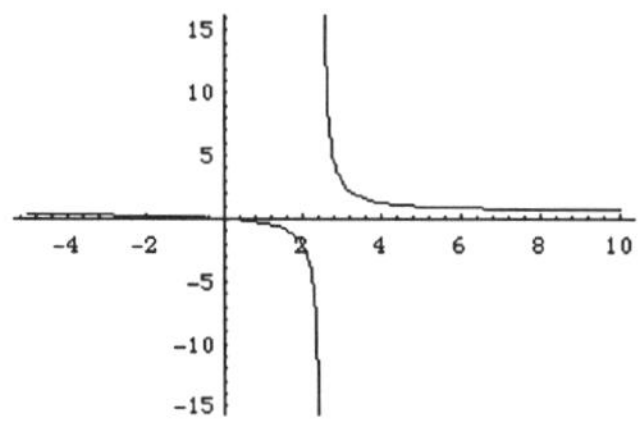

$$f(x) = \frac{x}{2x - 5}$$

(13) f has vertical asymptotes at $x = \pm 1$ and a horizontal one at $y = 1$.

(14) This function has a vertical asymptote at $x = 1$.

(15) $f(x) = \frac{2x^3}{\sqrt{3x^6+10}}$

$f(x) = \frac{2}{\sqrt{3+\frac{10}{x^6}}} \to 2/\sqrt{3}$ as $x \to \infty$ and $\to -2/\sqrt{3}$ as $x \to -\infty$.

(16) f has no vertical asymptotes but does have a horizontal one at $y = 2$.

(17) f has vertical asymptotes at $x = \pm 1$ and a horizontal asymptote at $y = 0 = \lim_{x \to -\infty} \frac{e^x}{x^2-1}$.

(18) f has a vertical asymptote at $x = 3$ and (using question 7) a vertical asymptote at $x = 0$ (on the right).

(19) We have vertical asymptotes at $x = \pm 1$ and horizontal asymptotes at $y = \lim_{x \to \pm\infty} \frac{e^{-x^2}}{x^2-1} = 0$.

(20) $f(x) = f(x) \times \frac{\sqrt{x^2+2}+\sqrt{x^2+1}}{\sqrt{x^2+2}+\sqrt{x^2+1}} = \frac{1}{\sqrt{x^2+2}+\sqrt{x^2+1}}$ and from this we can see that the denominator is never zero. So there are no vertical asymptotes. Since $f \to 0$ as $x \to \pm\infty$, there are horizontal asymptotes at $y = 0$.

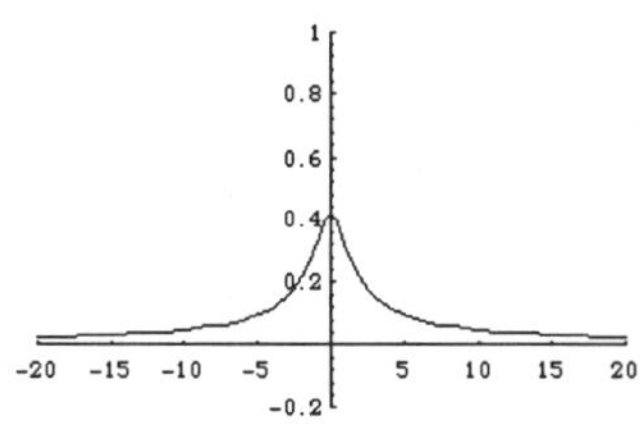

$$f(x) = \sqrt{x^2 + 2} - \sqrt{x^2 + 1}$$

(21) $\frac{\sin(x)}{x} = 0$ for $x = n\pi$ for all integers $n \neq 0$. Thus, the curve intersects the x–axis (line $y = 0$) infinitely often. Also, $\lim_{x\to\pm\infty} |\frac{\sin(x)}{x}| \leq \lim_{x\to\pm\infty} |\frac{1}{x}| = 0$ so $y = 0$ is a (wavelike) horizontal asymptote to the graph.

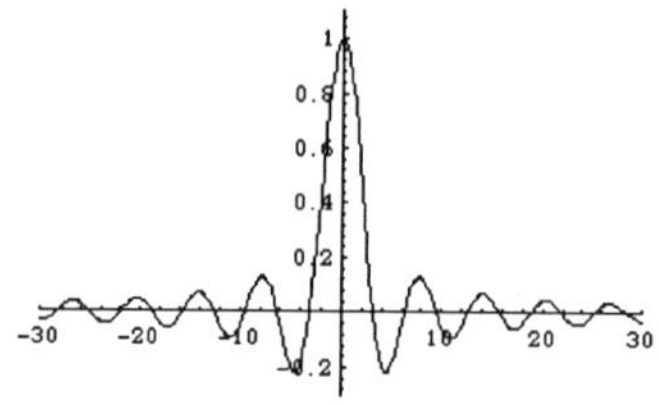

$$f(x) = \frac{\sin(x)}{x}$$

(22) We have $y = \frac{2x^2+3x+1}{x} = 2x + 3 + \frac{1}{x} \approx 2x + 3$ for large x.

(23) We have $y = \frac{x^4+3x^2+x-1}{x^2} = x^2 + 3 + \frac{1}{x} + \frac{-1}{x^2} \approx x^2 + 3$ as $x \to \pm\infty$.

Solutions for Additional Exercises for Chapter 12

(1) Vertical asymptotes at $x = \pm 1$. No horizontal asymptotes.

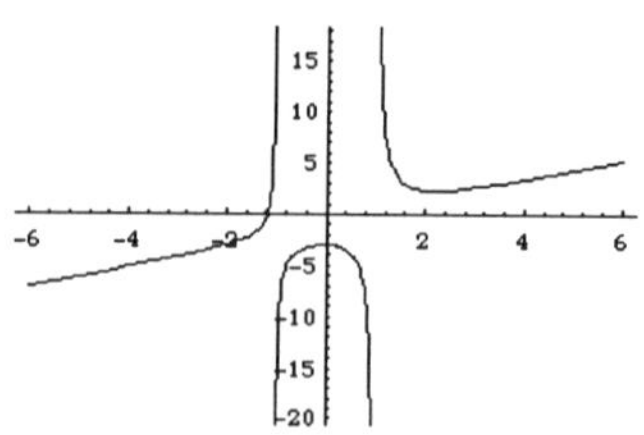

$$y = \frac{x^3 - x^2 + 3}{x^2 - 1}$$

(2) Vertical asymptotes at $x = \frac{\pi}{2} + n\pi$ with n a positive integer.

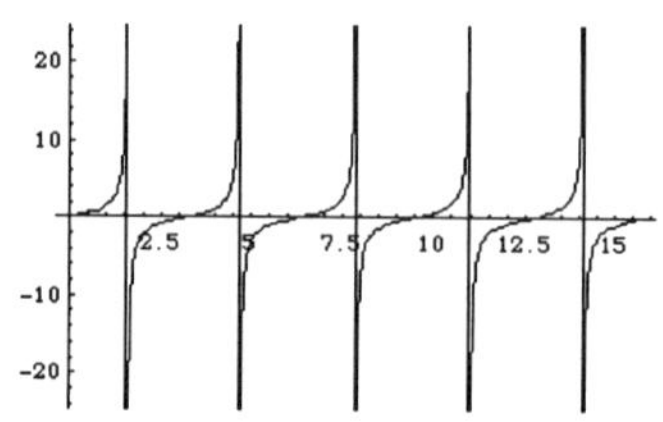

$$y = \tan(x)$$

(3) Horizontal asymptotes at $y = \pm 1$. No vertical asymptotes.

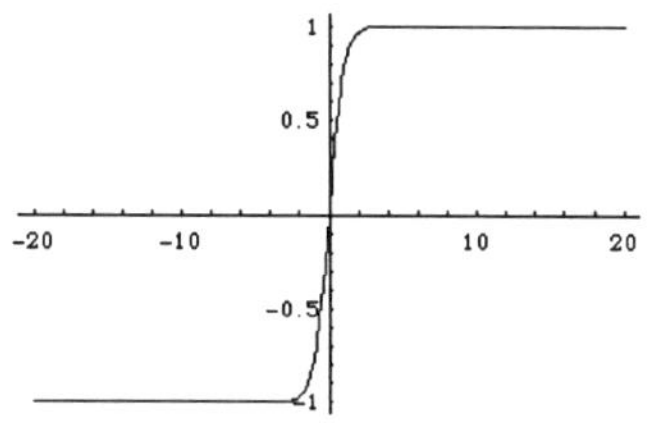

$$y = \tanh(x) = \frac{e^x - e^{-x}}{e^x + e^{-x}}$$

(4) Vertical asymptote at $x = 0$. No horizontal asymptotes.

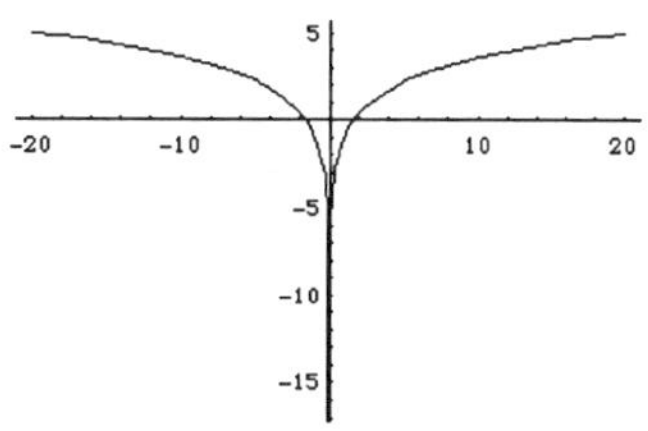

$$y = \ln(x^2) - 1$$

(5) No vertical asymptotes. No horizontal asymptotes.

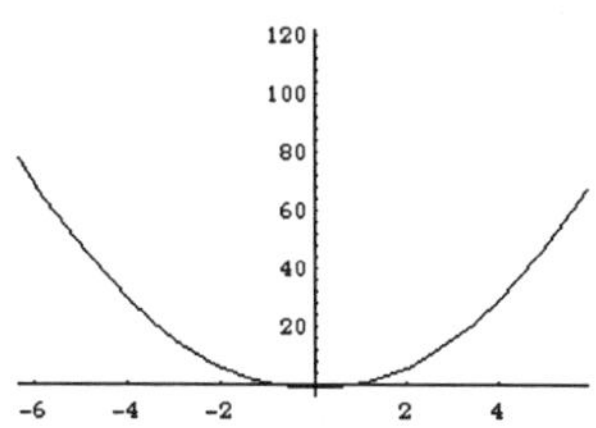

$$y = 2x^2 - x^{\frac{2}{3}}$$

(6) Vertical asymptotes when $\cos(x) = -1$, i.e., when $x = -\pi + 2\pi n$ with n an integer. There are no horizontal asymptotes.

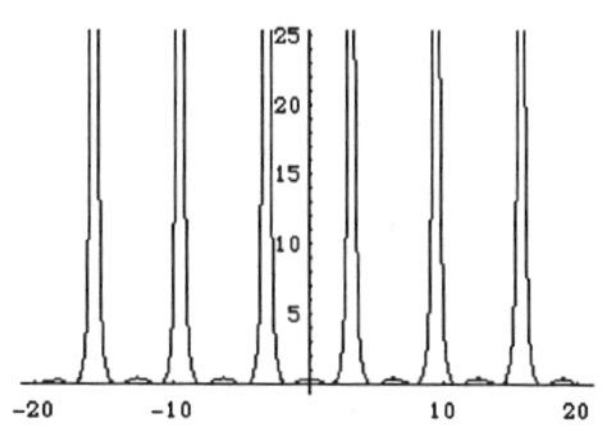

$$y = \frac{\cos^2(x)}{1 + \cos(x)}$$

(7) Vertical asymptotes when $x = -\pi + 2\pi n$ with n an integer. There are no horizontal asymptotes.

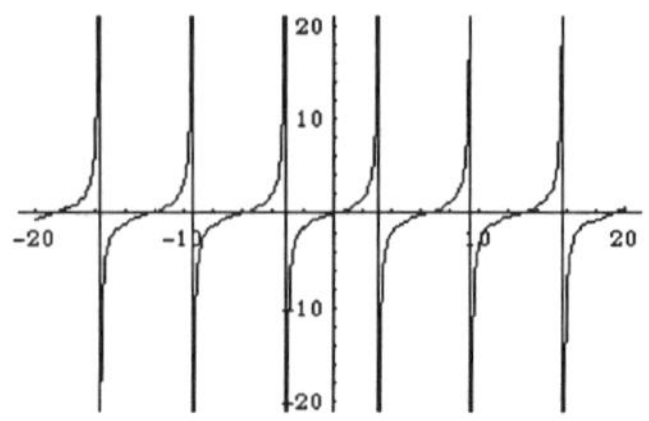

$$y = \frac{\sin(x)}{1 + \cos(x)}$$

(8) There are no vertical or horizontal asymptotes.

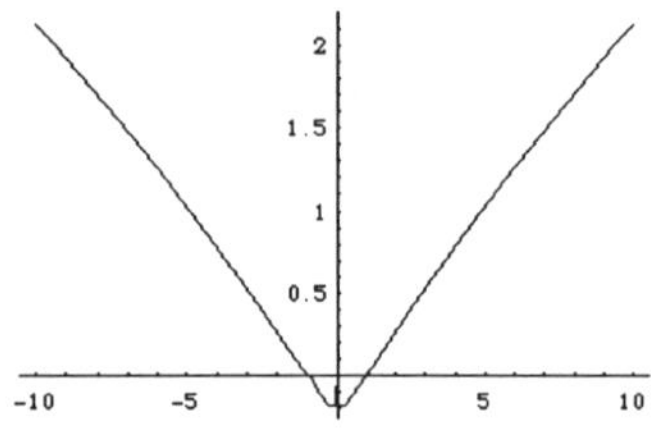

$$y = x^{\frac{2}{3}} - x^{\frac{2}{5}}$$

(9) Horizontal asymptotes at $y = 0$. There are no vertical asymptotes.

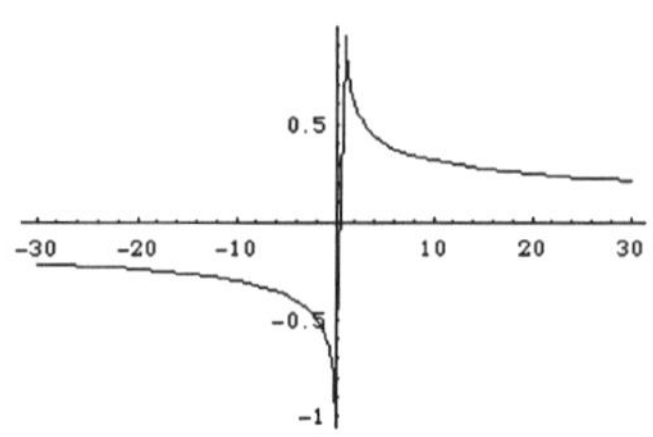

$$y = x^{\frac{2}{3}} - (1 - x)^{\frac{2}{3}}$$

(10) Vertical asymptote at $x \approx -2.24$. Horizontal asymptotics at $y = 0$.

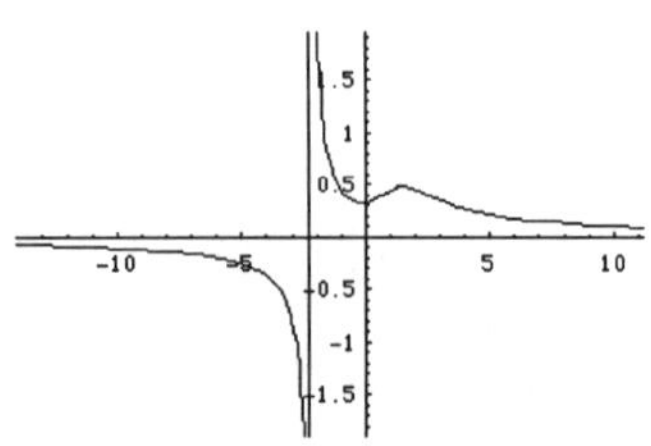

$$y = \frac{x^2 + 3}{x^3 - x + 9}$$

(11) If $\lim_{x\to\infty}(f(x)-(mx+b))=0$, then this means that the difference $|f(x)-(mx+b)|$ gets arbitrarily small as x gets large. Thus, the graph of the curve $y=f(x)$ gets arbitrarily close to the graph of the line $y=mx+b$ as x gets large. But this is just what one means by being asymptotic.

(12) We have $y=\frac{x^3}{x^2-3}=x+\frac{3x}{x^2-3}$. So the curve $y=\frac{x^3}{x^2-3}$ is asymptotic to the line $y=x$.

(13) The curve $y=7x-2+\frac{4x}{x^2-1}$ is asymptotic to the line $y=7x-2$.

(14) The curve $y=\frac{x}{\sqrt{x^2+1}}$ is asymptotic to the lines $y=1$ and $y=-1$.

(15) We have $y=\frac{2x^2-6x+4}{x-3}=2x+\frac{4}{x-3}$. So the curve $y=\frac{2x^2-6x+4}{x-3}$ is asymptotic to the line $y=2x$.

(16) We have $y=\frac{x^2}{3x+2}=\frac{x}{3}-\frac{2}{9}+\frac{4}{3x+2}$. So the curve $y=\frac{x^2}{3x+2}$ is asymptotic to the line $y=\frac{x}{3}-\frac{2}{9}$.

(17) The curve $y=x^3-2x+1$ cannot be asymptotic to any line.

(18) The hyperbola $(\frac{x}{a})^2-(\frac{y}{b})^2=1$ can be graphed as the union of the graphs of the two functions $y=\pm|b|\cdot\sqrt{(\frac{x}{a})^2-1}\approx\pm|b|\cdot\frac{|x|}{|a|}$ as $x\to\pm\infty$. So the hyperbola is asymptotic to the lines $y=\pm|b|\cdot\frac{|x|}{|a|}$.

Solutions for §13.1

(1) The integral $\int_0^5 3\,dx$ is the area of the rectangle in the figure below.

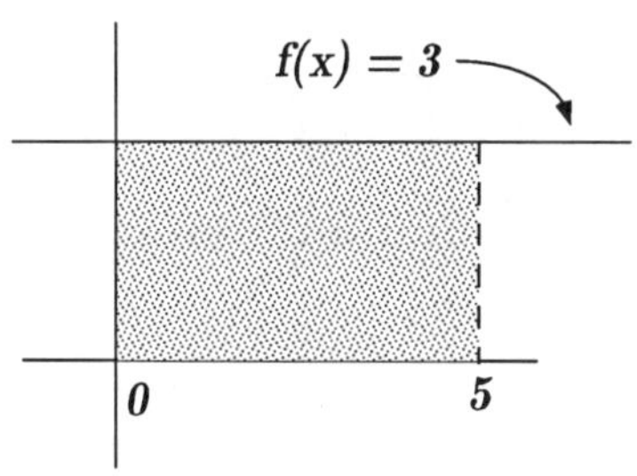

Clearly this area equals 15.

(2) The integral here is again the area of a rectangle and equals 15.

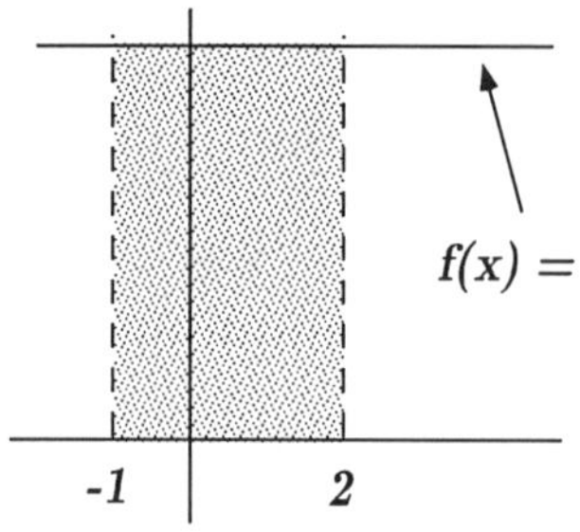

(3) The region under the graph of $f(x) = x$ where x ranges over the interval $0 \leq x \leq 3$ is the triangle below.

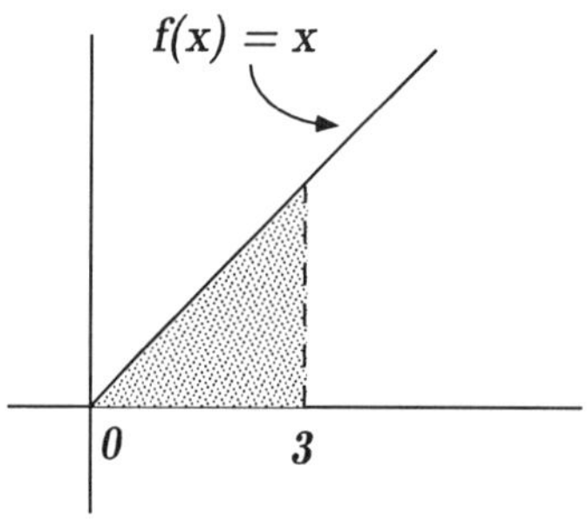

Thus the area is given $\frac{1}{2}(3 \cdot 3) = \frac{9}{2}$.

(4) Here the region under the line consists of a triangle and a rectangle.

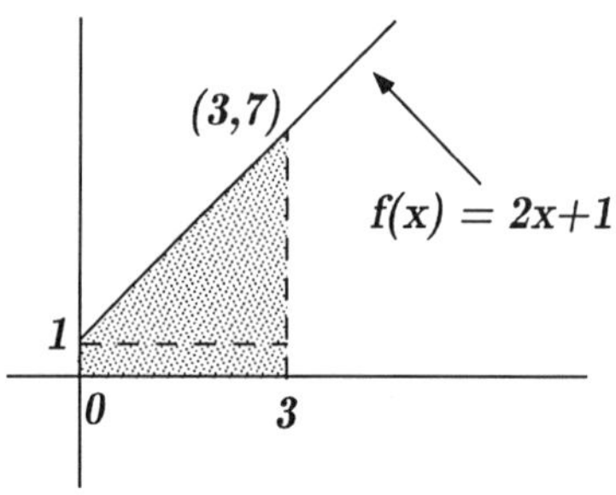

The integral is given by $\int_0^3 x\,dx = \frac{1}{2}(6 \cdot 3) + 3 \cdot 1 = 12$.

(5) Here again the region is the union of a triangle and a rectangle.

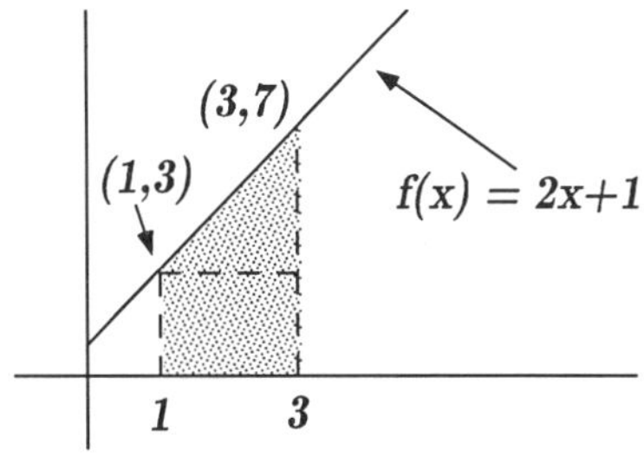

Thus the integral is given by $\int_1^3 (2x+1)\,dx = \frac{1}{2}(2\cdot 4) + (2\cdot 3) = 10.$

(6) Once again the region consists of a rectangle surmounted by a triangle,

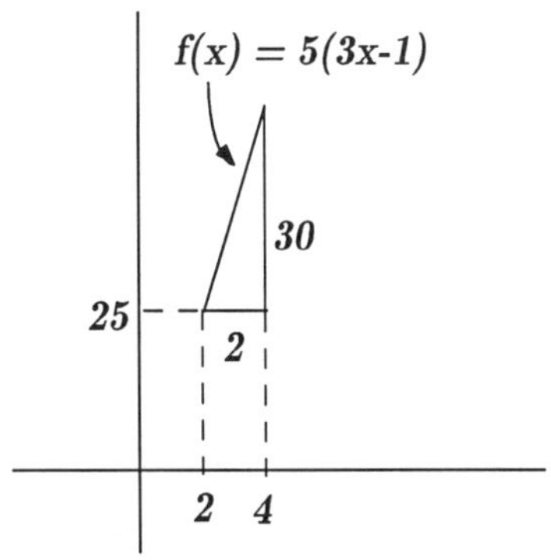

and the integral is $\int_2^4 5(3x-1)\,dx = (25\cdot 2) + \frac{1}{2}(2\cdot 30) = 80.$

(7) Dividing the interval $0 \le x \le 2$ in to 10 equal parts yields the points $\{0, .2, .4, .6, .8, 1, 1.2, 1.4, .1.6, 1.8, 2\}$ and the integral is approximated by

$$(.2)(.2) + (.4)(.2) + \cdots + (1.8)(.2) + (2)(.2) = 2.2.$$

The precise area of the region in question is 2.

(8) Dividing the interval up as we did in the previous problem approximates the integral by the sum

$$(1.2)(.2) + (1.4)(.2) + \cdots + (2.8)(.2) + (3)(.2) = 4.2.$$

Here the actual area is seen to be 4.

Solutions for §13.2

(1) From the figure below we see that the integral is given by $\int_1^3 (2x-4)\,dx = \frac{1}{2}(1\cdot 2) - \frac{1}{2}((1\cdot 2) = 0.$

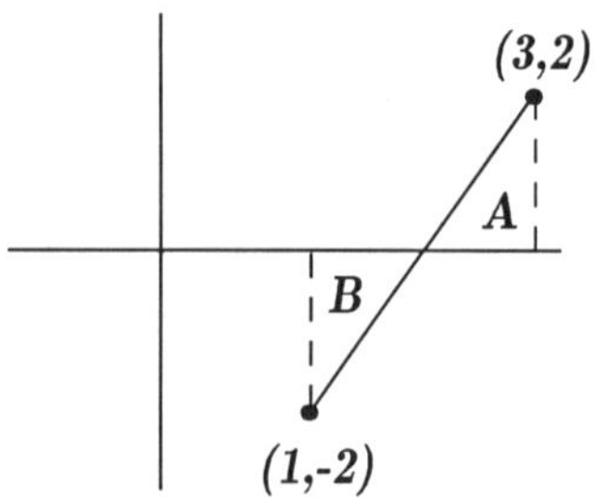

(2) The integral $\int_{-1}^{2} (3x-5)\,dx$ equals that area of the small triangle minus the area of the larger triangle in the figure below. Since $3x-5=0$ when $x=\frac{5}{3}$ we see that

$$\int_{-1}^{2} (3x-5)\,dx = \frac{1}{2}\left(\frac{1}{3}\cdot 1\right) - \frac{1}{2}\left(\frac{8}{3}\cdot 8\right) = -\frac{21}{2}.$$

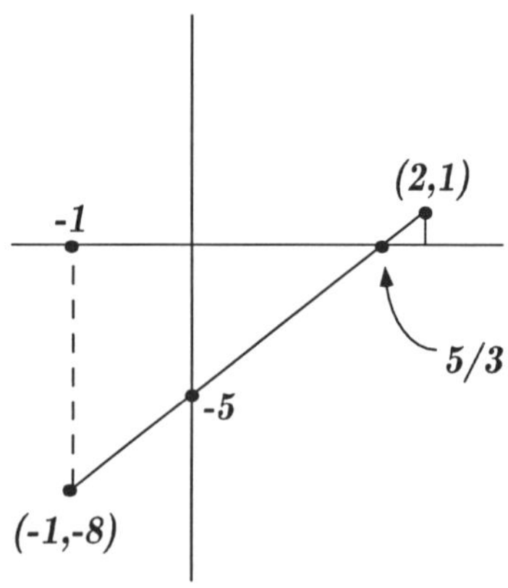

(3) Since the graph of the function $f(x)=x$ on the interval $-3\le x\le 3$ forms two identical triangles (one below the x–axis and one above the x–axis), we see that $\int_{-3}^{3} x\,dx = 0.$

(4) Referring to the figure below, $\int_{-2}^{3} (4x+1)\,dx$ equals the area of the triangle of height 13 minus the area of the triangle of height 7, i.e.,

$$\int_{-2}^{3} (4x+1)\,dx = \frac{1}{2}\left(13\cdot\frac{13}{4}\right) - \frac{1}{2}\left(\frac{7}{4}\cdot 7\right) = \frac{169}{8} - \frac{49}{8} = 15.$$

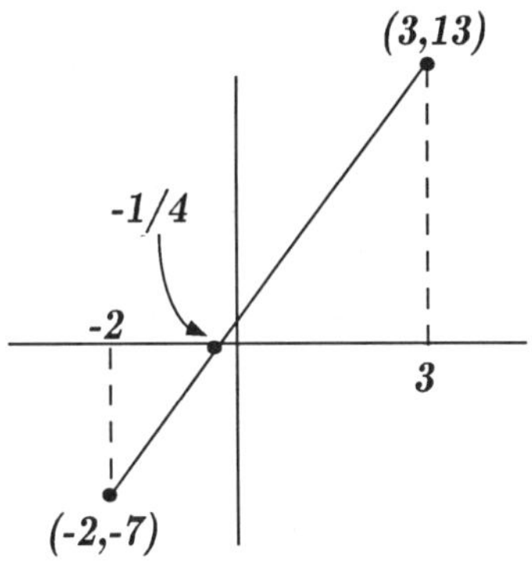

(5) The integral $\int_{-2}^{-1} (2x+1)\,dx$ must equal minus the area of the region in the figure below, i.e.,

$$\int_{-2}^{-1} (2x+1)\,dx = -\left(\frac{1}{2}(2\cdot 1)+1\right) = -2.$$

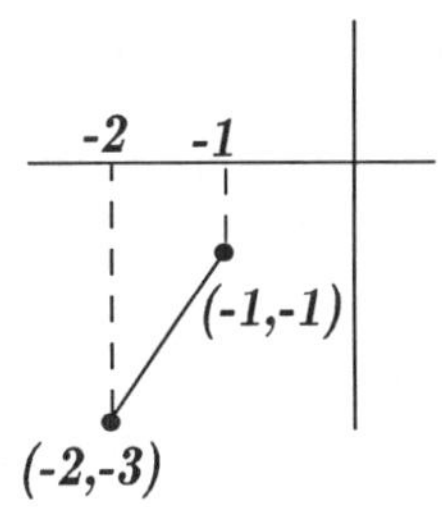

(6) Here again the graph of the function $f(x) = 3x-1$ falls below the x–axis on the interval $-1 \le x \le 0$ and thus

$$\int_{-1}^{0} (3x-1)\,dx = -\left(\frac{1}{2}\cdot 3\cdot 1 + 1\cdot 1\right) = -5/2.$$

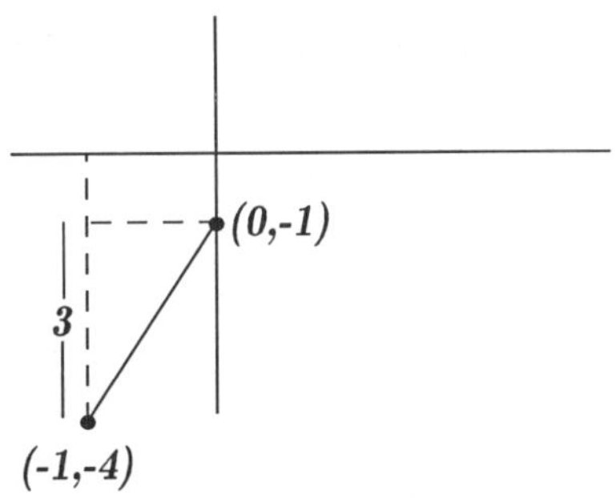

Solutions for §13.3

(1) Here $f(x) = x$, $a = 0$, and $b = 5$.

$$\begin{aligned}\int_0^5 x\,dx &= \lim_{N\to\infty}\Big(\frac{5}{N}+\frac{10}{N}+\cdots+\frac{N\cdot 5}{N}\Big)\frac{5}{N}\\ &= \lim_{N\to\infty} 25\Big(\frac{1+2+\cdots+N}{N^2}\Big)\\ &= \lim_{N\to\infty}\frac{25}{N^2}\Big(\frac{N(N+1)}{2}\Big)\\ &= \lim_{N\to\infty}\frac{25}{2}\left(\frac{N+1}{N}\right)\\ &= \lim_{N\to\infty}\left(\frac{25}{2}+\frac{25}{2N}\right)=\frac{25}{2}\end{aligned}$$

(2) As in problem (1), $f(x) = x$ and here $a = -1$ and $b = 2$.

$$\begin{aligned}\int_{-1}^2 x\,dx &= \lim_{N\to\infty}\left[\left(-1+\frac{3}{N}\right)+\left(-1+\frac{2\cdot 3}{N}\right)+\cdots+\left(-1+\frac{3N}{N}\right)\right]\frac{3}{N}\\ &= \lim_{N\to\infty}\frac{3}{N}\Big[-N+\frac{3}{N}+\frac{2\cdot 3}{N}+\cdots+\frac{3N}{N}\Big]\\ &= \lim_{N\to\infty}\frac{3}{N}\Big[-N+3\left(\frac{1+2+\cdots+N}{N}\right)\Big]\\ &= \lim_{N\to\infty}\left(-3+\frac{9}{2}\left(\frac{N+1}{N}\right)\right)\\ &= \lim_{N\to\infty}\left(-3+\frac{9}{2}+\frac{9}{2N}=-3+\frac{9}{2}\right)=\frac{3}{2}.\end{aligned}$$

(3) The function in question here is $f(x) = 4x - 1$ and $a = 1$, $b = 2$.

$$\begin{aligned}\int_1^2 (4x-1)\,dx &= \lim_{N\to\infty}\frac{1}{N}\Big[4\left(1+\frac{1}{N}\right)-1+4\left(1+\frac{2}{N}\right)-1+\cdots+4\left(1+\frac{N}{N}\right)-1\Big]\\ &= \lim_{N\to\infty}\frac{1}{N}\Big[\left(3+\frac{4}{N}\right)+\left(3+\frac{4\cdot 2}{N}\right)+\cdots+\left(3+\frac{4\cdot N}{N}\right)\Big]\\ &= \lim_{N\to\infty}\frac{1}{N}\Big[3N+4\left(\frac{1+2+\cdots+N}{N}\right)\Big]\\ &= \lim_{N\to\infty}\left(3+\frac{4}{N^2}\left(\frac{N(N+1)}{2}\right)\right)\\ &= \lim_{N\to\infty}\left(3+2\left(\frac{N^2+N}{N^2}\right)\right)\\ &= \lim_{N\to\infty}\left(3+2+\frac{2}{N}\right)=5.\end{aligned}$$

(4) Returning to the function $f(x) = x$ we consider the case $a = -3$ and $b = 3$.

$$\begin{aligned}\int_{-3}^{3} x\,dx &= \lim_{N\to\infty} \frac{6}{N}\Big[\Big(-3+\frac{6}{N}\Big)+\Big(-3+\frac{2\cdot 6}{N}\Big)+\cdots+\Big(-3+\frac{N\cdot 6}{N}\Big)\Big]\\ &= \lim_{N\to\infty} \frac{6}{N}\Big[-3N+\frac{6}{N}\cdot(1+2+\cdots+N)\Big]\\ &= \lim_{N\to\infty} \frac{6}{N}\Big[-3N+\frac{6}{N}\Big(\frac{N(N+1)}{2}\Big)\Big]\\ &= \lim_{N\to\infty} \frac{6}{N}(-3N+3N+3)\\ &= \lim_{N\to\infty} \frac{18}{N} = 0.\end{aligned}$$

(5) $f(x) = 2x + 1$, and $a = -2$, $b = -1$.

$$\begin{aligned}\int_{-2}^{-1} (2x+1)\,dx &= \lim_{N\to\infty} \frac{1}{N}\Big[f\Big(-2+\frac{1}{N}\Big)+f\Big(-2+\frac{2}{N}\Big)+\cdots+f\Big(-2+\frac{N}{N}\Big)\Big]\\ &= \lim_{N\to\infty} \frac{1}{N}\Big[2\Big(-2+\frac{1}{N}\Big)+1+2\Big(-2+\frac{2}{N}\Big)+1+\cdots+2\Big(-2+\frac{N}{N}\Big)+1\Big]\\ &= \lim_{N\to\infty} \frac{1}{N}\Big[N-4N+\Big(\frac{2}{N}+\frac{2\cdot 2}{N}+\cdots+\frac{2N}{N}\Big)\Big]\\ &= \lim_{N\to\infty} \frac{1}{N}\Big[-3N+\frac{2}{N}\Big(\frac{N^2+N}{2}\Big)\Big]\\ &= \lim_{N\to\infty} \Big(-3+\frac{N^2+N}{N^2}\Big)\\ &= \lim_{N\to\infty} \Big(-3+1+\frac{1}{N}\Big) = -2.\end{aligned}$$

(6) $f(x) = 3x - 1$ and $a = -1$, $b = 0$.

$$\begin{aligned}\int_{-1}^{0} (3x-1)\,dx &= \lim_{N\to\infty} \frac{1}{N}\Big[f\Big(-1+\frac{1}{N}\Big)+f\Big(-1+\frac{2}{N}\Big)+\cdots+f\Big(-1+\frac{N}{N}\Big)\Big]\\ &= \lim_{N\to\infty} \frac{1}{N}\Big[3\Big(-1+\frac{1}{N}\Big)-1+3\Big(-1+\frac{2}{N}\Big)-1+\cdots+3\Big(-1+\frac{N}{N}\Big)-1\Big]\\ &= \lim_{N\to\infty} \frac{1}{N}\Big(-4N+\frac{3}{N}(1+2+\cdots+N)\Big)\\ &= \lim_{N\to\infty} \Big(-4+\frac{3}{N^2}\Big(\frac{N^2+N}{2}\Big)\Big)\\ &= \lim_{N\to\infty} \Big(-4+\frac{3}{2}+\frac{3}{2N}\Big) = -4+\frac{3}{2} = -\frac{5}{2}.\end{aligned}$$

Solutions for Additional Exercises for Chapter 13

(1) Upper and lower bounds for f are given by $0 < f(x) < 1$. Furthermore, f is integrable, and the area under f is 2 as can be seen from the figure below.

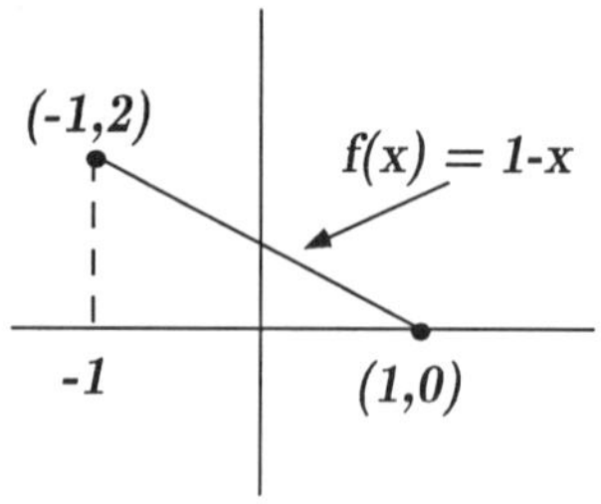

f is NOT differentiable at $x = -1$ and $x = 1$.

In the following problems we use the notation $\sum_{i=M}^{N} f(i) = f(M)+f(M+1)+f(M+2)+\cdots+f(N)$.

(2) Partition $[a, b]$ into n equal parts $a = x_0 < x_1 < \ldots < x_n = b$, with $x_i - x_{i-1} = h = \frac{b-a}{n}$. If $f(x) \leqslant g(x)$ on $[a, b]$ then $\sum_{i=1}^{n} f(x_i)h \leqslant \sum_{i=1}^{n} g(x_i)h$. Taking limits as $n \to \infty$, we obtain $\int_a^b f(x)\, dx \leqslant \int_a^b g(x)dx$.

(3a) In general, $\int_a^b s(x)\, dx = \sum_{i=1}^{n} f(x_{i-1})(x_i - x_{i-1})$. Therefore,

$\int_a^b s(x)\, dx = (0)(2-1) + (1)(3-2) + (2)(4-3) + (1)(5-4) + (1)(6-5) = 4.$

(3b) $\int_a^b s(x)\, dx = (-1)(\frac{1}{2} - 0) + (-2)(1 - \frac{1}{2}) + (-1)(\frac{3}{2} - 1) + (2)(2 - \frac{3}{2}) + (1)(\frac{5}{2} - 2) = -\frac{1}{2}.$

(3c) $\int_a^b s(x)\, dx = (\frac{1}{4})(0 - (-1)) + (0)(1 - 0) + (\frac{1}{2})(2 - 1) + (\frac{1}{2})(3 - 2) + (1)(4 - 3) = \frac{9}{4}.$

(4) $\int_0^b x^2\, dx = \lim_{n\to\infty} \sum_{i=1}^{n} \left(\frac{bi}{n}\right)^2 \frac{b}{n} = \lim_{n\to\infty} \frac{b^3}{n^3} \sum_{i=1}^{n} i^2 = \lim_{n\to\infty} \frac{b^3}{n^3} \left(\frac{1}{3}n^3 + \frac{1}{2}n^2 + \frac{1}{6}n\right) = \frac{b^3}{3}.$

(5) $\int_0^b x^3\, dx = \lim_{n\to\infty} \sum_{i=1}^{n} \left(\frac{bi}{n}\right)^3 \frac{b}{n} = \lim_{n\to\infty} \frac{b^4}{n^4} \sum_{i=1}^{n} i^3 = \lim_{n\to\infty} \frac{b^4}{n^4} \left(\frac{1}{4}n^4 + \frac{1}{2}n^3 + \frac{1}{4}n^2\right) = \frac{b^4}{4}.$

(6) In general we can express the power sum $\sum\limits_{i=1}^{n} x^m$ as an $(m+1)^{th}$ degree polynomial in the variable n, with leading coefficient $\frac{1}{m+1}$. Specifically, $\sum\limits_{i=1}^{n} x^m = \frac{1}{m+1}x^{m+1} +$ (terms of lower order). Therefore,

$$\int\limits_0^b x^m\, dx = \lim_{n\to\infty} \sum_{i=1}^{n} \left(\frac{bi}{n}\right)^m \frac{b}{n} = \lim_{n\to\infty} \frac{b^{m+1}}{n^{m+1}} \sum_{i=1}^{n} i^m$$

$$= \lim_{n\to\infty} \frac{b^{m+1}}{n^{m+1}} \left(\frac{1}{m+1} n^{m+1} + \ldots \text{ (terms of lower order)}\right) = \frac{b^{m+1}}{m+1}.$$

(7) Letting $f(x) = \exp - \left(\frac{1}{2}x^2\right)$, and $a = 0$, $b = 1$, we approximate $\int\limits_a^b f(x)\, dx$ using the finite sum $\sum\limits_{i=0}^{n-1} f\left(a + \frac{b-a}{n} i\right) \frac{b-a}{n}$. For $n = 10$ we get 0.8748. For $n = 20$ we get 0.8653. For $n = 30$ we get 0.8621.

(8) For $n = 10$ we get 0.2691. For $n = 20$ we get 0.2895. For $n = 30$ we get 0.2963.

(9) For $n = 10$ we get -1.6840. For $n = 20$ we get -2.3400. For $n = 30$ we get -2.7330.

(10) For $n = 10$ we get 0.1711. For $n = 20$ we get 0.1803. For $n = 30$ we get 0.1833.

(11) For n large, we have

$$\int_0^1 \frac{dx}{\left(1 + \frac{x^2}{3}\right)^2} \approx \frac{1}{n} \sum_{i=1}^{n} \left(1 + \frac{1}{3}\left(\frac{i}{n}\right)^2\right)^{-2}.$$

For $n = 10$, we get 0.806, for $n = 20$, we get 0.817, and for $n = 30$, we get 0.821.

(12) Note that $\frac{\sqrt{2\pi}}{2} \approx 1.2533$. Approximations are:

$\int\limits_0^1 \exp\left(-\frac{x^2}{2}\right)\, dx \approx 0.8556.$

$\int\limits_0^2 \exp\left(-\frac{x^2}{2}\right)\, dx \approx 1.1963.$

$\int\limits_0^3 \exp\left(-\frac{x^2}{2}\right)\, dx \approx 1.2499.$

$\int\limits_0^4 \exp\left(-\frac{x^2}{2}\right)\, dx \approx 1.2532.$

(13) For $n = 10$ we get 3.3045. For $n = 30$ we get 3.2011. Both of these values (roughly) estimate π.

Solutions for §14.1

(1) 3025. **(2)** 2585. **(3)** 2692. **(4)** $\frac{\sqrt{3}}{2}$. **(5)** $\frac{-1}{2}$. **(6)** ≈ 1.1975. **(7)**62. **(8)** $\frac{31}{32}$. **(9)** $\frac{3323}{128}$.

(10) The sums for $N = 10, 100, 1000$ are 1.5497, 1.6349, 1.6439 which are getting close to $\frac{\pi^2}{6}$ = 1.6449 (to 5 places).

(11) The sums in this case for $N = 10, 100, 1000$ are 2.9289, 5.1873, 7.4854. $\ln(N)$ is a good approximation to this sum since

$$\sum_{n=1}^{N} \frac{1}{n} \approx \int_{1}^{N} \frac{1}{x}\,dx = \ln(N)$$

Solutions for §14.2

(1) Letting $f(N) = \sum_{n=1}^{N} \frac{1}{n(n+1)}$ we see that $f(1) = \frac{1}{2}$, $f(2) = \frac{2}{3}$, $f(3) = \frac{3}{4}$ and so we guess that $f(N) = \frac{N}{N+1}$. So the initial cases are true, and it remains to check the induction step. Assuming that $f(N) = \frac{N}{N+1}$ we see that $f(N+1) = f(N) + \frac{1}{(N+1)(N+2)}$ which by assumption is $\frac{N}{N+1} + \frac{1}{(N+1)(N+2)} = \frac{N+1}{N+2}$, and the formula now holds for all positive integers N by induction.

(2) By experimentation we guess $f(N) = \sum_{n=1}^{N} \frac{1}{2^n} = \frac{(2^N-1)}{2^N}$ since this holds for small values of N. Assume it's true for N. Then $f(N+1) = f(N) + \frac{1}{2^{N+1}} = \frac{(2^N-1)}{2^N} + \frac{1}{2^{N+1}} = \frac{(2^{N+1}-1)}{2^{N+1}}$ as required and the formula is proven.

(3) By experimentation we guess $f(N) = \sum_{n=1}^{N} \frac{1}{3^n} = \frac{(3^N-1)}{2 \cdot 3^N}$ since this holds for small values of N. Assume it's true for N, then $f(N+1) = f(N) + \frac{1}{3^{N+1}} = \frac{(3^N-1)}{2\cdot 3^N} + \frac{1}{3^{N+1}} = \frac{(3^{N+1}-1)}{2\cdot 3^{N+1}}$ as required.

(4) Let $f(N) = \sum_{n=1}^{N} n^3$. Then the given formula holds for $N = 1$ and $f(N+1) = f(N) + (N+1)^3$, which by the induction hypothesis equals $\frac{N^2(N+1)^2}{4} + (N+1)^3 = \frac{(N+1)^2(N+2)^2}{4}$ as was to be shown.

(5) Let $f(N) = \sum_{n=1}^{N} n^4$. Then the given formula holds for $N = 1$. By the induction hypothesis, we have

$$\begin{aligned} f(N+1) = f(N) + (N+1)^4 &= \frac{N^5}{5} + \frac{N^4}{2} + \frac{N^3}{3} - \frac{N}{30} + (N+1)^3 \\ &= \frac{(N+1)^5}{5} + \frac{(N+1)^4}{2} + \frac{(N+1)^3}{3} - \frac{(N+1)}{30} \end{aligned}$$

as was to be shown.

(6) We want to show $f(N) = \sum_{n=1}^{N} a^n = \frac{(a^{N+1}-a)}{(a-1)}$ for $a \neq 1$. This formula is true for $N = 1$ and assuming it's true for N, $f(N+1) = f(N) + a^{N+1} = \frac{(a^{N+1}-a)}{(a-1)} + a^{N+1} = \frac{(a^{N+2}-a)}{(a-1)}$ as required.

Solutions for §14.3

(1)

$$\int_1^4 x^2\,dx = \lim_{N\to\infty}\left[\sum_{n=1}^{N}\left(1+\frac{3n}{N}\right)^2\right]\cdot\frac{3}{N}$$

$$= \lim_{N\to\infty}\frac{3}{N}\cdot\left[\left(\sum_{n=1}^{N}1\right)+\frac{6}{N}\left(\sum_{n=1}^{N}n\right)+\frac{9}{N^2}\left(\sum_{n=1}^{N}n^2\right)\right]$$

$$= \lim_{N\to\infty} 21+\frac{45}{2N}+\frac{9}{2N^2} \;=\; 21.$$

(2) In a similar manner to question (1) we get $\frac{16}{3}$.

(3) $\frac{3^4}{4}$. **(4)** $\frac{B^4}{4}$. **(5)** $\frac{B^3-A^3}{3}$. **(6)** $\frac{B^4-A^4}{4}$. **(7)** $\frac{B^5-A^5}{5}$.

(8)

$$\int_0^B x^k\,dx = \lim_{N\to\infty}\left[\sum_{n=1}^{N}\left(\frac{nB}{N}\right)^k\right]\cdot\frac{B}{N}$$

$$= \lim_{N\to\infty}\frac{B^{k+1}}{N^{k+1}}\left(\sum_{n=1}^{N}n^k\right)$$

$$= \lim_{N\to\infty}\frac{B^{k+1}}{N^{k+1}}\left(\frac{N^{k+1}}{k+1}+\frac{N^k}{2}+p(N)\right)$$

$$= \lim_{N\to\infty}\left(\frac{B^{k+1}}{k+1}+\frac{B^{k+1}}{2N}+\frac{B^{k+1}p(N)}{N^{k+1}}\right)=\frac{B^{k+1}}{k+1}.$$

Solutions for §14.4

(1) For this problem, we have $N=4$, $a=-1$, $b=3$, $x_0=-1$, $x_1=0$, $x_2=1$, $x_3=2$, $x_4=3$.

<u>Trapezoid Rule:</u> $\int\limits_{-1}^{3}(x^2+2x)\,dx = \frac{1}{2}\Big[\big((-1)^2-2\big)+\big(0^2+0\big)\Big]+\frac{1}{2}\Big[\big((0)^2+0\big)+\big(1^2+2\big)\Big]$

$+\frac{1}{2}\Big[\big(1^2+2\big)+\big(2^2+4\big)\Big]+\frac{1}{2}\Big[\big(2^2+4\big)+\big(3^2+6\big)\Big] \;=18.$

When $N=10$, we obtain 17.44, and when $N=50$, we obtain 17.3376.

<u>Simptson's Rule:</u> $\int\limits_{-1}^{3}(x^2+2x)\,dx = \frac{1}{3}\Big[\big((-1)^2-2\big)+4\,\big(0^2+0\big)+\big(1^2+2\big)\Big]$

$+\frac{1}{3}\Big[\big((1)^2+2\big)+4\,\big(2^2+4\big)+\big(3^2+6\big)\Big] \;=\; 17\frac{1}{3}$

When $N=10, 50$, we also obtain $17\frac{1}{3}$.

(2) For this problem, we have $N = 4$, $a = -2$, $b = 1$, $x_0 = -2$, $x_1 = -\frac{5}{4}$, $x_2 = -\frac{2}{4}$, $x_3 = \frac{1}{4}$, $x_4 = 1$.

Trapezoid Rule:

$$\int_{-2}^{3} (3x^2 + 4)\, dx = \frac{1}{2}\Big[(3(-2)^2 + 4) + \Big(3(\frac{-5}{4})^2 + 4\Big)\Big] \cdot \frac{3}{4} + \frac{1}{2}\Big[\Big(3(\frac{-5}{4})^2 + 4\Big) + \Big(3(\frac{-2}{4})^2 + 4\Big)\Big] \cdot \frac{3}{4}$$

$$+ \frac{1}{2}\Big[\Big(3(\frac{-2}{4})^2 + 4\Big) + \Big(3(\frac{1}{4})^2 + 4\Big)\Big] \cdot \frac{3}{4} + \frac{1}{2}\Big[\Big(3(\frac{1}{4})^2 + 4\Big) + (3 \cdot 1^2 + 4)\Big] \cdot \frac{3}{4}$$

$$= 21.8437.$$

When $N = 10$, we obtain 21.135, and when $N = 50$, we obtain 21.0054.

Simptson's Rule:

$$\int_{-2}^{1} (3x^2 + 4)\, dx = \frac{1}{3}\Big[(3(-2)^2 + 4) + 4\Big(3(\frac{-5}{4})^2 + 4\Big) + \Big(3(\frac{-2}{4})^2 + 4\Big)\Big] \cdot \frac{3}{4}$$

$$+ \frac{1}{3}\Big[\Big(3(\frac{-2}{4})^2 + 4\Big) + 4\Big(3(\frac{1}{4})^2 + 4\Big) + (3 \cdot 1^2 + 4)\Big] \cdot \frac{3}{4}$$

$$= 21.$$

When $N = 10, 50$, we also obtain 21 with Simpson's rule which is the exact answer.

For the following questions the approximations obtained for each integral using first the trapezoid rule (for N=4,10,50) and then Simpson's rule for these values of N are given.

(3) 3.4455, 3.4473, 3.4477; 3.4477, 3.4477, 3.4477.

(4) 0.7969, 0.7988, 0.7992; 0.7992, 0.7992, 0.7992.

(5) 1.0928, 1.0837, 1.0820; 1.0820, 1.0820, 1.0820.

(6) 1.1741, 1.1926, 1.1961; 1.1961, 1.1962, 1.1962.

(7) 3.1311, 3.1399, 3.1415; 3.1415, 3.1415, 3.1415.

(8) 1.8067, 1.8298, 1.8348; 1.8337, 1.8350, 1.8350.

(9) 1.5284, 1.5197, 1.5206; 1.5162, 1.5206, 1.5206.

(10) $0, -4.336 \times 10^{-20}, 7.3725 \times 10^{-20}$; $0, 1.4456 \times 10^{-20}, -6.0715 \times 10^{-20}$.

(11) 1.8314, 1.8450, 1.8475; 1.8476, 1.8476, 1.8476.

(12) 0.0924, 0.1063, 0.1025; 0.1118, 0.1023, 0.1023.

(13) Let $f(x) = ax^2 + bx + c$. Let $h > 0$ be arbitrary. Then

$$\int_0^{2h} \left(ax^2 + bx + c\right)\, dx = a\frac{(2h)^3}{3} + b\frac{(2h)^2}{2} + 2ch$$
$$= \frac{1}{3}h\big(f(0) + 4f(h) + f(2h)\big),$$

since $f(0) = c$, $f(h) = ah^2 + bh + c$, and $f(2h) = a(2h)^2 + b(2h) + c$. Setting $h = \frac{1}{2}$, we see that Simpson's rule (with one subdivision) gives the exact value of the integral. The same argument generalizes to several subdivisions.

Solutions for Additional Exercises for Chapter 14

(1) The left endpoint approximation to $\int_a^b f(x)\, dx$ is $\sum_{i=0}^{N-1} f\left(a + \frac{b-a}{N}i\right) \frac{b-a}{N}$. The right endpoint approximation to $\int_a^b f(x)\, dx$ is $\sum_{i=1}^{N} f\left(a + \frac{b-a}{N}i\right) \frac{b-a}{N}$. The trapeziodal approximation will always be equal to the average of the left endpoint and right endpoint approximations. We calculate these approximations for $N = 4$ and $N = 10$. For $N = 4$, the left endpoint approximation is 10, the right endpoint approximation is 26, and the trapezoidal approximation is 13. For $N = 10$, the left endpoint approximation is 14.24, the right endpoint approximation is 20.64, and the trapezoidal approximation is 17.44.

(2) We estimate π using the Simpson's rule approximation to the integral $\frac{4}{9}\int_0^3 \sqrt{9 - x^2}\, dx$. The estimate for π with $N = 4$ is 3.0836. The estimate for π with $N = 10$ is 3.1270. The estimate for π with $N = 50$ is 3.1403.

(3) We estimate π using the Simpson's rule approximation to the integral $4\int_0^1 \frac{1}{1+x^2}\, dx$. The estimate for π with $N = 10$ is 3.141592654.

(4a) The left endpoint approximation (lower bound) with $N = 4$ is $\frac{7}{4}$.

(4b) The right endpoint approximation (upper bound) with $N = 4$ is $\frac{15}{4}$.

(4c) The difference between the two estimates is $\frac{15}{4} - \frac{7}{4} = 2$.

(5) The left endpoint approximation to $\int_a^b f(x)\, dx$ is $\sum_{i=0}^{N-1} f\left(a + \frac{b-a}{N}i\right) \frac{b-a}{N}$. The right endpoint approximation to $\int_a^b f(x)\, dx$ is $\sum_{i=1}^{N} f\left(a + \frac{b-a}{N}i\right) \frac{b-a}{N}$. The difference between these two approximations is $\sum_{i=1}^{N} f\left(a + \frac{b-a}{N}i\right) \frac{b-a}{N} - \sum_{i=0}^{N-1} f\left(a + \frac{b-a}{N}i\right) \frac{b-a}{N} = \left(f(b) - f(a)\right) \frac{b-a}{N}$.

(6) $(\exp(9) - \exp(1))\, \frac{2}{N} < 50$ gives $N \geqslant 325$.

(7a) For the trapeziodal rule, the absolute value of the error is $< \frac{L}{150}$. For Simpson's rule, the absolute value of the error is $< \frac{M}{56250}$.

(7b) $\frac{M \cdot 32}{180N^4} < 0.00002$ gives $N \geqslant 9.71\sqrt[4]{M}$.

(8) With $f(x) = \frac{1}{1+x^2}$, we can compute the first four derivatives and obtain the inequalities $-2 \leqslant f''(x) \leqslant \frac{1}{2}$ and $-3 \leqslant f''''(x) \leqslant 24$, if $0 \leqslant x \leqslant 1$. So, if $0 \leqslant x \leqslant 1$ we have $|f''(x)| \leqslant 2$ and have $|f''''(x)| \leqslant 24$. We compute the value of N to insure the error is less that 10^{-3}: For the trapezoidal rule, $\frac{2}{12N^2} < 10^{-3}$ gives $N \geqslant 13$. For Simpson's rule, $\frac{24}{180N^4} < 10^{-3}$ gives $N \geqslant 4$.

(9a) Using Simpson's rule, the estimate of the integral with either $N = 10$ or $N = 40$ is -12.8333.

(9b) As long as N is divisible by 10, the Simpson's rule approximation of the integral will be exact.

In the following problems, we assume the formula holds for N, and then prove by induction that it holds for N + 1.

(10) $\sum_{n=1}^{N+1}(2n-1) = \sum_{n=1}^{N}(2n-1) + (2N+1) = N^2 + 2N + 1 = (N+1)^2.$

(11)
$$\sum_{n=1}^{N+1} n^5 = \sum_{n=1}^{N} n^5 + (N+1)^5 = \frac{1}{12}N^2(N+1)^2(2N^2+2N-1) + (N+1)^5$$
$$= \frac{1}{12}(N+1)^2(N+2)^2\big(2(N+1)^2 + 2(N+1) - 1\big).$$

(12) $\sum_{n=1}^{N+1}(2n-1)^2 = \frac{1}{3}N(4N^2-1) + (2N+1)^2 = \frac{1}{3}(N+1)\big(4(N+1)^2-1\big).$

(13) $\sum_{n=1}^{N+1} n(n+1)^2 = \frac{1}{12}N(N+1)(N+2)(3N+5) + (N+1)(N+2)^2$

$= \frac{1}{12}(N+1)(N+2)(N+3)\big(3(N+1)+5\big).$

(14) $\sum_{n=1}^{N+1} n! \cdot n = \Big[(N+1)! - 1\Big] + \Big[(N+1)! \cdot (N+2) - (N+1)!\Big] = (N+2)! - 1.$

(15) $\sum_{n=1}^{N+1} \frac{1}{n^2-1} = \left[\frac{3}{4} - \frac{2N+1}{2N(N+1)}\right] + \left[\frac{1}{N(N+2)}\right] = \frac{3}{4} - \frac{2(N+1)+1}{2(N+1)(N+2)}.$

Solutions for §15.1

(1) $\int_1^3 13\, dx = 13x\Big]_1^3 = 26.$

(2) $\int_1^5 \sqrt{x}\, dx = \frac{2}{3}x^{\frac{3}{2}}\Big]_1^5 = \frac{2}{3}(5\sqrt{5}-1).$

(3) $\int_1^7 \frac{1}{x^4}\, dx = -\frac{1}{3}x^{-3}\Big]_1^7 = \frac{114}{343}.$

(4) $\int_2^{10} \left(\frac{1}{x^4} - \frac{1}{x^3}\right) dx = \frac{-2+3x}{6x^3}\Big]_2^{10} = -\frac{59}{750}.$

(5) $\int\limits_0^3 (1+x)^3\,dx = x + \frac{3}{2}x^2 + x^3 + \frac{1}{4}x^4 \Big]_0^3 = \frac{255}{4}.$

(6) $\int\limits_{\frac{1}{2}}^4 (1+x)^4\,dx = x + 2x^2 + 2x^3 + x^4 + \frac{1}{5}x^5 \Big]_{\frac{1}{2}}^4 = \frac{99757}{160}.$

(7) $\int\limits_1^5 \frac{7}{\sqrt{x}}\,dx = 14\sqrt{x}\Big]_1^5 = 14(-1+\sqrt{5}).$

(8) $\int\limits_{-4}^{-2} \frac{7}{x^5}\,dx = \frac{-7}{4x^4}\Big]_{-4}^{-2} = \frac{-105}{1024}.$

(9) $\int\limits_0^1 x\sqrt[3]{x}\,dx = \frac{3x^{7/3}}{7}\Big]_0^1 = \frac{3}{7}.$

(10) $\int\limits_0^3 x(\sqrt{x}+x^3)\,dx = \frac{2}{5}x^{\frac{5}{2}} + \frac{1}{5}x^5\Big]_0^3 = \frac{9}{5}(27+2\sqrt{3}).$

(11) $\int\limits_1^2 \left(x+\frac{1}{x^2}\right)dx = -\frac{1}{x} + \frac{x^2}{2}\Big]_1^2 = 2.$

(12) $\int\limits_1^2 \left(x+\frac{1}{x^2}\right)^2 dx = -\frac{1}{3x^3} + \frac{1}{3}x^3 + 2\ln(x)\Big]_1^2 = \frac{21}{8} + 2\ln(2).$

(13) $\int\limits_1^{10} \frac{x-1}{\sqrt{x}}\,dx = \frac{2}{3}(-3+x)\sqrt{x}\Big]_1^{10} = \frac{2}{3}(2+7\sqrt{10}).$

(14) $\int\limits_0^1 x^{1.7}\,dx = \frac{10}{27}x^{2.7}\Big]_0^1 = \frac{10}{27}.$

(15) $\int\limits_2^4 \frac{x^2-1}{x-1}\,dx = x + \frac{1}{2}x^2\Big]_2^4 = 8.$

(16) $\int\limits_0^1 (x^7+x+3)\cdot(x^2-3)\,dx = -9x - \frac{3}{2}x^2 + x^3 + \frac{1}{4}x^4 - \frac{3}{8}x^8 + \frac{1}{10}x^{10}\Big]_0^1 = -\frac{381}{40}.$

(17) $\int\limits_0^2 (x^7+x+3)\cdot(\sqrt{x}-3)\,dx = -9x + 2x^{\frac{3}{2}} - \frac{3}{2}x^2 + \frac{2}{5}x^{\frac{2}{5}} - \frac{3}{8}x^8 + \frac{2}{17}x^{\frac{17}{2}}\Big]_0^2$

$= -120 + 3036\sqrt{2}/85.$

(18) $\int\limits_0^{\frac{\pi}{3}} \sin(x)\,dx = -\cos(x)\Big]_0^{\frac{\pi}{3}} = \frac{1}{2}.$

(19) $\int\limits_0^{\frac{\pi}{3}} \cos(4x)\,dx = \frac{1}{4}\sin(4x)\Big]_0^{\frac{\pi}{3}} = -\frac{1}{8}\sqrt{3}.$

(20) $\int_{0}^{\frac{\pi}{2}} (\cos(x) + \sin(x))\, dx = -\cos(x) + \sin(x)\Big]_{0}^{\pi/2} = 2.$

(21) $\int_{\frac{\pi}{4}}^{\pi} (2\sin(x) + \sec^2(x))\, dx = \sec(x)(-1 - \cos(2x) + \sin(x))\Big]_{\pi/4}^{\pi} = 1 + \sqrt{2}.$

(22) $\int_{\frac{2\pi}{3}}^{\pi} \sec(x)\tan(x)\, dx = \sec(x)\Big]_{\frac{2\pi}{3}}^{\pi} = 1.$

(23) By the fundamental theorem of calculus $\frac{d(\cos(x)+1)}{dx} = f(x) = -\sin(x)$. So, on integrating $-\sin(x)$ we see that $\cos(c) = -1$ and, therefore, one possible value for c is π.

(24) As before there exists an F with $\frac{dF(x)}{dx} = f(x)$ and $x^6 + 8 = F(x^2) - F(c) = \int_c^{x^2} f(t)\, dt$. Hence we see $6x^5 = F'(x^2)2x = f(x^2)2x \Rightarrow f(x) = 3x^2$. On integrating this we get $c = -2$.

(25) There exists F with $\frac{dF(t)}{dt} = \sqrt{t^6 - 1}$ and $F(x) - F(1) = h(x)$. Hence we see $\frac{dh(x)}{dx} = \frac{dF(x)}{dx} = \sqrt{x^6 - 1}$.

(26) There exists F with $\frac{dF(t)}{dt} = \sin^2(t)$ and $F(x) - F(1) = h(x)$. $\Rightarrow \frac{dh(x)}{dx} = \frac{dF(x)}{dx} = \sin^2(x)$.

(27) There exists F with $\frac{dF(t)}{dt} = \cos(t)$ and $F(1) - F(x) = h(x)$. $\Rightarrow \frac{dh(x)}{dx} = \frac{-dF(x)}{dx} = -\cos(x)$.

(28) There exists F with $\frac{dF(t)}{dt} = \frac{t^2}{t^7 - \sin(t)}$ and $F(x^2) - F(1) = h(x)$. $\Rightarrow \frac{dh(x)}{dx} = \frac{dF(x^2)}{dx}2x = \frac{2x^5}{x^14 - \sin(x^2)}$.

(29) There exists$F$ such that $\frac{dF(t)}{dt} = \sin(t^2)$ and $F(x^4) - F(\cos(x)) = h(x)$. These formulae imply that$\frac{dh(x)}{dx} = \frac{dF(x^4)}{dx}4x^3 - \frac{dF(\cos(x))}{dx}(-\sin(x)) = \sin(x^8)4x^3 + \sin(\cos(x))\sin(x)$.

(30) F exists with $\frac{dF(t)}{dt} = (3 - \sqrt[3]{t})^2$ and $F(10x) - F(2x) = h(x)$. It follows from these formulae that $\frac{dh(x)}{dx} = 10\frac{dF(10x)}{dx} - 2\frac{dF(2x)}{dx} = 10(3 - \sqrt[3]{10x})^2 - 2(3 - \sqrt[3]{2x})^2$.

(31) There exists F with $\frac{dF(t)}{dt} = f(t)$ and $F(g_2(x)) - F(g_1(x)) = h(x)$. It immediately follows that $h'(x) = F'(g_2(x))g_2'(x) - F'(g_1(x))g_1'(x) = f(g_2(x))g_2'(x) - f(g_1(x))g_1'(x)$.

Solutions for §15.2

(1) $\int \frac{x^2 - x - 1}{x}\, dx = -x + \frac{x^2}{2} - \ln(x) + C.$

(2) $\int (3 - \sqrt[3]{x})^2\, dx = 9x - \frac{9x^{\frac{4}{3}}}{2} + \frac{3x^{\frac{5}{3}}}{5} + C.$

(3) $\int (\sin(x) + 2\sec^2(x))\, dx = -\frac{1 + \cos(2x) - 4\sin(x)}{(2\cos(x))} + C.$

(4) $\int \left(\sqrt{x} + \frac{1}{\sqrt[3]{x}}\right)^2 dx = 3x^{\frac{1}{3}} + \frac{12x^{\frac{7}{6}}}{7} + \frac{x^2}{2} + C.$

(5) $\int (1-x)(1-x^2)(1-x^3)\,dx = x - \frac{x^2}{2} - \frac{x^3}{3} + \frac{x^5}{5} + \frac{x^6}{6} - \frac{x^7}{7} + C.$

(6) $\int \sec(x)\tan(x)\,dx = \sec(x) + C.$

(7) $\int \left(\frac{1}{x} + \frac{1}{x^2} + \frac{1}{x^3}\right)\,dx = -\frac{1}{2x^2} - \frac{1}{x} + \ln(x) + C.$

(8) $\int \sqrt[3]{x^{-2}}\,dx = 3x^{\frac{1}{3}} + C.$

(9) $\int \sqrt[4]{x^5}\,dx = \frac{5x^{\frac{9}{5}}}{9} + C.$

(10) $\int \sqrt{x}\left(x^3 - \frac{1}{\sqrt[3]{x}}\right)\,dx = -\frac{6x^{\frac{7}{6}}}{7} + \frac{2x^{\frac{9}{2}}}{9} + C.$

(11) $\int \sqrt{x}\left(x^3 - \frac{1}{\sqrt[3]{x}}\right)^2\,dx = -\frac{6x^{\frac{5}{6}}}{5} - \frac{12x^{\frac{25}{6}}}{25} + \frac{2x^{\frac{15}{2}}}{15} + C.$

(12) $\int (1-4)^4\,dx = x - 2x^2 + 2x^3 - x^4 + \frac{x^5}{5} + C.$

(13) $\int e^{-2x}\,dx = -\frac{1}{2e^{2x}} + C.$

(14) $\int \cos(5x)\,dx = \frac{\sin(5x)}{5} + C.$

(15) $\int \frac{2}{x}\,dx = 2\ln(x) + C.$

(16) $\int \sinh(x)\,dx = \cosh(x) + C.$

(17) $\int (e^{3x} + x^2)\,dx = \frac{e^{3x}}{3} + \frac{x^3}{3} + C.$

(18) $\int 2x\cos(x^2)\,dx = \sin(x^2) + C.$

(19) $\int 3x^2\sin(x^2)\,dx = -\cos(x^3) + C.$

(20) $\int 4x^3 e^{x^4}\,dx = e^{x^4} + C.$

Solutions for §15.3

(1) Let $u = 3x + 4$, $du = 3\,dx$. We obtain $\int (3x+4)^7\,dx = \frac{1}{3}\int u^7\,du = u^8/24 = \frac{(3x+4)^8}{24} + C.$

(2) Let $u = 2x$, $du = 2\,dx$. Then $\int_0^{\frac{\pi}{4}} \cos(2x)\,dx = \frac{1}{2}\int_0^{\frac{\pi}{2}} \cos(u)\,du = \sin(u)\big]_0^{\frac{\pi}{2}} = \frac{1}{2}.$

(3) Set $u = x^2 + 4x + 18$. Then $du = (2x+4)\,dx$ and $\int_{-1}^{2} \frac{2x+4}{x^2+4x+18}\,dx = \int_{15}^{30} \frac{du}{u} = \ln(30) - \ln(15) = \ln(2).$

(4) Let $u = 2x^2 - 3$, $du = 4x\,dx$. We compute:

$$\int_0^{\pi} x \cdot \sin(2x^2-3)\,dx = \int_{-3}^{2\pi^2-3} x \cdot \sin(2x^2-3)\,dx = \frac{1}{4}\int_{-3}^{2\pi^2-3} \sin(u)\,du = \frac{1}{4}\left(\cos(3) - \cos(3-2\pi^2)\right).$$

(5) Let $u = x^3$, $du = 3x^2\,dx$. Then $\int x^2 \sin(x^3)\,dx = \int \sin(u)\frac{du}{3} = -\frac{1}{3}\cos(x^3) + C$.

(6) Setting $u = x^2 - 1$, $du = 2x\,dx$, yields $\int \frac{x}{\sqrt{x^2-1}}\,dx = \int \frac{du}{2\sqrt{u}} = \sqrt{x^2-1} + C$.

(7) Letting $u = 2x^2$ we see that $du = 4x\,dx$ and $\int xe^{2x^2}\,dx = \int e^u \frac{du}{4} = \frac{e^{2x^2}}{4} + C$.

(8) Using the substitution $u = e^x$ and $du = e^x\,dx$, $\int e^{x+e^x}\,dx = \int e^x \cdot e^{e^x}\,dx = \int e^u\,du = e^u + C = e^{e^x} + C$.

(9) Let $u = 1 - e^x$ and $du = -e^x\,dx$. Then $\int_0^2 e^x(1-e^x)\,dx = -\int_0^{1-e^2} u\,du = \frac{u^2}{2}\Big]_0^{1-e^2} = -\frac{1}{2} + e^2 - \frac{e^4}{2}$.

(10) Set $u = \sin(x)$, $du = \cos(x)\,dx$. Then $\int_0^{\frac{1}{4}} e^{\sin(x)}\cos(x)\,dx = \int_0^{\frac{\sqrt{2}}{2}} e^u\,du = e^u\Big]_0^{\frac{\sqrt{2}}{2}} = e^{\frac{\sqrt{2}}{2}} - 1$.

(11) Setting $u = x - 1$, $du = dx$, we have

$$\int x\sqrt{x-1}\,dx = \int (u+1)\sqrt{u}\,du = \int (u^{\frac{3}{2}} + u^{\frac{1}{2}})\,du = \frac{2}{5}u^{\frac{5}{2}} + \frac{2}{3}u^{\frac{3}{2}} + C = \sqrt{u}\left(\frac{2}{5}u^2 + \frac{2}{3}u\right) + C$$
$$= \sqrt{x-1}\left(\frac{2}{5}(x-1)^2 + \frac{2}{3}(x-1)\right) + C = \sqrt{x-1}\left(\frac{2}{5}x^2 - \frac{2}{15}x - \frac{4}{15}\right) + C.$$

(12) In this problem we can evaluate the integral by simply dividing: $\int_{-2}^{-3} \frac{e^{2x}+1}{e^{2x}}\,dx = \int_{-2}^{-3} (1 + e^{-2x})\,dx = -1 + \frac{e^4}{2} - \frac{e^6}{2}$.

(13) Here again we compute directly without the method of substitution: since $\frac{x^3-1}{x^3-x^2} = \frac{(x-1)(x^2+x+1)}{x^2(x-1)} = 1 + \frac{1}{x} + \frac{1}{x^2}$, we have that $\int \frac{x^3-1}{x^3-x^2}\,dx = x + \ln(x) + \frac{1}{x} + C$.

(14) Notice that $(1+\sqrt{x})^2 = 1 + 2\sqrt{x} + x$, and hence, $\int (1+\sqrt{x})^2\,dx = x + \frac{4}{3}x^{\frac{3}{2}} + \frac{x^2}{2} + C$.

(15) Letting $u = \sqrt{x}$, $du = \frac{1}{2\sqrt{x}}\,dx$, we obtain $\int \frac{e^{\sqrt{x}}}{\sqrt{x}}\,dx = 2e^{\sqrt{x}} + C$.

(16) Substituting $u = x - 2$ yields $\int_3^4 \frac{x}{\sqrt{x-2}}\,dx = \int_1^2 \frac{u+2}{\sqrt{u}}\,du = \frac{16\sqrt{2}}{3} - \frac{14}{3}$.

(17) Substituting $u = 1 + \cos(x)$ and $du = -\sin(x)\,dx$ gives us $\int \frac{\sin(x)}{(1+\cos(x))^7}\,dx = (1+\cos(x))^{-6}/6 + C$.

(18) Using $u = \cos(x)$ we get

$$\int \sin^3(x)\,dx = \int \sin(x)(1-\cos^2(x))\,dx = -\int (1-u^2)\,du = \frac{\cos^3(x)}{3} - \cos(x) + C.$$

(19) Letting $u = 1+\sqrt{x}$ we get $du = \frac{1}{2}x^{-\frac{1}{2}}\,dx$ and $dx = 2\sqrt{x}\,du = 2(u-1)\,du$. Thus $\int \sqrt{1+\sqrt{x}}\,dx = 2\int \sqrt{u}(u-1)\,du = \sqrt{1+\sqrt{x}}(-8/15 + 4\sqrt{x}/15 + 4x/5) + C$.

(20) Let $u = x^2+a^2$, $du = 2x\,dx$. Then $\int x\sqrt{x^2+a^2}\,dx = \frac{1}{2}\int \sqrt{u}\,du = (a^2+x^2)^{\frac{3}{2}} + C$.

(21) Trying $u = x^2+2$, $du = 2x\,dx$ yields $\int_0^1 \frac{x}{\sqrt[3]{x^2+2}}\,dx = \frac{1}{2}\int_2^3 u^{-\frac{1}{3}}\,du = \frac{3}{4}\left(3^{\frac{2}{3}} - 2^{\frac{2}{3}}\right)$.

(22) Setting $u = \cos(x)$, $du = -\sin(x)\,dx$ gives $\int \frac{\sin(x)}{\cos^2(x)}\,dx = \int -u^{-2}\,du = \frac{1}{u} + C = \sec(x) + C$.

(23) Setting $u = x-4$, $du = dx$ gives $\int x\sqrt{x-4}\,dx = \sqrt{x-4}(-64/15 - 8x/15 + 2x^2/5) + C$.

(24) Letting $u = \sin(x)$, $du = \cos(x)\,dx$ yields $\int\limits_{-\frac{\pi}{2}}^{\frac{\pi}{2}} \frac{\cos(x)}{\sin^2(x)}\,dx = \int\limits_{-1}^{1} u^{-2}\,du = \frac{-1}{u}\Big]_{-1}^{1} = -2$.

(25) Since we can write $\tan(x) = \frac{\sin(x)}{\cos(x)}$, we see that letting $u = \cos(x)$, $du = -\sin(x)\,dx$ yields the formula $\int\limits_0^{\frac{\pi}{4}} \tan(x)\,dx = -\int\limits_1^{\frac{\sqrt{2}}{2}} \frac{1}{u}\,du = \ln(u)\Big]_1^{\frac{\sqrt{2}}{2}} = -\ln\left(\frac{\sqrt{2}}{2}\right)$.

(26) Let $u = \sqrt{x}$. We get $\int (x^3+x-1)^2 x^{-\frac{1}{2}}\,dx = \frac{2}{13}x^{\frac{13}{2}} + \frac{4}{9}x^{\frac{9}{2}} - \frac{2}{7}x^{\frac{7}{2}} + \frac{2}{5}x^{\frac{5}{2}} - \frac{4}{3}x^{\frac{3}{2}} + 2\sqrt{x} + C$.

(27) Using $u = 1+\sqrt{x}$, $du = \frac{1}{2\sqrt{x}}\,dx$ we obtain $\int\limits_1^3 \frac{(1+\sqrt{x})^6}{\sqrt{x}}\,dx = \frac{880}{7} + \frac{656\sqrt{3}}{7}$.

(28) Setting $u = x^2+1$, $du = 2x\,dx$ yields $\int\limits_0^2 x\sqrt{x^2+1}\,dx = \frac{1}{3}(x^2+1)^{\frac{3}{2}}\Big]_0^2 = -1/3 + 5\sqrt{5}/3$.

(29) This question uses proposition 17.1. Make the substitution $u = e^x/\sqrt{2}$ and get $\int \frac{e^x}{e^{2x}+2}\,dx = \frac{1}{\sqrt{2}}\int \frac{du}{u^2+1} = \arctan(e^x/\sqrt{2})/\sqrt{2} + C$.

(30) Let $u = e^x$, $du = e^x\,dx$. It follows that $\int e^x \sin(e^x)\,dx = -\cos(e^x) + C$.

(31) Use $u = \sin(e^x)$, $du = e^x\cos(e^x)\,dx$ to get $\int e^x\cos(e^x)\cos(\sin(e^x))\,dx = \sin(\sin(e^x)) + C$.

(32) Use $u = 5x$, $du = 5\,dx$ to obtain $\int \sec^2(5x)\,dx = \frac{1}{5}\tan(5x) + C$.

(33) Set $u = 1+x^3+x$, $du = (3x^2+1)\,dx$. Then $\int (x^2+1/3)\cos(1+x^3+x)\,dx = \frac{\sin(1+x+x^3)}{3} + C$.

(34) Let $u = 2x$ so $du = 2\,dx$. It follows that $\int\limits_0^1 \sinh(2x)\,dx = -1/2 + \cosh(2)/2 + C$.

(35) Let $u = \sinh(x)$, then $du = \cosh(x)\,dx$ and thus $\int \cosh(x)\sinh^2(x)\,dx = \frac{\sinh^3(x)}{3} + C$.

Solutions for §15.4

(1) Let $u = x$ and $v' = \cos(x)$. Then $u' = 1$, $v = \sin(x)$, and $\int x\cos(x)\,dx = x\sin(x) - \int \sin(x)\,dx = \cos(x) + x\sin(x) + C$.

(2) Let $u = (2x+1)$ and $v' = e^{-5x}$. We then have $u' = 2$ and $v = -\frac{1}{5}e^{-5x}$. It now follows that $\int (2x+1)e^{-5x}\,dx = (-7-10x)e^{-5x}/25 + C$.

(3) This problem requires us to integrate by parts twice. In the first stage let $u = x^2$ and $v' = e^{3x}$. Then $\int x^2 e^{3x}\,dx = \frac{x^2e^{3x}}{3} - \int 2x\cdot\frac{e^{3x}}{3}\,dx$. We now compute this new integral in a similar manner: let $u = 2x$ and $v' = \frac{e^{3x}}{3}$. We get $\int x^2e^{3x}\,dx = \frac{1}{27}(2 - 6x + 9x^2)e^{3x} + C$.

(4) Use first the substitution $z = 2x^3$, $dz = 6x^2\,dx = 6(\frac{z}{2})^{\frac{1}{3}}x\,dx$ to transform the original integral as follows:

$$\int x\cdot\ln(2x^3)\,dx = \int \frac{\ln(z)}{6}\cdot\left(\frac{z}{2}\right)^{-\frac{1}{3}}\,dz.$$

With this done we can integrate by parts: let $u = \ln(z)$ and $v' = \frac{1}{6(z/2)^{1/3}}$ to get $\frac{1}{4}x^2\left(2\ln(2x^3) - 3\right) + C$.

(5) By first substituting $u = \ln(x)$, $\int (\ln(x))^2\,dx = \int u^2e^u\,du$. We then integrate by parts twice integrate as in question (3) to get $x(2 - 2\ln(x) + \ln^2(x)) + C$.

(6) By substituting $u = \sqrt{x}$ and integrating the result the way we did in (1) above we see that $\int \cos(\sqrt{x})\,dx = 2\left(\cos(\sqrt{x}) + \sqrt{x}\sin(\sqrt{x})\right) + C$.

(7) Begin by substituting $w = \sqrt{x}$. Then $\int e^{\sqrt{x}}\,dx = \int 2we^w\,dw$. Now integrate by parts by letting $u = 2w$ and $v' = e^w$: $\int 2we^w\,dw = 2we^w - \int 2e^w\,dw = 2we^w - 2e^w + C = e^{\sqrt{x}}(2\sqrt{x} - 2) + C$.

(8) Letting $u = x$ and $v' = \sinh(x)$ integration by parts yields $\int x\sinh(x)\,dx = x\cosh(x) - \int \cosh(x)\,dx = x\cosh(x) - \sinh(x) + C$

(9) First integrate $\sin^2(x) = \sin(x)\sin(x)$ by parts by letting $u = \sin(x)$ and $v' = \sin(x)$. The result is $\int \sin^2(x)\,dx = -\sin(x)\cos(x) + \int \cos^2(x)\,dx$. Then since $\cos^2(x) = 1 - \sin^2(x)$ we can solve for the integral we want (as in example 15.16) to get $\int \sin^2(x)\,dx = \frac{1}{2}x - \frac{1}{4}\sin(2x) + C$.

(10) Integrate by parts with $u = \ln(x)$ and $v' = x^2$ we see that $\int x^2\ln(x)\,dx = \frac{x^3}{3}\ln(x) - \int\frac{x^3}{3}\ln(x) - \frac{x^3}{9} + C$.

(11) Set $u = x$ and $v' = \cosh(5x)$. Integrating by parts yields $\int x\cosh(5x)\,dx = \frac{x\sinh(5x)}{5} - \int\frac{\sinh(5x)}{5}\,dx = \frac{x\sinh(5x)}{5} - \frac{\cosh(5x)}{25} + C$

(12) This integral can be solved most easily using the identity $\cos(4x) = 1 - 2\sin^2(2x)$ and exercise 15.3.18 to get $\int \cos(4x)\sin(2x)\,dx = \frac{\cos(2x)}{2} - \frac{\cos^3(2x)}{3} + C$.

(13) Here, we let $u = \ln(3x)$ and $v' = (5x - 3)$. After integrating by parts we obtain $\int (5x-3)\ln(3x)\,dx = \frac{1}{4}x\left(12 - 5x - 12\ln(3x) + 10x\ln(3x)\right) + C$

(14) Begin by making the substitution $z = x^5$. Then $dz = 5x^4\,dx$ and $x^9\,dx = \frac{z}{5}\,dz$. Our original integral becomes $\int x^9 e^{x^5}\,dx = \int \frac{z}{5}e^z\,dz$. Letting $u = z$ and $v' = e^z$ we integrate the new integral by parts to complete the problem: $\int \frac{z}{5}e^z\,dz = \frac{1}{5}\left(ze^z - e^z\right) + C = \frac{e^{x^5}(x^5-1)}{5} + C$.

(15) Similarly let $z = 2x^{11}$ first. Then $\int x^{21}\cos\left(2x^{11}\right)\,dx = \frac{1}{44}\int z\cos(z)dz$. After integrating by parts, we obtain: $\int x^{21}\cos\left(2x^{11}\right)\,dx = \frac{1}{44}(\cos(2x^{11}) + 2x^{11}\sin(2x^{11})) + C$.

(16) The formula $\int x^n e^x\,dx = x^n e^x - n\int x^{n-1}e^x\,dx$ follows directly from one application of integration by parts with $u = x^n$ and $v' = e^x$.

(17) Let $S(n) = \int_0^{\pi/2}\sin^n(x)\,dx$. Integrate by parts with $u = \sin(x)$, $v' = \sin^{n-1}(x)$ to get

$$S(n) = \left(-\cos(x)\sin^{n-1}(x)\Big]_0^{\pi/2}\ (=0)\right) + (n-1)\int_0^{\pi/2}\cos^2(x)\sin^{n-2}(x)\,dx.$$

And since $\cos^2(x) = 1 - \sin^2(x)$ we can solve for $S(n)$ and get $S(n) = \frac{n-1}{n}S(n-2)$ when $n \geq 2$. This also provides the induction step in the proof of Wallis' formula (which is valid only for $n \geq 3$) and it only remains to verify that $S(1) = 1$ implying that $S(3) = 2/3$ and then by induction that $S(n) = \frac{2\cdot4\cdot6\cdots(n-1)}{3\cdot5\cdot7\cdots n}$ for n odd.

Solutions for §15.5

(1) $\int \frac{dy}{y^2+2y+1} = \int \frac{dy}{(y+1)^2} = -\frac{1}{y+1} = \int dx = x + C$. Thus $y = -\frac{1}{x+c} - 1$.

(2) $\int \frac{dy}{5y+2} = \frac{1}{5}\cdot\ln(5y+2) = \int dx = x + C$. We obtain $5y + 2 = e^{5(x+C)}$ which implies $y = \frac{1}{5}e^{5(x+C)} - \frac{2}{5}$. The initial condition $y(0) = 1$ implies that $C = \frac{1}{5}\ln(7)$.

(3) $y = \int dy = \int(1 + 2x + x^4)\,dx = x + x^2 + \frac{1}{5}x^5 + C$.

(4) $y = \int dy = \int\left(1 + \frac{x^3}{2}\right)\,dx = x + \frac{1}{8}x^4 + C$. The initial condition $y(0) = 1$ implies that $C = 1$.

(5) $\int y\,dy = \frac{1}{2}y^2 = \int(-2x)\,dx = -x^2 + C$. So $y = \sqrt{2C - 2x^2}$.

(6) $\int\sqrt{y+2}\,dy = \frac{2}{3}(y+2)^{\frac{3}{2}} = \int\sqrt{x}\,dx = \frac{2}{3}x^{\frac{3}{2}} + C$. Then $y(0) = 2$ implies that $C = \frac{2}{3}4^{\frac{3}{2}} = \frac{16}{3}$.

(7) $\int\frac{dy}{y} = \ln(y) = \int x^2\left(x^3+3\right)^{\frac{1}{4}}\,dx = \frac{4}{15}\left(x^3+3\right)^{\frac{5}{4}} + C$. So $y = \exp\left(\frac{4}{15}\left(x^3+3\right)^{\frac{1}{4}} + C\right)$.

(8) $\int\frac{dy}{y^2} = -y^{-1} = \int 2x\cos(3x)\,dx = \frac{2}{9}\Big(\cos(3x) + 3x\sin(3x) + C\Big)$. So $y = \frac{-9}{2(\cos(3x)+3x\sin(3x)+C)}$. The initial condition $y(0) = 1$ implies that $C = -\frac{11}{2}$.

(9) $\int\frac{y}{\sqrt{2y^2+1}}\,dy = \frac{1}{2}\sqrt{2y^2+1} = -\int(x^3+1)\,dx = -\frac{1}{4}x^4 - x + C$. Thus $2y^2 + 1 = \left(-\frac{1}{2}x^4 - 2x + 2C\right)^2$ and we finally get $y = \left[\frac{1}{2}\left(\left(-\frac{1}{2}x^4 - 2x + 2C\right)^2 - 1\right)\right]^{\frac{1}{2}}$.

(10) $\int\frac{dy}{y} = \ln(y) = \int 3x^2\,dx = x^3 + C$. The initial condition $y(0) = 3$ implies that $C = \ln(3)$. Thus, $y = \exp\left(x^3 + \ln(3)\right) = 3e^{x^3}$.

(11) $\int (y+3)\,dy = \frac{1}{2}y^2+3y = \int e^x\,dx = e^x+C$. So $y^2+6y-(2e^x+2C)=0$. Solving with the quadratic formula, we obtain: $y=-3\pm\sqrt{9+2e^x+2C}$.

(12) $\int y^2\,dy = \frac{1}{3}y^3 = \int \ln(x)\,dx = x\ln(x)-x+C$. The initial condition $y(1)=8$ implies that $\frac{1}{3}\cdot 8^3 = C-1$, which implies that $C = 1+\frac{8^3}{3}$. Thus, $y=\sqrt[3]{3\ln(x)-3x+515}$.

Solutions for §15.6

(1) Let the population (in millions) be $p(t)$, a function of time. Then $p(t) = Ce^{kt}$. The initial condition $p(0)=10$ implies that $C=10$. Since $p(15)=20=10e^{15k}$, we see that $k=\frac{1}{15}\ln(2)$ and $p(t)=10\cdot 2^{\frac{1}{15}t}$. Then $p(t)=40$ when $t=30$ and $p(t)=1000$ when $t=\frac{15\ln(100)}{\ln(2)}$.

(2) The amount of bacteria is $b(t)=10^4\cdot 3^{\frac{1}{2}t}$ giving a population of $10^4\cdot 3^6$ after 12 hours.

(3) Let $F(t)$ be the amount of carbon at time t. Then $\frac{1}{2}F(t)=F(t+5568)$ since carbon decays to $\frac{1}{2}$ its size after 5568 years. Thus, the solution takes the form $F(t)=Ce^{kt}$ with $k=\ln(\frac{1}{2})/5568$. If, in the stated problem, the time the carbon was formed was at $t=0$ then $F(t)=F(0)/10$ where t is the age of the site. So $t=5568\ln(\frac{1}{10})/\ln(\frac{1}{2}) \approx 18\frac{1}{2}$ thousand years.

(4) The number of millions of gallons of oil left in the well is $3e^{t\ln(\frac{5}{6})/5} \Rightarrow$ (a) $5\ln(\frac{1}{3})/\ln(\frac{5}{6}) \approx 30$ years. (b) $\ln(\frac{5}{6})/10 \approx 18.2$ thousand gallons per year.

(5) $T'(t) = k(T(t)-T_a) \Rightarrow$ (a) $T(t)=e^{k(t+C)}+T_a$ and (b) the tea reaches 120°F after time $t=\ln(\frac{50}{142})/\ln(\frac{90}{142}) \approx 2.28$ minutes.

(6) We have

(a) $L'(t)=k(100-L(t)) \Rightarrow L(t)=100+Ce^{kt}$.

(b) $L(0)=0 \Rightarrow C=-100$ and $k=\ln(\frac{75}{100})/3 \Rightarrow 90\%$ will be mastered in $3\ln(\frac{1}{10})/\ln(\frac{75}{100}) \approx 24$ hours.

(c) In this case $L=100-70e^{t\ln(\frac{5}{7})/2}$ giving 90% learned in $2\ln(\frac{1}{7})/\ln(\frac{5}{7}) \approx 11\frac{1}{2}$ hours.

(d) For large t, L gets closer and closer to 100% but according to the model the program can *never* be completely mastered which is unrealistic.

Solutions for Additional Exercises for Chapter 15

(1a) By the fundamental theorem of calculus, if $F(x)=\int_0^x f(t)\,dt$, then $F'(x)=f(x)$. So, for the function $f(x)=\frac{1}{2}-e^{-x^2}$, we see that $F'(x)=\frac{1}{2}-e^{-x^2}$ is positive for $x>\sqrt{\ln(x)}$ and negative for $x<\sqrt{\ln(2)}$. Thus, F is increasing for $x>\sqrt{\ln(2)}$ and decreasing for $x<\sqrt{\ln(2)}$.

(1b) $F''(x)=2xe^{-x^2}$ which is positive for $x>0$ and negative for $x<0$. Consequently, F is concave up for $x>0$ and concave down for $x<0$.

(1c) $F'(x)=\frac{1}{2}-e^{-x^2}=0$ for $x=\sqrt{\ln(2)}$.

(2) Anti-derivatives must differ by a constant on intervals where they are continuous. This is true in this example. For $x>0$, we have $G(x)-F(x)\equiv 2$ and for $x<0$, we have $G(x)-F(x)\equiv 4$.

(3) We have $y' = (x+2)e^y$ or equivalently, $y'e^{-y} = x+2$. Integrating both sides yields:

$$\int e^{-y}\,dy = \int (x+2)\,dx \qquad \Longrightarrow \qquad -e^{-y} = \frac{1}{2}x^2 + 2x + C,$$

for some constant C. The initial condition $y(0) = 1$, implies that $C = -\frac{1}{e}$, and we then obtain the solution $y = -\log\left(-\frac{1}{2}x^2 - 2x + \frac{1}{e}\right)$.

(4) Let $y = f(x)$ be a curve whose normal at (x, y) has slope $-\frac{2y}{3x^2}$. Then $y' = \frac{3x^2}{2y}$ since if a line has slope m, a line normal to it has slope $-\frac{1}{m}$. Integrating, gives

$$\int 2y\,dy = \int 3x^2\,dx \qquad \Longrightarrow \qquad y^2 = x^3 + C,$$

for some constant C. Since the curve passes through $(1, 1)$, we see that $C = 0$ and $y^2 = x^3$.

(5) We know that $\int dx = x + C$ and $\int x\,dx = \frac{x^2}{2} + C$. Assume $\int x^{n-1}\,dx = \frac{x^n}{n} + C$. Integrating $\int x^n dx$ by parts by letting $u = x$ and $v' = x^{n-1}$ we see that

$$\int x^n\,dx = x \cdot \frac{x^n}{n} - \int \frac{x^n}{n}\,dx \Longrightarrow \int x^n\,dx + \frac{1}{n}\int x^n\,dx = \frac{x^{n+1}}{n}.$$

Hence $\left(1 + \frac{1}{n}\right)\int x^n\,dx = \frac{x^{n+1}}{n} \Longrightarrow \int x^n\,dx = \frac{x^n}{n+1}$.

(6) We see from the diagram that

$$f(t) = \begin{cases} -2t, & 0 \le t \le 1 \\ 2t - 4, & 1 < t \le 3 \\ -2t + 8, & 3 < t \le 4. \end{cases}$$

It follows that $F(1) = \int_0^1 (-2t)\,dt = -1$, $F(2) = -1 + \int_1^2 (2t-4)\,dt = -2$, $F(3) = -2 + \int_2^3 (2t-4)\,dt = -1$, and $F(4) = 0$.

(7) Since

$$F'(x) = \begin{cases} -2x, & 0 \le x \le 1 \\ 2x - 4, & 1 < x \le 3 \\ -2x + 8, & 3 < x \le 4, \end{cases}$$

we see that $F'(x) = 0$ when $x = 0, 2, 4$. There are additional critical points at $x = 1, 3$, since the derivative of $F(x)$ is not well defined at these points.

(8) F is increasing for $2 < x < 4$ and decreasing for $0 < x < 2$.

(9) For $0 \le x \le 1$, $F(x) = -x^2$. For $1 < x \le 3$, we have $F(x) = -x^2 + \int_1^x (2t-4)\,dt = 3 - 4x$, and for $3 < x \le 4$, we have $F(x) = 3 - 4x + \int_3^x (-2t+8)\,dt = -x^2 + 4x + 18$.

(10) Consider first the initial condition $y(0) = 1$. We have $y'(0) = (y(0) - 2)(y(0) - 4) = 3 > 0$. So $y(x)$ is increasing for x near 0. Further, $y' > 0$ for $y < 2$ and $y' < 0$ for $2 < y < 4$. So y is increasing for $y < 2$ and decreasing for $2 < y < 4$. We see that y cannot get larger than 2.

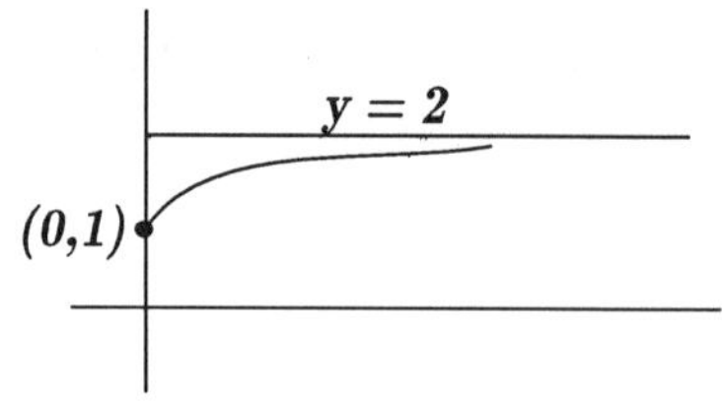

$$y' = (y - 2)(y - 4), \quad y(0) = 1$$

For the initial condition, $y(0) = 6$, we have $y'(0) = (y(0) - 2)(y(0) - 4) = 8 > 0$. So again $y(x)$ is increasing for $y > 4$.

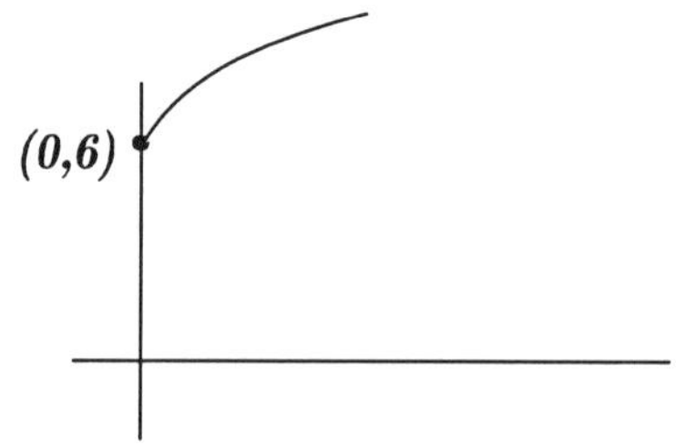

$$y' = (y - 2)(y - 4), \quad y(0) = 6$$

(11)

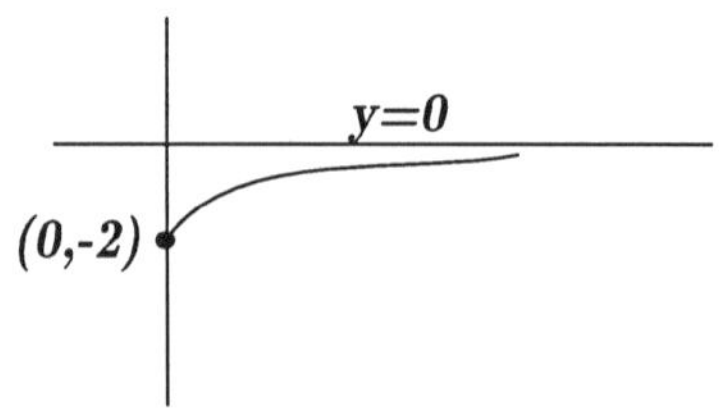

$$y' = -y(1 - y), \qquad y(0) = -2$$

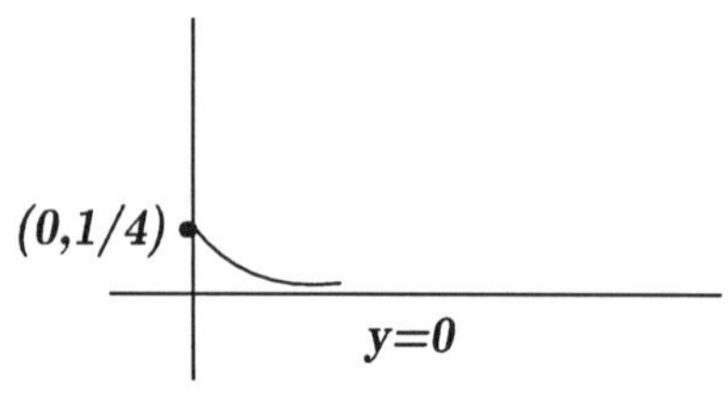

$$y' = -y(1 - y), \qquad y(0) = \frac{1}{4}$$

(12)

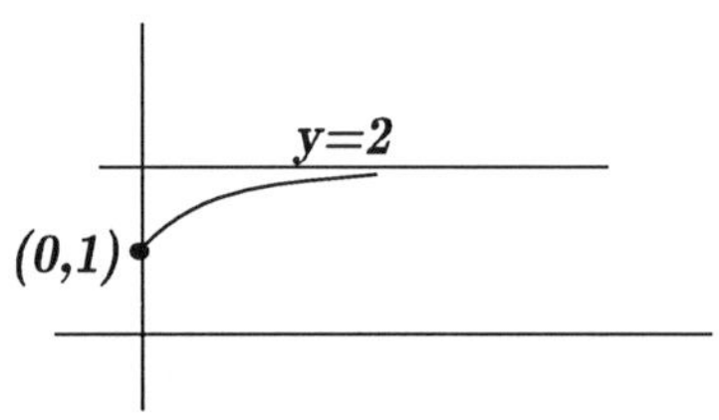

$$y' = y(y - 2)(y - 5)(y + 3), \qquad y(0) = 1$$

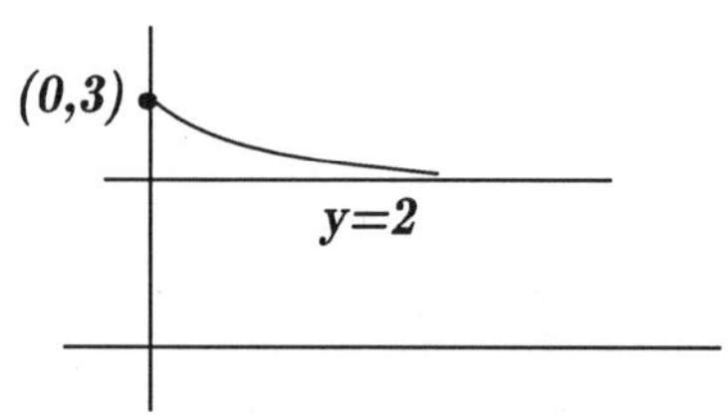

$$y' = y(y - 2)(y - 5)(y + 3), \qquad y(0) = 3$$

Solutions for §16.1

(1) Average Value $= \frac{1}{5}\int_{10}^{15} x^3\,dx = \frac{8125}{4}$.

(2) Average Value $= \frac{1}{21}\int_{100}^{121} \sqrt{x}\,dx = \frac{662}{63}$.

(3) Average Value $= \frac{3}{2\pi}\int_{-\frac{\pi}{3}}^{\frac{\pi}{3}} \cos(x)\,dx = \frac{3\sqrt{3}}{2\pi}$.

(4) Average Value $= \frac{1}{5}\int_{2}^{7} e^x\,dx = \frac{e^7-e^2}{5}$.

(5) Average Value $= \frac{1}{2}\int_{1}^{3} x\cdot e^{-x^2}\,dx = \frac{1}{4}\left(e^{-1}-e^{-9}\right)$.

(6) Average Value $= \frac{1}{2}\int_{0}^{2} x\cdot\sin^2(\pi x)\,dx = \frac{1}{2}$.

(7) Average Value $= \frac{1}{e^2-e}\int_{e}^{e^2} x^{-1}\,dx = \frac{1}{e^2-e}$.

(8) Average Value $= \frac{1}{4}\int_{1}^{5} 10^x\,dx = \frac{49995}{2\ln(10)}$.

(9) Average Value $= \frac{1}{\pi}\int_{0}^{\pi} \sin(x)\,dx = \frac{2}{\pi}$.

(10) Average Value $= \frac{1}{100}\int_{100}^{200} x^{\frac{1}{4}}\,dx = 4\sqrt{\frac{2}{5}}\left(2^{\frac{5}{4}}-1\right)$.

Solutions for §16.2

(1) The functions $f(x) = x^2+3$ and $g(x) = 1-x$, do not cross on the interval $0 \le x \le 1$, so the area between them is $\left|\int_0^1 f(x)-g(x)\,dx\right| = \left|\int_0^1 x^2+x+2\,dx\right| = \left|2x+\frac{1}{2}x^2+\frac{1}{3}x^3\Big]_0^1\right| = \frac{17}{6}$.

(2) f and g do not cross on this interval either so the area between them is $\left|\int_0^2 (f(x)-g(x))\,dx\right| = \frac{26}{3}$.

(3) Solving $f(x) = g(x)$ we see that $x = 0$ or 1. So the area between the graphs on the interval $-1 \le x \le 2$ is $\left|\int_{-1}^{0} (f(x)-g(x))\,dx\right| + \left|\int_0^1 (f(x)-g(x))\,dx\right| + \left|\int_1^2 (f(x)-g(x))\,dx\right| = \frac{25}{12}$.

(4) $f(x) = g(x) \Rightarrow x = -2, -1$ or 1. So the area between f and g is

$$\left|\int_{-1}^{1} f(x) - g(x)\,dx\right| + \left|\int_{1}^{2} f(x) - g(x)\,dx\right| = \frac{8}{3} + \frac{59}{12} = \frac{91}{12}.$$

(5) $x^4 = 6x^2 - 4 \Rightarrow x = -\sqrt{3+\sqrt{5}} = a, \; -\sqrt{3-\sqrt{5}} = b, \; \sqrt{3-\sqrt{5}} = c$ or $\sqrt{3+\sqrt{5}} = d$. So we integrate $f - g$ separately on the intervals $[-3, a]\,, [a, b]\,, [b, c]\,, [c, d]$ and $[d, 3]$ (taking the absolute values) to get $8.86 + 4.52 + 4.52 + 4.52 + 4.52 \approx 26.96$.

(6) f is strictly greater than g on this interval so the area is $\left|\int_2^{10} f(x) - g(x)\,dx\right| = \left.\frac{1}{2}e^{2x} - \ln(x)\right]_2^{10} \approx 2.42 \times 10^8$.

(7) The graphs of f and g cross at $x = \frac{4}{3}$ and the area we want is $\left|\int_0^{\frac{4}{3}} (f(x) - g(x))\,dx\right| + \left|\int_{\frac{4}{3}}^{2} (f(x) - g(x))\,dx\right|$. To integrate $f - g$ we in fact need results from chapter 17, the answer being ≈ 2.9525.

(8) See example 16.2.

(9) It is clear from the graph that f intersects g only for $x = 0, \frac{1}{2}, 1$. Then

$$\left|\int_{-1}^{0} (f(x) - g(x))\,dx\right| + \left|\int_{0}^{\frac{1}{2}} (f(x) - g(x))\,dx\right| + \left|\int_{\frac{1}{2}}^{1} (f(x) - g(x))\,dx\right| = \frac{3}{2} + \frac{2}{\pi}$$

since $\int (f(x) - g(x))\,dx = -x - x^2 + \frac{1}{\pi}\sin(\pi x) + C$.

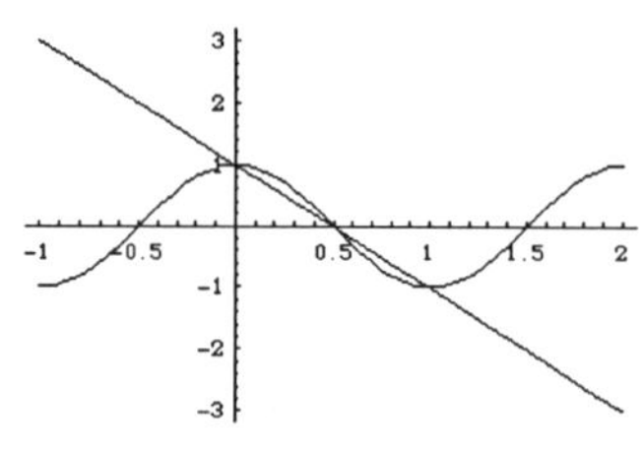

$f(x) = \cos(\pi x), \qquad g(x) = 1 - 2x$

(10) The graphs of f and g intersect at the origin so we must evaluate $\left|\int_{-\frac{1}{4}}^{0} f - g\right| + \left|\int_{0}^{\frac{1}{4}} f - g\right|$ which is $\int \tan(\theta)\,d\theta = \ln(\sec(\theta)) + C) \approx 0.2206$ (since $\int \tan(\theta)\,d\theta = \ln(\sec(\theta)) + C$).

(11) On the interval $0 \le x \le \frac{\pi}{2}$, $\cos(x)$ is just the reflection of $\sin(x)$ in the line $x = \frac{\pi}{4}$ and so their areas on this interval will be equal.

(12) Since f and g pass through $(1,1)$ and $(2,1)$ we see that $a=-\frac{1}{2},\ \ b=\frac{3}{2},\ \ c=-3$ and $d=3$. Then $\left|\int_1^2 f-g\right| = \left|\int_1^2 \left(-\frac{3}{2}x^2+\frac{9}{2}x-3\right)\,dx\right| = \frac{1}{4}$.

(13) The three points of intersection (from left to right) are $x=0, y=1,\quad x=1, y=1$, and $x=2, y=\frac{1}{3}$. Then the area we want is

$$\left|\int_0^1 \left(1-(1+4x)^{-\frac{1}{2}}\right)dx\right| + \left|\int_1^2 \left(-\frac{2}{3}x+\frac{5}{3}-(1+4x)^{-\frac{1}{2}}\right)dx\right| \approx 0.666667.$$

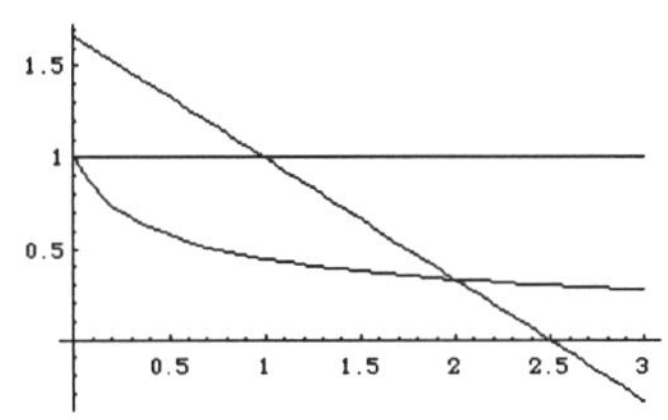

$$y=1,\qquad y=(1+4x)^{-\frac{1}{2}},\qquad y=-\frac{2}{3}x+\frac{5}{3}$$

(14) We see that $|\int_0^1 x-x^2\,dx| = \frac{1}{6} = |\int_0^1 \sqrt{x}-x\,dx|$.

Solutions for §16.3

(1) Here we consider the part of the curve in the first quadrant with $1\le x\le 2$. Since $y=+3\sqrt{x^3}$, $y'=\frac{9}{2}x^{\frac{1}{2}}$, and arclength $=\int_1^2 \sqrt{1+\frac{81}{4}x}\,dx = \left(\frac{4}{243}+\frac{1}{3}x\right)\sqrt{4+81x}\Big]_1^2 \approx 5.576$.

(2) Since $y=\cosh(x)$, $y'=\sinh(x)$ and

$$\text{arclength } = \int_0^3 \sqrt{1+\sinh^2(x)}\,dx = \int_0^3 \cosh(x)\,dx = \sinh(x)\Big]_0^3 = \sinh(3).$$

(3), (4) See question 11.

(5) Since $y=\sqrt{\ln(x)}$ we have $y'=\frac{1}{2x\sqrt{\ln(x)}}$ and arclength $=\int_1^e \sqrt{1+\left(\frac{1}{2x\sqrt{\ln(x)}}\right)^2}\,dx \approx 2.127$.

(6) The arclength is $\int_0^2 \sqrt{1+4x^2}\,dx \approx 4.646$.

(7) Arclength is $\int_1^{\frac{\pi}{2}} \sqrt{1+\cos^2(x)}\,dx \approx 0.598$.

(8) $y = \ln(\cos(x))$ so $y' = -\tan(x)$, and arclength $= \int_0^{\frac{\pi}{4}} \sqrt{1+\tan^2(x)}\,dx = \int_0^{\frac{\pi}{4}} \sec(x) \approx 0.8813$.

(9) The arclength is $\int_0^{10} \sqrt{1+e^{2x}}\,dx \approx 22025.9$.

(10) Here $y = \sqrt{\sin(x)e^{\frac{1}{x}}}$ and so $y' = \frac{1}{2\sqrt{\sin(x)e^{\frac{1}{x}}}}\left(e^{\frac{1}{x}}\cos(x) - \frac{e^{\frac{1}{x}}\sin(x)}{x^2}\right)$. The arclength is given by $\int_{\frac{1}{10}}^{1} \sqrt{1+(y')^2}\,dx \approx 45.68$.

(11) Here we have $y' = x^p - x^{-p}/4 \Rightarrow (y')^2 = x^{2p} - 1/2 + 1/(16px^2)$ and hence

$$(y')^2 + 1 = x^{2p} + 1/2 + 1/(16px^2) = (x^p + x^{-p}/4)^2.$$

Thus the arclength $= \int_1^2 (x^p + x^{-p}/4)\,dx = 2^{p+1}/(p+1) + 2^{-p-1}/(1-p) - 1/(p+1) - 1/(4-4p)$.

Solutions for §16.4

(1) Let $A(y) = x^2$ be the area of the cross sectional square at a distance y from the top of the pyramid.

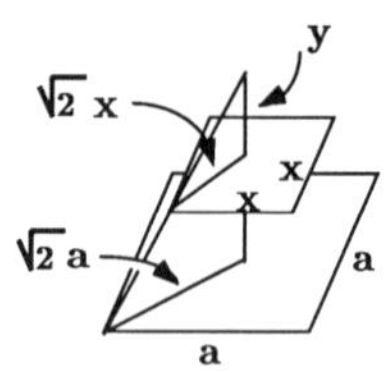

By similar triangles, $\frac{y}{\sqrt{2}x} = \frac{h}{\sqrt{2}a}$ and thus $x = \frac{ay}{h}$ and $A(y) = \left(\frac{ay}{h}\right)^2$. The volume of the pyramid is

$$\int_0^h A(y)\,dy = \int_0^h \left(\frac{ay}{h}\right)^2 dy = \frac{1}{3}a^2h.$$

(2) By similar triangles, $\frac{y}{x} = \frac{h}{r}$ and hence $y = \frac{hx}{r}$ and $x = \frac{ry}{h}$. Then Volume $= \int_0^h A(y)\,dx = \int_0^h \pi x^2\,dy = \int_0^h \frac{\pi r^2 y^2}{h^2}\,dy = \frac{1}{3}\pi r^2 h$.

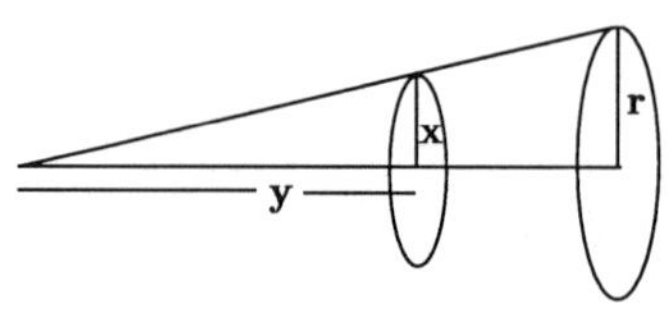

(3) Let $A(y) = \pi x^2$ be the area of the cross sectional slice at a distance y from the top of the sphere. By the Pythagorean theorem we have $x^2 + (r - y)^2 = r^2$ i.e., $x^2 = 2yr - y^2$.

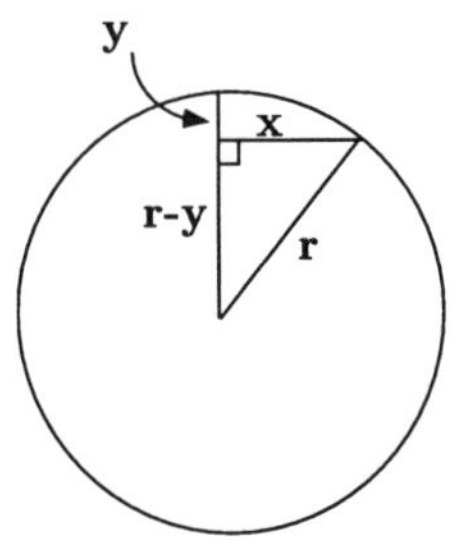

Then $A(y) = \pi x^2 = \pi(2yr - y^2)$. We have

$$\text{Volume} = \int_0^a \pi(2yr - y^2)\, dy = \pi r a^2 - \frac{\pi}{3} a^3.$$

(4) Volume $= \int_{-1}^{1} \pi \frac{1-x^2}{4}\, dx = \frac{\pi}{3}$.

(5) An Equilateral triangle of side b has area $\frac{\sqrt{3}}{4} b^2$. Thus, the area of the cross sectional slice is given by $\frac{\sqrt{3}}{4} L(x)^2 = \frac{\sqrt{3}}{64}(1 - x)^4$ and Volume $= \int_0^1 \frac{\sqrt{3}}{64}(1 - x)^4\, dx = \frac{\sqrt{3}}{320}$.

(6) Volume $= \int_0^2 A(x)\, dx = \int_0^2 (2x^3 + x^2 - 1)\, dx = \frac{26}{3}$.

Solutions for §16.5

(1) Volume $= \int_0^3 \pi x^8\, dx = \pi \cdot 3^7$.

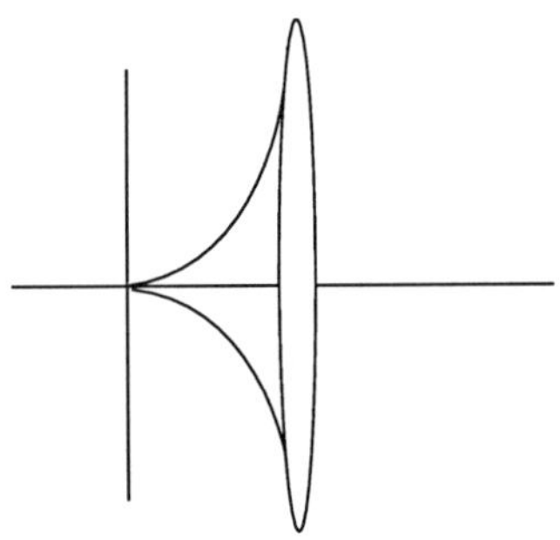

(2) Volume $= \int_1^2 \pi(3x+1)^2\,dx = 31\pi$.

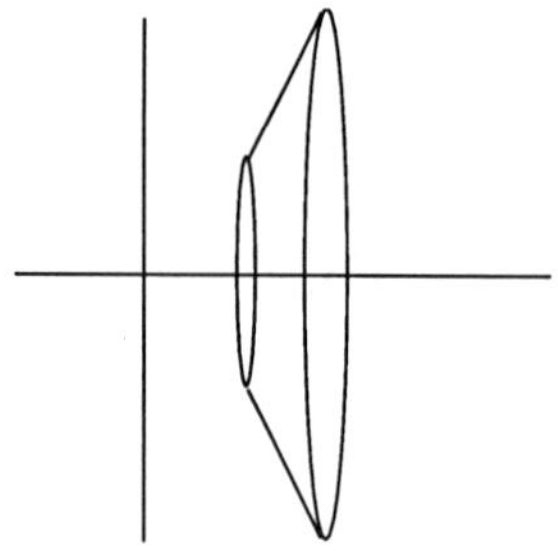

$y = 3x + 1 \qquad (1 \le x \le 2)$

(3) Volume $= \int_0^2 2\pi x^4\,dx = \frac{1}{5}\pi \cdot 2^6$.

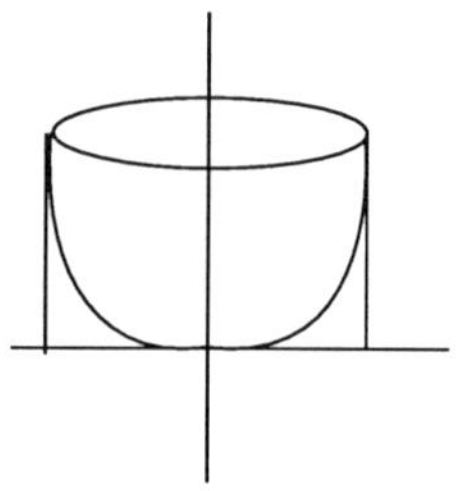

$y = x^3 \qquad (0 \le x \le 2)$

(4) Volume $= \int_{-\frac{1}{2}}^{\frac{1}{2}} \pi(1-x)^2\,dx = \frac{13}{12}\pi$.

(5) In this example we need only rotate half the region around the y–axis: Volume $= \int_0^{\frac{1}{2}} 2\pi x|1-x|\,dx = \frac{\pi}{6}$.

(6) Volume $= \int_0^{10} \pi x^{\frac{4}{3}}\,dx = \frac{300}{7}10^{\frac{1}{3}}$.

(7) Volume $= \int_{\frac{\pi}{6}}^{\frac{\pi}{2}} 2\pi x\sin(x)\,dx = \frac{\pi(6+\sqrt{3}\pi)}{6}$.

(8) Volume $= \int_{-1}^{1} \pi(1-x^2)\,dx = \frac{4}{3}\pi$.

(9) Volume $= \int_0^1 2\pi x\sqrt{1-x^2}\,dx = \frac{2}{3}\pi$.

(10) Volume $= \int_1^{10} \pi e^{2x}\,dx = \frac{\pi}{2}\left(e^{20} - e^2\right)$.

(11) Consider the semicircle of radius R at a distance D from the origin.

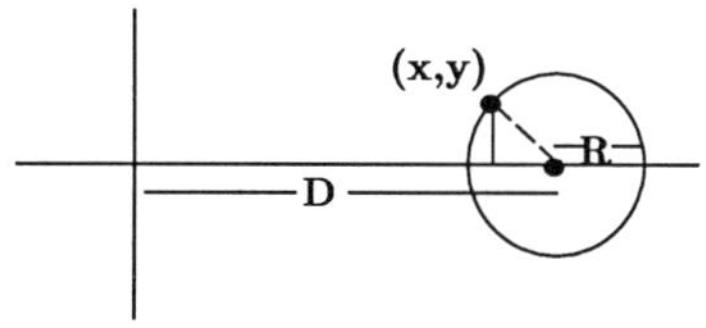

We can write the equation for this semicircle as $(D-x)^2 + y^2 = R^2$ or $y = \sqrt{R^2 - (D-x)^2}$. If we rotate this semicircle about the y–axis, we obtain $\frac{1}{2}$ of the torus. Thus, the volume of the torus is given by

$$\text{Volume} = 2\int_{D-R}^{R} 2\pi x\sqrt{R^2-(D-x)^2}\,dx.$$

Note that we have defined D to be the distance from the semicircle to the origin. The diameter of the torus is normally defined to be $2(D+R)$.

Solutions for §17.1

(1) $\ln(x-r)+C$

(2) Let $u = x^4+2$, $du = 4x^3\,dx$. Then

$$\int \frac{x^3}{x^4+2}\,dx = \frac{1}{4}\int \frac{du}{u} = \frac{1}{4}\ln(u)+C = \frac{1}{4}\ln(x^4+2)+C.$$

(3) Let $u=\sin(x)$, $du = \cos(x)\,dx$. Then $\int \frac{\cos(x)}{\sin(x)}\,dx = \int \frac{du}{u} = \ln(\sin(x)) = \ln(\sin(x))+C$.

(4) Let $u = \ln(x)$, $du = \frac{dx}{x}$. Then $\int \frac{\ln(x)}{x}\,dx = \int u\,du = \frac{1}{2}(\ln(x))^2 + C$.

(5) $u = 1+x^2$, $du = 2x\,dx$. Then

$$\int \frac{x}{1+x^2}\,dx = \frac{1}{2}\int \frac{du}{u} = \frac{1}{2}\ln(u)+C = \frac{1}{2}\ln(1+x^2)+C.$$

(6) $u = \tan(x)$, $du = \sec^2(x)\,dx$. Then $\int \frac{\sec^2(x)}{\tan(x)}\,dx = \frac{du}{u} = \ln(u)+C = \ln(\tan(x))+C$.

(7) Notice first that $\ln(\sqrt{x}) = \frac{1}{2}\ln(x)$. Thus $\int \ln(\sqrt{x})\,dx = \frac{1}{2}\int \ln(x)\,dx$. This latter integral can be done by integration by parts setting $u=\ln(x)$ and $v'=1$. Then $\int \ln(x)\,dx = x\ln(x) - x + C$.

(8) Let $u = \ln(x)$, $du = \frac{1}{x}\,dx = e^{-u}\,dx$. So $dx = e^u\,du$ and $\sin(\ln(x))\,dx = \int \sin(u)\cdot e^u\,du$, which can be evaluated as in example 15.16 to yield

$$\frac{1}{2}\left(x\sin(\ln(x)) - \cos(\ln(x))\right)+C.$$

(9) $\int \frac{x-1}{x^2-1}\,dx = \int \frac{1}{x+1}\,dx = \ln(x+1) + C.$

(10) Use $u = 1 + e^x$, $du = e^x\,dx$. Then $\int \frac{e^x}{1+e^x}\,dx \int \frac{du}{u} = \ln(u) + C = \ln(1+e^x) + C.$

(11) Let $u = \sin(x) - \cos(x) \Rightarrow du = (\cos(x) + \sin(x))dx$ and $\int \frac{du}{u} = \ln(\sin(x) - \cos(x)) + C.$

(12) Use the identity $1 - \cos^2(x) = \sin^2(x)$ and refer back to question 3.

(13) Since $\frac{1}{x} \leq 1$ for $x \geq 1 \Rightarrow \ln(X) = \int_1^X \frac{dx}{x} \leq \int_1^X 1\,dx = x]_1^X = X - 1.$

(14) Because $\frac{1}{x} > 0$ for $0 \leq x \leq 1 \Rightarrow$ when $0 \leq X \leq 1$ that $\int_X^1 \frac{dx}{x} > 0$ and so $\int_1^X \frac{dx}{x} < 0.$

(15) $\int_1^2 \frac{dt}{t} < 1$ since it's area is less than the trapezoid on the x-axis with sides of length 1 and $\frac{1}{2}$ and, therefore, area $\frac{3}{4}$. It can be shown that $\int_1^3 \frac{dt}{t} > 1$ by getting good approximations to the integral, e.g., using the trapezoidal rule. Therefore, $\ln(3) > 1$ and $\ln(2) < 1$ and so $2 < e < 3$ since the function e^x is increasing.

(16) Approximation with Simpson's rule yields 0.693.

Solutions for §17.2

(1) $\int_0^1 \frac{1}{1+x^2} = \arctan(x)\Big]_0^1 = \frac{\pi}{4}.$

(2) With the substitution $u = \frac{x}{2}$ we get $\int_0^2 \frac{dx}{4+x^2} = \frac{1}{2}\int_0^1 \frac{du}{1+u^2} = \frac{1}{2}\arctan(\frac{x}{2})\Big]_0^2 = \frac{\pi}{8}.$

(3) $\int = -\arccos(x)\Big]_{-\frac{1}{2}}^0 = \frac{\pi}{6}.$

(4) Let $u = \frac{x}{3} \Rightarrow dx = 3\,du$. Then $\int_0^{\frac{3}{4}} \frac{dx}{9+x^2} = \int_0^{\frac{1}{4}} \frac{3\,du}{9+9u^2} = \frac{1}{3}\int_0^{\frac{1}{4}} \frac{du}{1+u^2} = \frac{1}{3}\left(\arctan(\frac{1}{4}) - \arctan(0)\right) = .0816.$

(5) After the substitution $u = \cos(x)$ we get $-\int \frac{du}{1+u^2} = -\arctan(\cos(x)) + C.$

(6) Let $u = e^{\frac{1}{2}x}$, $du = \frac{1}{2}e^{\frac{1}{2}x}\,dx$. Then

$$\int \frac{e^{\frac{1}{2}x}}{1+e^x}\,dx = \int \frac{2}{1+u^2}\,du = 2\arctan(e^{\frac{x}{2}}) + C.$$

(7) Let $u = x^5$ then $du = 5x^4\,dx$ and $\int \frac{x^4}{\sqrt{1-x^{10}}}\,dx = \frac{1}{5}\int \frac{du}{\sqrt{1-u^2}} = \frac{1}{5}\arcsin(x^5) + C.$

(8) $u = x^{\frac{3}{2}} \Rightarrow \frac{2}{3}\,du = x^{\frac{1}{2}}dx$ and $\int = \frac{8}{3}\int \frac{du}{1+u^2} = \frac{8}{3}\arctan(x)]_0^1 = \frac{2\pi}{3}.$

(9) $\int \frac{x}{x^2+1}dx + \int \frac{dx}{x^2+1} = \frac{1}{2}\int \frac{du}{u} + \arctan(x)$ where $u = x^2+1$. Therefore we get $\frac{\ln(x^2+1)}{2} + \arctan(x) + C$.

(10) As in question 9 this is $\frac{\ln(x^2+1)}{2} - \arctan(x)\Big]_0^{\frac{1}{\sqrt{2}}}$.

Solutions for §17.3

(1) Already in form required.

(2) $\frac{x^3+2}{x^2-4} = \frac{x^3-4x+4x+2}{x^2-4} = x + \frac{4x+2}{x^2-4}$.

(3) We get after division $x^6 + 2x^5 + 4x^4 + 8x^3 + 16x^2 + 32x + 64 + \frac{131}{x-2}$.

(4) $6x - 2 + \frac{15-12x}{x^3+2}$

(5) $3x^{16} + 6x^{14} - 3x^{13} + 12x^{12} - 12x^{11} + 27x^{10} - 36x^9 + 66x^8 - 99x^7 + 168x^6 - 264x^5 + 435x^4 - 696x^3 + 1132x^2 - 1827x + 2960 + \frac{-4786x^2+7747x-2959}{x^3-2x+1}$

(6) $x^7 - 2x^4 - 3x^3 + 4x + 12 + \frac{-4x^3-8x^2-34x-39}{x^4+2x+3}$

(7) Let $\frac{x+2}{x^2-9x} = \frac{a}{x} + \frac{b}{x-9}$. Multiplying both sides by $x^2 - 9x$ yields $x + 2 = a(x-9) + bx$. Set $x = 0 \Rightarrow a = -\frac{2}{9}$. Set $x = 9 \Rightarrow b = \frac{11}{9}$. Then $\int \frac{x+2}{x^2-9x}\,dx = -\frac{2}{9}\int \frac{dx}{x} + \frac{11}{9}\int \frac{dx}{x-9} = -\frac{2}{9}\ln(x) + \frac{11}{9}\ln(x-9) + C$.

(8) Notice first that $\frac{x^4+2}{x^2-2x-3} = x^2 + 2x + 7 + \frac{23+20x}{x^2-2x-3}$. Now we write

$$\frac{23+20x}{x^2-2x-3} = \frac{a}{x-3} + \frac{b}{x+1} \Rightarrow 23 + 20x = a(x+1) + b(x-3).$$

When $x = 3 \Rightarrow 83 = 4a \Rightarrow a = \frac{83}{4}$. When $x = -1 \Rightarrow 13 = -4b \implies b = -\frac{13}{4}$. Putting all this together we have:

$$\begin{aligned}\int \frac{x^4+2}{x^2-2x-3}\,dx &= \int \left(x^2 + 2x + 7 + \frac{83}{4}\cdot\frac{1}{x-3} - \frac{13}{4}\cdot\frac{1}{x+1}\right)dx \\ &= \frac{1}{3}x^3 + x^2 + 7x + \frac{83}{4}\ln(x-3) - \frac{13}{4}\ln(x+1) + C.\end{aligned}$$

(9) Let $\frac{3x+1}{x^3-9x} = \frac{a}{x} + \frac{b}{x-3} + \frac{c}{x+3}$. Then $3x + 1 = a(x-3)(x+3) + bx(x+3) + cx(x-3)$. Setting $x = 0 \implies a = -\frac{1}{9}$. Setting $x = 3 \implies b = \frac{10}{18} = \frac{5}{6}$. Finally, setting $x = -3 \implies c = -\frac{8}{18} = -\frac{4}{9}$. In conclusion,

$$\int \frac{3x+1}{x^3-9x}\,dx = -\frac{1}{9}\ln(x) + \frac{5}{6}\ln(x-3) - \frac{4}{9}\ln(x+3) + C.$$

(10) Here $\frac{x^5}{x^4-5x^2+4} = x + \frac{5x^3-4x}{x^4-5x^2+4}$, and happily we are able to factor the expression $x^4 - 5x^2 + 4 = (x^2-4)(x^2-1) = (x+2)(x-2)(x+1)(x-1)$. Thus we set

$$\frac{5x^3-4x}{x^4-5x^2+4} = \frac{a}{(x+2)} + \frac{b}{(x-2)} + \frac{c}{(x+1)} + \frac{d}{(x-1)},$$

which yields

$$5x^3-4x = a(x-2)(x+1)(x-1)+b(x+2)(x+1)(x-1)+c(x+2)(x-2)(x-1)+d(x+2)(x-2)(x+1).$$

Plugging in $x = \pm 1, \pm 2$, we obtain

$$a = \frac{8}{3}, \quad b = \frac{8}{3}, \quad c = -\frac{1}{6}, \quad d = -\frac{1}{6},$$

and

$$\int \frac{x^5}{x^4-5x^2+4}\,dx = \frac{1}{2}x^2 + \frac{8}{3}\ln(x+2) + \frac{8}{3}\ln(x-2) - \frac{1}{6}\ln(x+1) - \frac{1}{6}\ln(x-1) + C.$$

(11) $\frac{8\ln(x-3)}{15} - \frac{ln(x)}{6} + \frac{3\ln(2+x)}{10} + \frac{\ln(x+3)}{3} + C$

(12) $\frac{x^4}{4} + 3x^2 + \frac{113\ln(x-2)}{8} + \frac{\ln(x)}{4} + \frac{113\ln(x+2)}{8} - \frac{5\ln(x+\sqrt{2})}{4} - \frac{5\ln(x-\sqrt{2})}{4} + C$

(13) Cross multiplying yields

$$3x+1 = A(x+3)^2 + B_1(x-1)(x+3) + B_2(x-1).$$

Setting $x = 1 \implies 4 = 16A \implies A = \frac{1}{4}$. Setting $x = -3 \implies -8 = -4B_2 \implies B_2 = 2$. Thus $3x+1 = \frac{1}{4}(x+3)^2 + B_1(x-1)(x+3) + 2(x-1)$. Setting $x = 0 \implies 1 = \frac{9}{4} - 3B_1 - 2 \implies B_1 = -\frac{1}{4}$. In conclusion

$$\int \frac{3x+1}{(x-1)(x+3)^2}\,dx = \frac{1}{4}\ln(x-1) - \frac{1}{4}\ln(x+3) - \frac{2}{x+3} + C.$$

(14) $A_1 = -\frac{1}{54}$, $A_2 = \frac{3}{2}$, $B_1 = -\frac{40}{27}$, $B_2 = -\frac{13}{9}$ and $B_3 = -\frac{4}{3} \Rightarrow$

$$\int \frac{x-2}{(x^2-1)(x+2)^3}\,dx = -\frac{\ln(x-1)}{54} + \frac{3\ln(x+1)}{2} - \frac{40\ln(x+2)}{27} + \frac{13}{(9x+18)} + \frac{4}{2(x+2)^2} + C.$$

(15) $A_1 = \frac{1}{8}$, $A_2 = \frac{1}{8}$, $B = 0$ and $C = -\frac{1}{2} \Rightarrow \int \frac{x^2}{16-x^4}\,dx = -\frac{\ln(2-x)}{8} + \frac{\ln(x+2)}{8} - \frac{\arctan(\frac{x}{2})}{4} + C$ since $\frac{-1}{2}\int \frac{dx}{x^2+4} = \frac{-1}{4}\int \frac{du}{u^2+1}$ when $x = 2u$.

(16) $A = 0, B = 1, C = 0, D = -4$. Now as before $\int \frac{dx}{x^2+3} = \frac{\arctan(\frac{x}{\sqrt{3}})}{\sqrt{3}}$ and to compute $-4\int \frac{dx}{(x^2+3)^2}$ we use the substitution $x = \sqrt{3}\tan(u)$ to get $\frac{-4}{9}\int \cos^2(u)\,du = -\frac{2u}{9} + \frac{\sin(2u)}{9}$ (see question 3 in §17.4). The final answer is

$$\arctan\left(\frac{x}{\sqrt{3}}\right)\left(\frac{1}{\sqrt{3}} - \frac{2}{9}\right) + \frac{\sin(2\arctan(\frac{x}{\sqrt{3}}))}{9} + C.$$

Solutions for §17.4

(1) Let $u = \cos(x)$, $du = -\sin(x)dx$. We have

$$\int \sin^3(x)\,dx = \int (1-\cos^2(x))\sin(x)\,dx = -\int (1-u^2)\,du = -\cos(x) + \frac{1}{3}\cos^3(x) + C.$$

(2) Let $u = \cos(x)$, $du = -\sin(x)dx$. We compute

$$\int \sin^3(x)\cos^2(x)\,dx =$$
$$= \int (1-\cos^2(x))\cos^2(x)\sin(x)\,dx = -\int (u^2-u^4)\,du = -\frac{1}{3}\cos^3(x) + \frac{1}{5}\cos^5(x) + C.$$

(3) $\int \cos^2(x)\,dx = \frac{1}{2}\int 1+\cos(2x)\,dx = \frac{1}{2}x + \frac{1}{4}\sin(2x) + C.$

(4) $\int \sin^4(x)\,dx = \frac{3}{8}x - \frac{1}{4}\sin(2x) + \frac{1}{32}\sin(4x).$

(5) Let $u = \sin(x)$ and $du = \cos(x)dx$. We have

$$\int \cos^5(x)\,dx = \int (1-\sin^2(x))^2\cos(x)\,dx = \int (1-2u^2+u^4)\,du = \sin(x) - \frac{2}{3}\sin^3(x) + \frac{1}{5}\sin^5(x) + C.$$

(6)
$$\begin{aligned}\int \cos^2(x)\cdot\Big(\sin(x)\cos(x)\Big)^2 dx &= \int \left(\frac{1+\cos(2x)}{2}\right)\cdot\left(\frac{\sin(2x)}{2}\right)^2 dx\\ &= \frac{1}{16}\int \sin^2(u)\,du + \frac{1}{16}\int \sin^2(u)\cos(u)\,du\\ &= \frac{1}{16}x - \frac{1}{64}\sin(4x) + \frac{1}{48}\sin^3(2x) + C.\end{aligned}$$

(7) Let $x = \sin(u)$ then $\int \frac{1}{x\sqrt{1-x^2}}\,dx = \int \frac{du}{\sin(u)} = \ln\left(\tan(\frac{u}{2})\right) = \ln\left(\tan\left(\frac{\arcsin(x)}{2}\right)\right) + C$ after an application of the Weierstrass substitution.

(8) Let $u = \sqrt{1-x^2}$. Then $du = -x(1-x^2)^{-\frac{1}{2}}dx = -\frac{x}{u}dx$. We have

$$\begin{aligned}\int \frac{\sqrt{1-x^2}}{x}\,dx = -\int \frac{u^2du}{1-u^2} &= -\int -du - \int \frac{1}{1-u^2}\,du = u + \frac{1}{2}\ln(1-u) - \frac{1}{2}\ln(1+u) + C\\ &= \sqrt{1-x^2} + \frac{1}{2}\ln(1-\sqrt{1-x^2}) - \frac{1}{2}\ln(1+\sqrt{1-x^2}) + C.\end{aligned}$$

(9) Use $u = \sqrt{x}$ to get $\int \frac{1}{x+\sqrt{x}}\,dx = \int \frac{2u}{u^2+u}\,du = \int \frac{2du}{u+1} = 2\ln(\sqrt{x}+1) + C.$

(10) $u = 1 + x^{\frac{3}{2}} \Rightarrow \int \frac{\sqrt{x}}{1+x^{\frac{3}{2}}}\,dx = \frac{2}{3}\int \frac{du}{u} = \frac{2}{3}\ln(1+x^{\frac{3}{2}}) + C.$

(11) Let $x = a\sin(u)$. Then $dx = \cos(u)du$. We have

$$\int \frac{1}{x^2\sqrt{a^2-x^2}}\,dx = \frac{1}{a^2}\int \frac{du}{\sin^2(u)} = \frac{-1}{a^2\tan(u)} = \frac{-1}{a^2\tan(\arcsin(\frac{x}{a}))} = \frac{-\sqrt{a^2-x^2}}{a^2x} + C.$$

(12) Let $x = \tan(u)$. Then $dx = \sec^2(u)du$. We get

$$\int \frac{1}{(x^2+1)^2}\,dx = \int \cos^2(u)\,du = \frac{1}{2}u + \frac{1}{4}\sin(2u) = \frac{1}{2}\arctan(x) + \frac{1}{4}\sin(2\arctan(x)) + C.$$

(13) With the given substitution $u = e^x$, $du = e^x\,dx = u\,dx$, we obtain

$$\begin{aligned}\int \frac{e^{3x}}{e^x+1}\,dx = \int \frac{u^2\,du}{u+1} &= \int u\,du - \int \frac{u}{u+1}\,du = \frac{1}{2}u^2 + u + 1 - \ln(u+1) + C\\ &= \frac{1}{2}e^{2x} + e^x - \ln(e^x+1) + C.\end{aligned}$$

(14) See question 19 in §15.3.

(15) Using the hint we get

$$\begin{aligned}\int \frac{\sqrt{x}}{\sqrt{1-x}}\,dx = 2\int \sin^2(u)\,du = \int (1-\cos(2u))\,du &= u - \frac{1}{2}\sin(2u)\\ &= \arcsin(\sqrt{x}) - \frac{1}{2}\sin(2\arcsin(\sqrt{x})) + C.\end{aligned}$$

(16) Letting $x = a\sinh(u)$ gives $\int x^2\sqrt{x^2+a^2}\,dx = a^4\int \sinh^2(u)\cosh^2(u)\,du$. Now use the identity $\sinh(u)\cosh(u) = \frac{1}{2}\sinh(2u)$ to get $\frac{a^4}{4}\int \sinh^2(2u)\,du = -\frac{1}{8}a^4u + \frac{1}{32}a^4\sinh(4u) + C$.

(17) After the substitution $x = \sin(u)$ we get $\int x^3\sqrt{1-x^2}\,dx = \int \sin^3(u)\cos^2(u)\,du$. See question(2).

(18) Let $x = \sinh(u)$, $dx = \cosh(u)\,du$. Then

$$\begin{aligned}\int \ln\left(x+\sqrt{x^2+1}\right)\,dx &= \int \ln(\sinh(u)+\cosh(u))\cosh(u)\,du\\ &= \int \ln(e^u)\cosh(u)\,du = \int u\cosh(u)\,du\\ &= x\sinh^{-1}(x) - \cosh(\sinh^{-1}(x)) + C = x\sinh^{-1}(x) - \sqrt{1+x^2} + C,\end{aligned}$$

after integration by parts.

(19) $x = \sinh(t) = \frac{1}{2}(u - \frac{1}{u}) \Rightarrow u^2 - 2ux - 1 = 0 \Rightarrow u = x \pm \sqrt{x^2+1}$ and, therefore, $t = \ln(u) = \ln(x+\sqrt{x^2+1})$, (we must take the log of the positive answer).

(20) $x = \cosh(t) = \frac{1}{2}(u + \frac{1}{u}) \Rightarrow u^2 - 2ux + 1 = 0 \Rightarrow u = x \pm \sqrt{x^2-1}$ and, therefore, $t = \ln(u) = \ln(x+\sqrt{x^2-1})$, (again we take the log of the positive answer).

(21) Let $x = \sinh(u)$. Then

$$\int \frac{x^2}{\sqrt{x^2+a^2}}\,dx = \int \sinh^2(u)\,du = \frac{1}{4}\int (e^{2u} + e^{-2u} - 2)\,du = \frac{1}{4}\sinh(2u) - \frac{1}{2}u$$
$$= \frac{1}{4}\sinh\left(2\sinh^{-1}(x)\right) - \frac{1}{2}\sinh^{-1}(x)/2 + C.$$

(22) Use $x = \cosh(u)$ to get $\frac{1}{4}\sinh\left(2\cosh^{-1}(x)\right) - \frac{1}{2}\cosh^{-1}(x) + C$.

(23) We begin by integrating by parts to obtain $\frac{1}{2}x^2\arcsin(x) - \frac{1}{2}\int \frac{x^2}{\sqrt{1-x^2}}\,dx$. To evaluate the latter integral we use the substitution $x = \sin(u)$ to get $\int \sin^2(u)\,du = \frac{1}{2}(u - \frac{1}{2}\sin(2u))$, and the final answer is then given by $\frac{1}{4}(2x^2-1)\arcsin(x) + \frac{1}{8}\sin(2\arcsin(x)) + C$.

(24) Use $x = 5u$ and refer back to question 8.

(25) Using the same method as question 23 we get $\int x\arccos(x)\,dx = \frac{1}{2}x^2\arccos(x) + \frac{1}{2}\int \frac{x^2}{\sqrt{1-x^2}}\,dx$...

(26) Let $t = \tan(\frac{x}{2})$. We have $\int \frac{1}{2+\tan(x)}\,dx = \int \frac{(1-t^2)dt}{(1+t-t^2)(1+t^2)}$. Now $1+t-t^2 = (t-\alpha)(t-\beta)$ with $\alpha, \beta = \frac{1}{2}(1 \pm \sqrt{5})$. Partial fraction decomposition yields $\frac{(1-t^2)}{(1+t-t^2)(1+t^2)} = \frac{-1}{5(t-\alpha)} + \frac{-1}{5(t-\beta)} + \frac{\frac{2t}{5} - \frac{4}{5}}{1+t^2}$ and on integrating we get: $\frac{1}{5}\ln(1+\tan(\frac{x}{2})) - \ln\left(1 + \tan(\frac{x}{2}) - \frac{1}{5}\tan^2(\frac{x}{2})\right) - \frac{2}{5}x + C$.

(27) $\int x^2\ln(x)\,dx = \frac{1}{3}x^3\ln(x) - \frac{1}{3}\int x^2\,dx = \frac{1}{3}x^3\ln(x) - \frac{1}{9}x^3 + C$.

(28) $\int x^2(\ln(x))^2\,dx = \frac{1}{3}x^3(\ln(x))^2 - \int \frac{2}{3}x^2\ln(x)\,dx$ which using the previous question is equal to $\frac{1}{3}x^3(\ln(x)^2 - \frac{2}{9}x^3\ln(x) + \frac{2}{27}x^3 + C$.

(29) Let $x = \frac{1}{z}$, $dx = -\frac{1}{z^2}\,dz$. We obtain: $\int \frac{\sqrt{x-x^2}}{x^4}\,dx = -\int z\sqrt{z-1}\,dz = -\int (u^{3/2} + u^{1/2})\,du$ where $u = z-1$ and so the answer is $-\frac{2}{5}\left(\frac{1}{x}-1\right)^{\frac{5}{2}} - \frac{2}{3}\left(\frac{1}{x}-1\right)^{\frac{3}{2}} + C$.

(30) Use a Weierstrass substitution to get $2\int \frac{dt}{1+5t^2}$. Then $\int \frac{dx}{3-2\cos(x)} = \frac{2}{\sqrt{5}}\arctan\left(\sqrt{5}\tan(\frac{x}{2})\right) + C$.

(31) Following the hint we obtain $-\int \frac{dz}{\sqrt{-z^2+2z+3}} = -\int \frac{dz}{\sqrt{4-(1-z)^2}} = \int \frac{du}{\sqrt{1-u^2}}$ where $u = \frac{1}{2}(1-z)$ and the answer is $\arcsin((x-1)/2x) + C$.

(32) $\frac{1}{4}\int \frac{dx}{\sin(x)} = \frac{1}{4}\int \frac{1+t^2}{2t}\frac{2dt}{1+t^2} = \frac{1}{4}\int \frac{dt}{t} = \frac{1}{4}\ln\left(\tan(\frac{x}{2})\right)$.

(33) Use the partial fraction decomposition method to get $\frac{1}{x^2(4+x^2)} = \frac{1}{4x^2} - \frac{1}{4(4+x^2)}$. We now have $\int \frac{1}{x^2(4+x^2)}\,dx = -\frac{1}{4x} - \frac{1}{8}\arctan(\frac{x}{2}) + C$.

(34) Let $u = \sqrt{x}$ then $\int \frac{\sqrt{x}}{1+x}\,dx = \int \frac{u^2 du}{1+u^2} = \int du - \int \frac{du}{1+u^2} = \sqrt{x} - \arctan(\sqrt{x}) + C$.

(35) Let $u = \sqrt[4]{x}$ then $\int \frac{1}{\sqrt{x}+\sqrt[4]{x}}\,dx = \int \frac{u^3 du}{u+u^2} = \int \frac{(v-1)^2}{v}\,dv = 2(\sqrt[4]{x}+1)^2 - 8(\sqrt[4]{x}+1) + 4\ln(\sqrt[4]{x}+1) + C$.

(36) Integrate by parts to get $-\frac{\arctan(x)}{x} + \int \frac{dx}{x^3+x^2}$. This last integral can be solved using partial fractions $(\frac{1}{x^3+x} = \frac{1}{x} - \frac{x}{1+x^2})$ to get $\ln(x) - \frac{1}{2}\ln(1+x^2) + C$.

Solutions for §17.5

(1) $\int\limits_2^\infty \frac{dx}{x^3} = \lim\limits_{B\to\infty} \int\limits_2^B \frac{dx}{x^3} = \lim\limits_{B\to\infty} -\frac{1}{4}x^{-4}\Big]_2^B = \lim\limits_{B\to\infty}\left(-\frac{1}{4}B^{-4} + \frac{2^{-4}}{4}\right) = 2^{-6}$.

(2) $\int\limits_2^\infty \frac{dx}{x^{\frac{3}{4}}} = \lim\limits_{B\to\infty} \int\limits_2^B \frac{dx}{x^{\frac{3}{4}}} = \lim\limits_{B\to\infty} 4x^{\frac{1}{4}}\Big]_2^B = \lim\limits_{B\to\infty}\left(4B^{\frac{1}{4}} - 4\cdot 2^{\frac{1}{4}}\right) = \infty$, so this integral is infinite.

(3) Here we have

$$\int\limits_1^3 \frac{dx}{(x-2)^{\frac{1}{3}}} = \lim_{\substack{\beta\to 2\\ \beta<2}} \int\limits_1^\beta \frac{dx}{(x-2)^{\frac{1}{3}}} + \lim_{\substack{\beta\to 2\\ \beta>2}} \int\limits_\beta^3 \frac{dx}{(x-2)^{\frac{1}{3}}}$$

$$= \lim_{\substack{\beta\to 2\\ \beta<2}} \left(\frac{3}{2}(\beta-2)^{\frac{2}{3}} - \frac{3}{2}(1-2)^{\frac{2}{3}}\right) + \lim_{\substack{\beta\to 2\\ \beta>2}} \left(\frac{3}{2}(3-2)^{\frac{2}{3}} - \frac{3}{2}(\beta-2)^{\frac{2}{3}}\right) = 0.$$

(4) $\int\limits_2^4 \frac{dx}{(x-3)^3}$ diverges since $\int\limits_2^3 \frac{dx}{(x-3)^3} = \lim\limits_{\substack{\beta\to 3\\ \beta<3}} -\frac{1}{2}(x-3)^{-2}\Big]_2^\beta = \infty$.

(5) $\int\limits_1^\infty xe^{-x}\,dx = \lim\limits_{B\to\infty} \int\limits_1^B xe^{-x}\,dx = \lim\limits_{B\to\infty} -xe^{-x} - e^{-x}\Big]_1^B = \frac{2}{e}$.

(6) This integral diverges.

(7) This integral diverges because $\int\limits_1^\infty \frac{e^{2-x}}{(x-2)^2}\,dx \ge \int\limits_1^2 \frac{e^{2-x}}{(x-2)^2}\,dx \ge \int\limits_1^2 \frac{dx}{(x-2)^2} = \infty$.

(8) $\int\limits_1^\infty \frac{dx}{x} = \lim\limits_{B\to\infty} \int\limits_1^B \frac{dx}{x} = \lim\limits_{B\to\infty} \ln(B) = \infty$, so this integral diverges.

(9) $\int\limits_0^1 \frac{dx}{x} = \lim\limits_{\substack{\beta\to 0\\ \beta>0}} \int\limits_\beta^1 \frac{dx}{x} = \lim\limits_{\substack{\beta\to 0\\ \beta>0}} -\ln(\beta) = \infty.$

(10) Observe that $\frac{1}{x^2+2x} = \frac{1}{2}\left(\frac{1}{x} - \frac{1}{x+2}\right)$. Thus

$$\lim_{B\to\infty} \int\limits_2^B \frac{dx}{x^2+2x} = \lim_{B\to\infty} \int\limits_2^B \frac{1}{2}\left(\frac{1}{x} - \frac{1}{x+2}\right) dx$$

$$= \frac{1}{2} \lim_{B\to\infty} \ln(x) - \ln(x+2)\Big]_2^B$$

$$= \frac{1}{2} \lim_{B\to\infty} \left(\ln\left(\frac{B}{B+2}\right) - \ln\left(\frac{2}{4}\right)\right)$$

$$= \frac{\ln(2)}{2}.$$

(11) $\int\limits_1^3 \frac{dx}{\sqrt{x^2+1}} = \arcsin(3) - \arcsin(1) \approx .937.$

(12) $\int\limits_1^3 \frac{dx}{(x-1)^{\frac{3}{4}}} = \lim\limits_{\substack{\beta\to 1\\ \beta>1}} \frac{1}{4}(x-1)^{\frac{1}{4}}\Big]_\beta^3 = \frac{2^{\frac{1}{4}}}{4}.$

(13) $\int\limits_1^3 \frac{dx}{(x-1)} = \lim\limits_{\substack{\beta\to 1\\ \beta>1}} \int\limits_\beta^3 \frac{dx}{x-1} = \lim\limits_{\substack{\beta\to 1\\ \beta>1}} \ln(x-1)\Big]_\beta^3 = \infty.$

(14) This integral does not converge.

(15) $\int\limits_{10}^{20} \frac{\cos(3x)}{\sqrt{x-10}}\, dx = \int\limits_{10}^{10.01} \frac{\cos(3x)}{\sqrt{x-10}}\, dx + \int\limits_{10.01}^{20} \frac{\cos(3x)}{\sqrt{x-10}}\, dx.$ The latter of these integrals is clearly finite. That the original integral is convergent can be seen by observing that

$$\int\limits_{10}^{10.01} \frac{\cos(3x)}{\sqrt{x-10}}\, dx \le \Big|\operatorname*{Max}_{10\le x\le 10.01}(\cos(3x))\Big| \cdot \int\limits_{10}^{10.01} \frac{dx}{\sqrt{x-10}} \le \frac{2}{10}.$$

(16) $\int\limits_1^5 \frac{x^{50}}{(x-1)^{\frac{3}{2}}}\, dx \ge \int\limits_1^5 \frac{dx}{(x-1)^{\frac{3}{2}}} = \infty.$

(17) $\int\limits_3^6 \frac{\ln(x)e^{\frac{1}{x}}}{(x-3)^5}\, dx \ge \ln(3)e^{\frac{1}{6}} \int\limits_3^6 \frac{dx}{(x-3)^5} = \infty$, so the original integral is also infinite.

(18) Since $\arcsin(x) \approx x$ for $x \approx 0$, we see that $\frac{\arcsin(x)}{\sqrt[3]{x}} \approx 0$ for $x \approx 0$. Thus, this integral does not have a singularity and must be finite.

(19) One checks directly that $e^x > x^2$ for $1 \le x \le 2$. Let $f(x) = \frac{e^x}{x^2}$. Then $f'(x) = \frac{e^x}{x^3}(x-2)$ and $f'(x) > 0$ for $x > 2$. This implies that $f(x)$ is increasing for $x > 2$.

(20) Let $f(x) = \frac{e^x}{x^3}$. Then $f'(x) = \frac{e^x}{x^4}(x-3)$ and $f(x)$ is increasing for $x > 3$.

(22) Integral is infinite.

(23) Integral is finite.

(24) Integral is finite.

(25) Integral is infinite.

(26) $\Gamma(1) = \int_0^\infty e^{-u}\,du = 1$. By integration by parts,

$$\Gamma(x+1) = \int_0^\infty e^{-u}u^x\,du = -e^{-u}u^x\Big]_0^\infty + x\int_0^\infty e^{-u}u^{x-1}\,du = x\Gamma(x).$$

Solutions for Additional Exercises for Chapter 17

(1) Using the trigonometric identity:

$$\sin(x-3)\cos(x) = \frac{1}{2}\Big[\sin(2x-3) - \sin(3)\Big],$$

we have:

$$\int \frac{1}{\sin(x-3)\cos(x)}dx = \sec(3)\cdot\ln\frac{-\sin(3)\tan(x-\frac{3}{2}) + 1 - \cos(3)}{-\sin(3)\tan(x-\frac{3}{2}) + 1 + \cos(3)} + C.$$

(2) $\int \frac{1}{\sin(x)+\tan(x)}dx = \frac{1}{4}\Big[\ln\big(\tan^2\big(\frac{x}{2}\big)\big) - \tan^2\big(\frac{x}{2}\big)\Big] + C.$

(3a) $a = \sqrt{2}, \quad b = -\sqrt{2}, \quad$ so that $\quad x^4 + 1 = (x^2 + \sqrt{2}x + 1)(x^2 - \sqrt{2}x + 1).$

(3b)

$$\begin{aligned}
\int_0^1 \frac{x}{x^4+1}dx &= \int_0^1 \frac{x}{(x^2+\sqrt{2}x+1)(x^2-\sqrt{2}x+1)}dx \\
&= \frac{1}{2\sqrt{2}}\Big[\int_0^1 \frac{1}{x^2-\sqrt{2}x+1}dx - \int_0^1 \frac{1}{x^2+\sqrt{2}x+1}dx\Big] \\
&= \frac{1}{2\sqrt{2}}\Big[\sqrt{2}\tan^{-1}(\sqrt{2}x-1)\Big|_0^1 - \sqrt{2}\tan^{-1}(\sqrt{2}x+1)\Big|_0^1\Big] \\
&= \frac{1}{2}\Big[\sqrt{2}\tan^{-1}(\sqrt{2}-1) - \tan^{-1}(\sqrt{2}+1) + 2\tan^{-1}(1)\Big] \\
&= \frac{1}{2}\Big[-\frac{\pi}{4} + 2\big(\frac{\pi}{4}\big)\Big] = \frac{\pi}{8}.
\end{aligned}$$

(4)$A_1 = 1, \quad B_1 = 2, \quad C_1 = 0, \quad B_2 = 4, \quad C_1 = 0,$ so that

$$\frac{3x^4 + 4x^3 + 16x^2 + 20x + 9}{(x+2)(x^2+3)^2} = \frac{1}{x+2} + \frac{2x}{x^2+3} + \frac{4x}{(x^2+3)^2}$$

and

$$\int \frac{3x^4 + 4x^3 + 16x^2 + 20x + 9}{(x+2)(x^2+3)^2} dx = \ln|(x+2)(x^2+3)| - \frac{2}{x^2+3} + C.$$

(5) $\int\limits_0^1 \arctan(x)dx = \frac{\pi}{4} - \int\limits_0^{\frac{\pi}{4}} \tan(y)dy = \frac{\pi}{4} + \ln\cos(y)\Big]_0^{\frac{\pi}{4}} = \frac{\pi}{4} - \ln\sqrt{2}.$

(6) Let $\quad u^2 = 1 + \exp(x), \quad \frac{2u}{u^2-1}du = dx.$ We compute:

$$\begin{aligned}\int \sqrt{1+\exp(x)}dx &= \int u \cdot \frac{2u}{u^2-1}du = \int \frac{u}{u-1}du + \int \frac{u}{u+1}du \\ &= 2u + \ln\left|\frac{u-1}{u+1}\right| \\ &= 2\sqrt{1+\exp(x)} + \ln\left|\frac{\sqrt{1+\exp(x)}-1}{\sqrt{1+\exp(x)}+1}\right| + C.\end{aligned}$$

(7) With $\quad x = 3\cos\theta, \quad dx = -3\sin\theta d\theta,$ and

$$\begin{aligned}\int\limits_1^3 \frac{\sqrt{9-x^2}}{x}dx &= -3\int \frac{\sin^2\theta}{\cos\theta}d\theta = -3\int \frac{1-\cos^2\theta}{\cos\theta}d\theta \\ &= 3\int \cos\theta d\theta - 3\int \sec\theta d\theta \\ &= 3\sin\theta - 3\ln|\sec\theta + \tan\theta| \\ &= \sqrt{9-x^2}\Big]_1^3 - 3\ln\left|\frac{3+\sqrt{9-x^2}}{x}\right|\Big]_1^3 \\ &= 3\ln(3+2\sqrt{2}) - 2\sqrt{2}.\end{aligned}$$

(8) $\delta(x) = 500 - \sqrt{x}, \quad$ so

$$\begin{aligned}\text{Population} &= \lim_{\Delta x \to 0} \sum_{i=1}^{N} 300\delta(x)\Delta x \\ &= \int\limits_0^{800} 300\delta(x)dx = 300\int\limits_0^{800} (500 - \sqrt{x})\, dx \\ &= 300(500x - \frac{2}{3}x^{\frac{3}{2}})\Big]_0^{800} \\ &= 1600000(75 - 2\sqrt{2}) \approx 115{,}474{,}516.6\end{aligned}$$

(9) Area $= \int_1^e \ln(x)dx + \int_e^{2e} \left(-\frac{1}{e}x + 2\right) dx = (x\ln(x) - x)\Big]_1^e + (2x - \frac{1}{2e}x^2)\Big]_e^{2e} = 1 + \frac{e}{2}.$

(10a) $\int_1^3 \ln(x)\, dx = x\ln(x) - x\Big]_1^3 = 3\ln(3) - 2.$

(10b) With $x = \exp(y), \quad dx = \exp(y)\, dy,$ and

$$\begin{aligned}\int_0^3 \ln(x)dx &= \int_{-\infty}^{\ln(3)} y\exp(y)dy \\ &= ye^y - e^y\Big]_{-\infty}^{\ln(3)} \\ &= 3\ln(3) - 2.\end{aligned}$$

(11) $\dfrac{x^4 + x^2 - 2}{x^3 + 3x + 1} = x - \dfrac{2x^2 + x + 2}{x^3 + 3x + 1}.$

(12) $\dfrac{3x^4 + 12x - 2}{x^2 + 1} = 3x^2 - 3 + \dfrac{12x + 1}{x^2 + 1}.$

(13)

$$\frac{x^{14} - x^{12} + 2x^{10} + x^8 + 2x^6 - x^4 + 1}{3x^3 + 2x + 1} =$$

$$= \frac{1}{3}x^{11} - \frac{5}{9}x^9 - \frac{1}{9}x^8 + \frac{28}{27}x^7 + \frac{7}{27}x^6$$

$$- \frac{26}{81}x^5 - \frac{42}{81}x^4 + \frac{193}{243}x^3 + \frac{29}{243}x^2$$

$$- \frac{503}{729}x - \frac{58}{729} + \frac{\frac{919}{729}x^2 + \frac{619}{729}x + \frac{787}{729}}{3x^3 + 2x + 1}.$$

(14) Here

$$\frac{x^{16} + 6x + 2}{x^3 - 6} = x^{13} + 6x^{10} + 36x^7 + 216x^4 + 1296x + \frac{7782x + 2}{x^3 - 6}.$$

(15) Since

$$\frac{x^5 + 2x^3 + x^2 - 3x - 1}{x^3 + 3x + 1} = x^2 - 1,$$

we have that

$$\int \frac{x^5 + 2x^3 + x^2 - 3x - 1}{x^3 + 3x + 1}\, dx = \int (x^2 - 1)\, dx = \frac{x^3}{3} - x + C.$$

(16)

$$\int \frac{3x^4+12x-2}{x^2+1}dx = x^3-3x+\int \frac{12x+1}{x^2+1}dx$$
$$= x^3-3x+6\ln|(x^2+1)|+\arctan(x)+C.$$

(17) Since $x^4-3x^2-4=(x^2-4)(x^2+1)$ we have that

$$\frac{1}{x^4-3x^2-4}=\frac{1}{20(x-2)}-\frac{1}{20(x+2)}-\frac{1}{5(x^2+1)},$$

and thus

$$\int \frac{dx}{x^4-3x^2-4}=-\frac{1}{5}\arctan(x)+\frac{\ln(x-2)}{20}-\frac{\ln(x+2)}{20}.$$

Solutions for §18.1

(1) $\sum\limits_{n=0}^{\infty} 2\cdot 3^{-n}=\frac{2}{\left(1-\frac{1}{3}\right)}=3.$

(2) $\sum\limits_{n=0}^{\infty} 2\cdot 3^{-2n}=\frac{2}{\left(1-\frac{1}{9}\right)}=\frac{9}{4}.$

(3) $\sum\limits_{n=2}^{\infty} 5\cdot 4^{-n}=5\cdot(4^{-2}+4^{-3}+\cdots)=5\cdot 4^{-2}(1+4^{-1}+4^{-2}+\cdots)=\frac{5\cdot 4^{-2}}{1-\frac{1}{4}}=\frac{5}{12}.$

(4) $\sum\limits_{n=0}^{\infty} \left(\frac{-3}{4}\right)^n=\frac{1}{1+\frac{3}{4}}=\frac{4}{7}.$

(5) $\sum\limits_{n=1}^{\infty} 2\cdot\left(\frac{2}{3}\right)^n=2\cdot\frac{2}{3}\sum\limits_{n=0}^{\infty}\left(\frac{2}{3}\right)^n=2\cdot\frac{2}{3}\cdot\frac{1}{1-\frac{2}{3}}=4.$

(6) $\sum\limits_{n=1}^{\infty}(-2)^{-3n}=(-2)^{-3}\cdot\sum\limits_{n=0}^{\infty}(-2)^{-3n}=\frac{-1}{8}\cdot\frac{1}{1+\frac{1}{8}}=-\frac{1}{9}.$

(7) $\sum\limits_{n=0}^{\infty} 4\cdot 5^{-2n+2}=4\cdot 25\sum\limits_{n=0}^{\infty}(5^2)^{-n}=100\cdot\frac{1}{1-\frac{1}{25}}=\frac{625}{6}.$

(8) $\sum\limits_{n=1}^{\infty}\left(\frac{-5}{7}\right)^{n-2}=\left(\frac{-5}{7}\right)^{-2}\cdot\left(\frac{-5}{7}\right)\sum\limits_{n=0}^{\infty}\left(\frac{-5}{7}\right)^n=-\frac{7}{5}\cdot\frac{1}{\frac{5}{7}}=-\frac{49}{60}.$

(9) $\sum\limits_{n=4}^{\infty}\left(\frac{3}{5}\right)^{n-2}=\left(\left(\frac{3}{5}\right)^2+\left(\frac{3}{5}\right)^3+\left(\frac{3}{5}\right)^4+\cdots\right)=\left(\frac{3}{5}\right)^2\cdot\left(\frac{1}{1-\frac{3}{5}}\right)=\frac{9}{10}.$

(10) $0.\overline{35}=35\cdot\left(\frac{1}{100}+\frac{1}{10{,}000}+\cdots\right)=35\cdot\sum\limits_{n=1}^{\infty}10^{-2n}=35\cdot 10^{-2}\cdot\sum\limits_{n=0}^{\infty}10^{-2n}=35\cdot 10^{-2}\cdot\frac{1}{1-10^{-2}}=\frac{35}{99}.$

(11) $0.\overline{129}=129\cdot\sum\limits_{n=1}^{\infty}10^{-3n}=\frac{129}{999}.$

(12) $2.\overline{17} = 2 + 17 \cdot \sum_{n=1}^{\infty} 10^{-2n} = 2 + \frac{17}{99}$.

(13) $5.\overline{621} = 5 + 621/999$.

(14) The series converges if $|\cos(x)| < 1$, that is, $x \neq \pi n$, n any integer. If the series converges, the answer is $\frac{1}{(1-\cos(x))}$.

(15) The series converges if $e^x < 1$, or $x < 0$. The answer is $\frac{1}{1-e^x}$.

(16) The series converges if $|\ln(x)| < 1$, or $\frac{1}{e} < x < e$. The answer is $\frac{1}{1-\ln(x)}$.

(17) The total distance is given by $15 + 15 \cdot 2 \cdot \frac{4}{5} + 15 \cdot 2 \cdot \left(\frac{4}{5}\right)^2 + \ldots = 15 + \frac{24}{\left(1-\frac{4}{5}\right)} = 135$ feet. The total distance traveled by the 50th bounce is given by the finite series

$$15 + 15 \cdot 2 \cdot \frac{4}{5} + 15 \cdot 2 \cdot \left(\frac{4}{5}\right)^2 + \ldots + 15 \cdot 2 \cdot \left(\frac{4}{5}\right)^{49} = 15 + 24 \cdot \frac{\left(1 - \left(\frac{4}{5}\right)^{50}\right)}{\left(1 - \frac{4}{5}\right)}$$

$$= 15 + 120 \cdot \left(1 - \left(\frac{4}{5}\right)^{50}\right).$$

Solutions for §18.3

(1) The series diverges, because $\int_1^{\infty} \frac{dx}{x^{\frac{2}{3}}} = \lim_{N\to\infty} \int_1^N \frac{dx}{x^{\frac{2}{3}}} = \lim_{N\to\infty} 3x^{\frac{1}{3}}\Big]_1^N = \infty$.

(2) The series converges, because $\int_1^{\infty} \frac{15\,dx}{x^{1.01}} = \lim_{N\to\infty} \int_1^N \frac{15\,dx}{x^{1.01}} = \lim_{N\to\infty} -\frac{15}{.01}x^{-.01}\Big]_1^N = 1500$.

(3) The series diverges, because $\int_1^{\infty} \frac{2\,dx}{\sqrt{x}} = \lim_{N\to\infty} \int_1^N \frac{2\,dx}{\sqrt{x}} = \lim_{N\to\infty} 4\sqrt{x}\Big]_1^N = \infty$.

(4) The series diverges, because $\int_1^{\infty} \frac{2\,dx}{5x+1} = \lim_{N\to\infty} \int_1^N \frac{2\,dx}{5x+1} = \lim_{N\to\infty} \frac{2}{-5}\ln(5x+1)\Big]_1^N = \infty$.

(5) The series converges, because $\int_1^{\infty} \frac{x\,dx}{x^3+1} < \int_1^{\infty} \frac{dx}{x^2}$.

(6) The series diverges, because $\int_1^{\infty} \frac{x\,dx}{2x^2+3} > \int_1^{\infty} \frac{dx}{2x+3}$.

(7) The series converges, because so does $\int_1^{\infty} xe^{-x}\,dx$.

(8) The series diverges, because so does $\int\limits_1^\infty \ln x\,dx > \int\limits_e^\infty dx$. (One can apply the integral test to an increasing series as well. This illustrates formally that an increasing series is divergent.)

(9) The series converges, because so does $\int\limits_1^\infty \frac{dx}{1+x^2}$.

(10) The series converges, because so does $\int\limits_1^\infty \frac{dx}{x^2+3x+2} < \int\limits_1^\infty \frac{dx}{x^2}$.

(11) The series diverges, because so does $\int\limits_1^\infty \frac{2x+3}{x^2+3x+2}dx > \int\limits_1^\infty \frac{dx}{3x}$. Note that the integral test is applicable because the corresponding function is decreasing from some (possibly, large) value of x.

(12) The integral test is not applicable here, because the corresponding function $f(x) = \sin^2(x)$ is not monotone (even for large values of x).

(13) If $f(x)$ is a positive increasing integrable function then we have $\sum\limits_{n=2}^{N} f(n) < \int\limits_1^N f(x)\,dx$ and, furthermore, $\sum\limits_{n=1}^{N} f(n) > \int\limits_1^{N+1} f(x)\,dx$. Applying these inequalities to $f(x) = \frac{1}{x}$ yields

$$\ln(N+1) = \int_1^{N+1} \frac{dx}{x} < \sum_{n=1}^{N} \frac{1}{n} < 1 + \sum_{n=2}^{N} \frac{1}{n} < 1 + \int\limits_1^N \frac{dx}{x} = 1 + \ln(N).$$

(14) Using problem (13), one can write: $\sum\limits_{n=1}^{10^{10}} \frac{1}{n} < 1 + \ln\left(10^{10}\right) = 1 + 10 \cdot \ln(10) = 1 + 23.02585\ldots < 25$.

Solutions for §18.4

(1) The series converges by the comparison test , because it is bounded by $2 \sum\limits_{n=1}^{\infty} \frac{1}{n^4}$, which converges.

(2) The series diverges, because it is greater than the series $\frac{1}{3} \sum\limits_{n=1}^{\infty} \frac{1}{\sqrt{n}}$, which is divergent.

(3) The series converges, because it is bounded by the series $\sum\limits_{n=1}^{\infty} \frac{1}{n^2+1}$, which is convergent.

(4) The series converges, because it is bounded by the series $\sum\limits_{n=1}^{\infty} \left(\frac{3}{4}\right)^n$, which is convergent.

(5) The series diverges, because for $n > 1$ it bounds the series $\frac{1}{2} \sum\limits_{n=1}^{\infty} \frac{1}{\sqrt{n}}$, which is divergent.

(6) The series is divergent, because $\frac{a_{n+1}}{a_n} = 4 \cdot \left(\frac{n}{n+1}\right)^{100}$ tends to 4 as n tends to infinity (the ratio test).

(7) The series is convergent, because $\frac{a_{n+1}}{a_n} = \frac{1}{4} \cdot \left(\frac{n+1}{n}\right)^{100}$ tends to $\frac{1}{4}$ as n tends to infinity (the ratio test).

(8) The series is convergent, because $\frac{a_{n+1}}{a_n} = \frac{1}{n+1} \cdot \left(\frac{n+1}{n}\right)^3$ tends to 0 as n tends to infinity (the ratio test).

(9) The series is divergent, because $\frac{a_{n+1}}{a_n} = \frac{n+1}{5}$ tends to infinity as n tends to infinity.

(10) The series is convergent, because $\frac{a_{n+1}}{a_n} = \frac{5}{n+1}$ tends to 0 as n tends to infinity (the ratio test).

(11) The series is convergent, because $\frac{a_{n+1}}{a_n} = \frac{2}{n+1} \cdot \left(\frac{n+1}{n}\right)^3$ tends to 0 as n tends to infinity (the ratio test).

(12) The series is convergent, because it is bounded by $\sum_{n=1}^{\infty} \frac{3^n}{n!}$, which is convergent by the ratio test.

(13) $\frac{a_{n+1}}{a_n} = 2$, so the series is divergent due to the ratio test. Or, one can note that $a_n = 2^n$.

(14) $a_n = \frac{n^n}{n!} > 1$, therefore, the divergence of the series is provided by the comparison test.

(15) The series is convergent, because $\frac{a_{n+1}}{a_n} = \frac{1}{1.01} \cdot \left(\frac{n+1}{n}\right)^{100}$ tends to $\frac{1}{1.01}$ as $n \to \infty$ (the ratio test).

Solutions for §18.5

(1) The series is absolutely convergent since $\sum_{n=1}^{\infty} 2\left(\frac{5}{8}\right)^n = \frac{10}{13}$.

(2) Since $1 > \frac{1}{2^{\frac{1}{4}}} > \frac{1}{3^{\frac{1}{4}}} > \ldots \to 0$, the series is conditionally convergent, as an alternating series with terms approaching zero.

(3) The series diverges.

(4) The series converges absolutely: $\sum_{n=1}^{\infty} \frac{n^5}{3^n}$ converges by the ratio test since $\frac{a_{n+1}}{a_n} = \left(\frac{n+1}{n}\right)^5 \cdot \frac{1}{3} \to \frac{1}{3}$ as $n \to \infty$.

(5) The series converges absolutely: $\frac{a_{n+1}}{a_n} = \frac{3}{n+1} \to 0$ as $n \to \infty$ thus $\sum_{n=1}^{\infty} \frac{3^n}{n!}$ converges by the ratio test.

(6) The series is divergent, because $\frac{a_{n+1}}{a_n} = e \cdot \left(\frac{n}{n+1}\right)^{100}$ tends to e as n tends to infinity, and therefore, $|a_n| \to \infty$ (the ratio test).

(7) We have $\sum_{n=1}^{\infty} \frac{15n^{\frac{1}{4}}}{n^2+1} < 15 \sum_{n=1}^{\infty} \frac{15n^{\frac{1}{4}}}{n^2} < 15 \sum_{n=1}^{\infty} \frac{1}{n^{\frac{7}{4}}}$ which converges. By the comparison test, the series converges absolutely.

(8) The series is conditionally convergent, because $a_n = \frac{n}{n^{\frac{3}{2}}+2}$ tends to 0 as n tends to infinity. The series does not converge absolutely (apply the comparison test).

(9) The ratio test implies that the series is divergent. It does not converge conditionally.

(10) The series is conditionally convergent as an alternating series with a_n approaching zero.

(11) The series is absolutely convergent (one can use the comparison test).

(12) The series is absolutely convergent (one can use the ratio test).

(13) The series converges absolutely as a geometric series with a term $\left(-\frac{5}{9}\right)^n$.

(14) One has: $a_n = \frac{1}{2^{\ln n}} = \frac{1}{n^{\ln 2}}$, and, as $\ln(2) < 1$, the series is divergent (integral test).

(15) The series is divergent, because $|a_n|$ does not tend to zero for large n.

Solutions for §18.6

(1) $\sum_{n=0}^{\infty} (2x)^n = \sum_{n=0}^{\infty} 2^n \cdot x^n = \sum_{n=0}^{\infty} a_n x^n$ with $a_n = 2^n$. By the ratio test, if $\left|\frac{a_{n+1}}{a_n} \cdot \frac{x^{n+1}}{x^n}\right| = |2x| < 1$ then the series converges absolutely. Thus, the radius of convergence is $R = \frac{1}{2}$.

(2) $\sum_{n=0}^{\infty} \frac{x^{3n}}{n+1}$ converges absolutely if

$$\lim_{n\to\infty} \left| \frac{\frac{x^{3(n+1)}}{n+2}}{\frac{x^{3n}}{n+1}} \right| = \lim \left| \frac{n+1}{n+2} x^3 \right| = \left|x^3\right| < 1.$$

Thus $R = 1$.

(3) $\sum_{n=1}^{\infty} n^2 x^n$ converges absolutely if

$$\lim_{n\to\infty} \left| \frac{n+1}{n} \right|^2 \cdot |x| = |x| < 1.$$

Thus $R = 1$.

(4) $\sum_{n=0}^{\infty} \frac{x^n}{n!}$ converges absolutely if

$$\lim_{n\to\infty} \left| \frac{n!}{(n+1)!} x \right| = \lim_{n\to\infty} \left| \frac{x}{n+1} \right| = 0 < 1.$$

So the series converges absolutely if $|x| < R = \infty$.

(5) $R = 1$.

(6) $R = \frac{1}{3}$.

(7) $\sum_{n=1}^{\infty} \frac{(x-4)^n}{\sqrt{n}}$ converges absolutely if

$$\lim_{n\to\infty} \frac{\frac{|(x-4)^{n+1}|}{\sqrt{n+1}}}{\frac{|(x-4)^n|}{\sqrt{n}}} = \lim_{n\to\infty} \sqrt{\frac{n}{n+1}} \cdot |x-4| = |x-4| < 1.$$

Thus the series converges absolutely if $|x-4| < R = 1$

(8) The series is absolutely convergent for $|2x-1|<R=1$.

(9) The ratio test tells us that the series is divergent for any value of x, except for $x=0$. Therefore, $R=0$.

(10) Both $\sum_{n=1}^{\infty}\frac{1}{n^2}$ and $\sum_{n=1}^{\infty}\frac{(-1)^2}{n^2}$ converge absolutely, so that the series converges absolutely at $x=\pm 1$.

(11) $\sum_{n=1}^{\infty}\frac{1}{\sqrt{n}}$ diverges, but $\sum_{n=1}^{\infty}\frac{(-1)^n}{\sqrt{n}}$ is conditionally convergent. So the series diverges at $x=1$ and converges conditionally at $x=-1$.

(12) The series diverges for $x=\pm 1$.

(13) The series is absolutely convergent for $x=\pm 1$ (apply the ratio test).

(14) The series is absolutely convergent for $x=\pm 1$ (apply the integral test).

(15) The series only converges conditionally for $x=-1$, as an alternating series with $a_n>a_{n+1}>\cdots\to 0$. It diverges for $x=+1$.

(16) $\frac{d}{dx}\frac{1}{1-x^2}=\frac{2x}{(1-x^2)^2}=\sum_{n=1}^{\infty}2nx^{2n-1}$. Integrating both sides of this identity yields:

$$\int\frac{x}{1-x^2}dx=-\frac{1}{2}\ln(1-x^2)+C=\int\sum_{n=0}^{\infty}x^{2n+1}dx=\sum_{n=0}^{\infty}\frac{x^{2n+2}}{2n+2}+C.$$

(17) From **(16)** one can conclude that $\frac{1}{1+x^2}=\sum_{n=0}^{\infty}(-1)^n x^{2n}$. $\frac{d}{dx}\frac{1}{1+x^2}=\frac{2x}{(1+x^2)^2}=\sum_{n=1}^{\infty}2n(-1)^n x^{2n-1}$. $\arctan x=\int\frac{1}{1+x^2}dx=\int\sum_{n=0}^{\infty}(-1)^n x^{2n}dx=\sum_{n=0}^{\infty}(-1)^n\frac{x^{2n+1}}{2n+1}$. Note that we take $C=0$ for the constant of integration, so that $\arctan(0)=0$.

(18) From **(16)** one can conclude that $\frac{1}{1-x^4}=\sum_{n=0}^{\infty}x^{4n}$. For the derivative one has: $\frac{d}{dx}\frac{1}{1-x^4}=\frac{4x^3}{(1-x^4)^2}=\sum_{n=1}^{\infty}4nx^{4n-1}$. The power series for the indefinite integral is $\int\frac{x}{1-x^4}dx=\int\sum_{n=0}^{\infty}x^{4n}dx=\sum_{n=0}^{\infty}\frac{x^{4n+1}}{4n+1}+C$.

Solutions for Additional Exercises for Chapter 18

(1) $\lim_{N\to\infty}\sum_{n=1}^{N}(a_n+b_n)=\lim_{N\to\infty}\left(\sum_{n=1}^{N}a_n+\sum_{n=1}^{N}b_n\right)=,\ \lim_{N\to\infty}\sum_{n=1}^{N}a_n+\lim_{N\to\infty}\sum_{n=1}^{N}b_n.$

(2) If $\sum_{n=1}^{\infty}a_n$ is convergent, then there is some S such that for every $\epsilon>0$ there is an N so that $|a_n-S|<\frac{\epsilon}{2}$ if $n\geqslant N$. So, if $n,m\geqslant N$ then $|a_n-a_m|\leqslant|a_n-S|+|a_m-S|<\epsilon$. This proves than every convergent series is Cauchy. The converse is also true.

(3) The series is NOT Cauchy.

(4) The series is Cauchy.

(5) The series is NOT Cauchy.

(6) The series is Cauchy (use the alternating series test).

(7) The series is Cauchy (use the alternating series test).

(8a) If $\prod_{i=1}^{\infty}(1+a_i) = p < \infty$ and $p \neq 0$, then $\lim\limits_{n\to\infty} p_n = p$, where $p_n = \prod_{i=1}^{n}(1+a_i)$. Therefore $\ln(p_n) = \sum\limits_{i=1}^{n} \ln(1+a_i)$ and $\lim\limits_{n\to\infty} \ln(p_n) = \ln(p) < \infty$, since $p \neq 0$. Therefore $\sum\limits_{i=1}^{\infty} \ln(1+a_i) = \ln(p) < \infty$. Conversely, if $\sum\limits_{i=1}^{\infty} \ln(1+a_i) < \infty$ then $\prod_{i=1}^{\infty}(1+a_i) < \infty$.

(8b) If $a_i \geqslant 0$, then $\ln(1+a_i) \leqslant a_i$. Therefore, if $\sum\limits_{i=1}^{\infty} a_i < \infty$ then $\sum\limits_{i=1}^{\infty} \ln(1+a_i) < \infty$ and, by part (a), $\prod\limits_{i=1}^{\infty}(1+a_i) < \infty$. Conversely, if $\prod\limits_{i=1}^{\infty}(1+a_i) < \infty$, then $\sum\limits_{i=1}^{\infty} a_i \leqslant \prod\limits_{i=1}^{\infty}(1+a_i) < \infty$.

(9) Yes.

(10) The series $\sum\limits_{n=1}^{\infty} (\cos(x))^{2n}$ converges if $|\cos(x)| < 1$, i.e., if $x \neq n\pi$ (with n an integer). In this case

$$\sum_{n=1}^{\infty} (\cos(x))^{2n} = \frac{\cos^2(x)}{1-\cos^2(x)}.$$

(11) $\sum\limits_{n=1}^{\infty} \frac{1}{(2x)^n} = \frac{1}{2x} \sum\limits_{n=0}^{\infty} \frac{1}{(2x)^n} = \frac{1}{2x}\left(1-\frac{1}{2x}\right)^{-1}$ provided $2x > 1$ or $x > \frac{1}{2}$. The series converges conditionally if $x = -\frac{1}{2}$ and diverges if $x = \frac{1}{2}$.

(12) $\sum\limits_{n=1}^{\infty} (-1)^{n+1}\frac{x^n}{n}$ converges absolutely if $|x| < 1$. It converges conditionally for $x = 1$ and diverges for $x = -1$. For $|x| < 1$, we have $\sum\limits_{n=1}^{\infty} (-1)^{n+1}\frac{x^n}{n} = -\ln(1-x)$ as can be verified by differentiating both sides.

(13) $\sum\limits_{n=0}^{\infty} 3^n \cdot n \cdot \cdot x^{n-1} = \frac{d}{dx} \sum\limits_{n=0}^{\infty} (3x)^n = \frac{d}{dx}\left(3x \sum\limits_{n=0}^{\infty} (3x)^n\right) = \frac{d}{dx}\left(3x\left(1-\frac{1}{3x}\right)^{-1}\right) = \frac{27x^2-18x}{(3x-1)^2}$ provided $|x| < \frac{1}{3}$.

(14) The series converges absolutely if $x \neq \pm n\pi$, (n a natural number).

(15) The series converges absolutely for every x. Its sum is $\cos(x) - 1$.

(16) The series converges absolutely for every x. Its sum is $\sin(x)$.

(17) The series converges if $x \neq \pm n\pi$, where n is a natural number.

(18) The series converges absolutely for every $x > 0$. Its sum is x.

Solutions for §19.1

(1) $\sqrt{15.5} = \sqrt{16 - .5} \approx \sqrt{16} - \frac{1}{2\sqrt{16}}0.5 = \frac{63}{16}$.

(2) $67^{\frac{1}{3}} = (64+3)^{\frac{1}{3}} \approx 4 + \frac{1}{3 \cdot 16}3 = \frac{65}{16}$.

(3) $1.01^{\frac{1}{5}} \approx 1 + \frac{1}{5}0.01 = 1.002$.

(4) $\cos(62^\circ) = \cos(60^\circ + 2^\circ) \approx \frac{1}{2} - \frac{\sqrt{3}}{2}2\frac{\pi}{180}$.

(5) $\tan(34^\circ) = \tan(30^\circ + 4^\circ) \approx \frac{1}{\sqrt{3}} + \frac{4}{3}4\frac{\pi}{180}$.

(6) $\ln(x+h) \approx \ln(x) + \frac{h}{x} \implies \ln(0.9) \approx \ln(1) - 0.1 = -\frac{1}{10}$.

(7) $e^{x+h} \approx e^x + he^x \implies e^{3.2} \approx e^3(1+0.2) = 1.2e^3$.

(8) $\sin(x+h) \approx \sin(x) + h\cos(x) \implies \sin(87^\circ) \approx \sin(90^\circ) - \frac{\pi}{60}\cos(90^\circ) = 1$.

(9) $\cos(x+h) \approx \cos(x) - h\sin(x) \implies \cos(112^\circ) = \cos(120^\circ - 8^\circ) \approx -\frac{1}{2} + \frac{\sqrt{3}}{2}8\frac{\pi}{180}$.

(10) $(x+h)^{100} \approx x^{100} + 100 \cdot h \cdot x^{99} \implies 1.01^{100} \approx 1 + 100 \cdot 0.01 = 2$.

(11) $\ln(x+h) = \ln(x) + \frac{h}{x} \implies \ln 3 = \ln(e + (3-e)) \approx 1 + \frac{3-e}{e} = \frac{3}{e}$.

(12) $\cosh(x+h) \approx \cosh(x) + h\sinh(x) \implies \cosh\left(\frac{1}{4}\right) = \cosh\left(0 + \frac{1}{4}\right) \approx 1$.

Solutions for §19.2

(1) $1 + 3x + \frac{9x^2}{2} + \frac{9x^3}{2} + \frac{27x^4}{8}$.

(2) $1 - \frac{x^2}{2} + \frac{x^4}{4!}$.

(3) $x - \frac{x^3}{6}$.

(4) $-x - \frac{x^2}{2} - \frac{x^3}{3} - \frac{x^4}{4}$.

(5) $x + \frac{x^3}{3}$.

(6) $\frac{1}{2}x - \frac{1}{4}\frac{x^2}{2} + \frac{3}{8}\frac{x^3}{3!} - \frac{15}{16}\frac{x^4}{4!}$.

(7) $\frac{1}{3}2x - \frac{2}{9}\frac{4x^2}{2} + \frac{10}{27}\frac{8x^3}{3!} - \frac{80}{81}\frac{16x^4}{4!}$.

(8) $\cosh x \approx 1 + \frac{x^2}{2} + \frac{x^4}{4!}$.

(9) $\sinh x \approx x + \frac{x^3}{3!}$.

(10) Use the polynomial $1 + x + \frac{x^2}{2} + \frac{x^3}{3!} + \frac{x^4}{4!}$ for $x = 1$.

(11) Use the polynomial $\frac{1}{2} - \frac{\sqrt{3}}{2}x - \frac{1}{2}\frac{x^2}{2} + \frac{\sqrt{3}}{2}\frac{x^3}{3!} + \frac{1}{2}\frac{x^4}{4!}$ for $x = -2\frac{\pi}{180}$.

(12) Use the polynomial $\frac{\sqrt{2}}{2} + \frac{\sqrt{2}}{2}x - \frac{\sqrt{2}}{2}\frac{x^2}{2} - \frac{\sqrt{2}}{2}\frac{x^3}{3!} + \frac{\sqrt{2}}{2}\frac{x^4}{4!}$ for $x = 3\frac{\pi}{180}$.

(13) Use the polynomial from **(4)** for $x = -0.1$.

(14) Use the polynomial from **(5)** for $x = -5\frac{\pi}{180}$. (Note that $\tan 175° = \tan -5°$.)

(15) Use the polynomial from **(6)** for $x = 0.1$.

(16) As $\sqrt[3]{9} = 2\sqrt[3]{1 + 1/8}$, one can use the polynomial $2\sqrt[3]{1+x} = 2(\frac{1}{3}x - \frac{2}{9}\frac{x^2}{2} + \frac{10}{27}\frac{x^3}{3!} - \frac{80}{81}\frac{x^4}{4!})$ for $x = \frac{1}{8}$.

(17) Use the polynomial $\ln(1 + x) = x - \frac{x^2}{2} + \frac{x^3}{3} - \frac{x^4}{4}$ for $x = 2$, or $\ln(e(1 + x)) = 1 + \ln(1 + x) = 1 + x - \frac{x^2}{2} + \frac{x^3}{3} - \frac{x^4}{4}$ for $x = \frac{3}{e} - 1$.

(18) Use the polynomial from **(8)** for $x = \frac{1}{4}$.

Solutions for §19.3

(1) $\sum\limits_{n=0}^{\infty} \frac{3^n x^n}{n!}$, $R = \infty$.

(2) $\sum\limits_{n=0}^{\infty} (-1)^n \frac{x^n}{n!}$, $R = \infty$.

(3) $\sum\limits_{n=0}^{\infty} (-1)^n \frac{2^{2n} x^{2n}}{(2n)!}$, $R = \infty$.

(4) $\ln(e + x) = \ln e + \ln\left(1 + \frac{x}{e}\right) = 1 + \sum\limits_{n=1}^{\infty} (-1)^{n+1} \frac{x^n}{ne^n}$, $R = e$.

(5) $\sum\limits_{n=0}^{\infty} (-1)^n 2^n x^n$, $R = \frac{1}{2}$.

(6) $\sum\limits_{n=0}^{\infty} 3^n x^n$, $R = \frac{1}{3}$.

(7) $\sum\limits_{n=0}^{\infty} \frac{x^{2n}}{(2n)!}$, $R = \infty$.

(8) $\sum\limits_{n=0}^{\infty} \frac{x^{2n+1}}{(2n+1)!}$, $R = \infty$.

(9) $\sum\limits_{n=0}^{\infty} \frac{x^{2n}}{n!}$, $R = \infty$.

(10) $10^x = e^{x\ln 10} = \sum\limits_{n=0}^{\infty} \frac{(x\ln 10)^n}{n!}$, $R = \infty$.

(11) $\sum\limits_{n=0}^{\infty} 2^n x^{2n}$, $R = 1/\sqrt{2}$.

(12) $-1 + 3x - 3x^2 + x^3$, $R = \infty$.

Solutions for §19.4

(1) $\binom{5}{2} = \frac{5\cdot 4}{1\cdot 2} = 10$.

(2) $\binom{2}{8} = \frac{2\cdot 1\cdot 0\cdot(-1)\cdot(-2)\cdot(-3)\cdot(-4)\cdot(-5)}{1\cdot 2\cdot 3\cdot 4\cdot 5\cdot 6\cdot 7\cdot 8} = 0$.

(3) $\binom{\frac{1}{3}}{2} = \frac{(\frac{1}{3})\cdot(-\frac{2}{3})}{1\cdot 2} = -\frac{1}{9}$.

(4) $\binom{\frac{1}{2}}{0} = 1$.

(5) $\binom{-\frac{1}{2}}{4} = \frac{(-\frac{1}{2})\cdot(-\frac{3}{2})\cdot(-\frac{5}{2})\cdot(-\frac{7}{2})}{1\cdot 2\cdot 3\cdot 4} = \frac{35}{128}$.

(6) $\binom{-\frac{1}{3}}{3} = \frac{(-\frac{1}{3})\cdot(-\frac{4}{3})\cdot(-\frac{7}{3})}{1\cdot 2\cdot 3} = -\frac{14}{81}$.

(7) $\binom{0.21}{2} = \frac{(0.21)\cdot(-0.79)}{1\cdot 2} = -0.08295$.

(8) $\binom{-3.21}{2} = \frac{(-3.21)\cdot(-4.21)}{1\cdot 2} = 6.757075$.

(9) $(1+x)^{\frac{1}{3}} = \sum\limits_{n=0}^{\infty} \binom{\frac{1}{3}}{n} x^n$, $(1+x)^{-\frac{1}{3}} = \sum\limits_{n=0}^{\infty} \binom{-\frac{1}{3}}{n} x^n$, $(1+x)^{\frac{2}{3}} = \sum\limits_{n=0}^{\infty} \binom{\frac{2}{3}}{n} x^n$, and $(1+x)^{-\frac{2}{3}} = \sum\limits_{n=0}^{\infty} \binom{\frac{2}{3}}{n} x^n$.

(10) $1 + 0.23\,x - 0.08855\,x^2 + 0.0522445\,x^3 - 0.0361793\,x^4$.

(11) 1.2428; 1.25992....

(12) 8.82121; 8.82496....

Solutions for §19.5

(1) $e^{\frac{1}{3}} = 1 + \frac{1}{3} + \frac{1}{2!\cdot 3^2} + \frac{1}{3!\cdot 3^3} + \cdots + \frac{1}{m!\cdot 3^m} + \text{Error}$ where

$$\text{Error} = \int_0^{\frac{1}{3}} \frac{\left(\frac{1}{3} - t\right)^m}{m!} e^t\,dt < \int_0^{\frac{1}{3}} \frac{\left(\frac{1}{3}\right)^m \cdot 2}{m!}\,dt < \frac{2}{3^{m+1}\cdot m!}.$$

If we choose $m = 5$ we see that Error $< 10^{-4}$. Thus $e^{\frac{1}{3}} \approx 1.395$ after summing the first 5 terms of the above series.

(2) If we choose $m = 10$ (in problem (1)), we see that Error $< 10^{-11}$ and $e^{\frac{1}{3}} \approx 1.39561242508608952863$.

(3) $\cos\left(\frac{\pi}{7}\right) = 1 - \frac{\left(\frac{\pi}{7}\right)^2}{2!} + \frac{\left(\frac{\pi}{7}\right)^4}{4!} - \cdots + (-1)^m \frac{\left(\frac{\pi}{7}\right)^{2m}}{(2m)!} + \text{Error}$, where

$$|\text{Error}| = \left| \int_0^{\frac{\pi}{7}} (-1)^m \sin(t) \frac{\left(\frac{\pi}{7} - t\right)^{2m}}{(2m)!}\, dt \right| < \int_0^{\frac{\pi}{7}} \frac{\left(\frac{\pi}{7}\right)^{2m}}{(2m)!}\, dt < \left(\frac{\pi}{7}\right)^{2m+1} \cdot \frac{1}{(2m)!}.$$

If we choose $m = 3$, we have $|\text{Error}| < 10^{-4}$. When we sum the first 3 terms of the above series, we obtain $\cos\left(\frac{\pi}{7}\right) \approx 0.901$.

(4) Let $f(x) = \sqrt{4+x} = 2\left(1 + \frac{x}{4}\right)^{\frac{1}{2}} = 2\sum_{n=0}^{\infty} \binom{\frac{1}{2}}{n} \cdot \left(\frac{x}{4}\right)^n$. Then setting $x = 1$,

$$f(1) = \sqrt{5} = 2\sum_{n=0}^{m} \binom{\frac{1}{2}}{n} \cdot 4^{-n} + \text{Error},$$

with

$$\text{Error} = \int_0^1 f^{m+1}(t) \cdot \frac{(1-t)^m}{m!}\, dt.$$

Now $f'(x) = \frac{1}{2}(4+x)^{-\frac{1}{2}}$, $f''(x) = \frac{1}{2} \cdot \left(-\frac{1}{2}\right)(4+x)^{-\frac{3}{2}}$, $f'''(x) = \frac{1}{2} \cdot \left(-\frac{1}{2}\right)\left(-\frac{3}{2}\right)(4+x)^{-\frac{5}{2}}, \ldots$ We see that

$$f^{m+1}(t) = \frac{1}{2} \cdot \left(-\frac{1}{2}\right)\left(-\frac{3}{2}\right) \cdots \left(-\frac{2m-1}{2}\right)(4+t)^{-\frac{2m+1}{2}},$$

and

$$\left|f^{m+1}\right| < 1 \cdot 1 \cdot 2 \cdot 3 \cdots m \cdot (4+t)^{-m-\frac{1}{2}} = m!(4+t)^{-m-\frac{1}{2}}.$$

With this estimate:

$$\text{Error} < \int_0^1 m!(4+t)^{-m-\frac{1}{2}} \frac{(1-t)^m}{m!}\, dt < 4^{-m-\frac{1}{2}}.$$

So if $m = 18$, we have Error$< 10^{-11}$. Summing the first 18 terms of the Taylor series for $f(x)$ gives $\sqrt{5} \approx 2.23606797749978969641$.

(5) 2.23607....

(6) 1.25992....

(7) 0.69314....

(8) 1.09861....

(9) 3.76220....

(10) 11.68172....

Solutions for §19.6

(1) $e^{2x-1} \approx 1 + 2\left(-\frac{1}{2} + x\right) + 2\left(-\frac{1}{2} + x\right)^2 + \frac{4}{3}\left(-\frac{1}{2} + x\right)^3 + \frac{2}{3}\left(-\frac{1}{2} + x\right)^4.$

(2) $\cos(3x - 2) \approx 1 - \frac{9}{2}\left(-\frac{2}{3} + x\right)^2 + \frac{27}{8}\left(-\frac{2}{3} + x\right)^4.$

(3) $\sin(x) \approx \frac{\sqrt{3}}{2} - \frac{1}{2}\left(\frac{-2\pi}{3} + x\right) - \frac{\sqrt{3}}{4}\left(\frac{-2\pi}{3} + x\right)^2 + \frac{1}{12}\left(\frac{-2\pi}{3} + x\right)^3 + \frac{1}{16\sqrt{3}}\left(\frac{-2\pi}{3} + x\right)^4.$

(4) $\cos(x) \approx \frac{\sqrt{3}}{2} + \frac{1}{2}\left(\frac{-11\pi}{6} + x\right) - \frac{\sqrt{3}}{4}\left(\frac{-11\pi}{6} + x\right)^2 - \frac{1}{12}\left(\frac{-11\pi}{6} + x\right)^3 + \frac{1}{16\sqrt{3}}\left(\frac{-11\pi}{6} + x\right)^4.$

(5) $\sqrt{x} \approx 4 + \frac{-16+x}{8} - \frac{(-16+x)^2}{512} + \frac{(-16+x)^3}{16384} - \frac{5\,(-16+x)^4}{2097152}.$

(6) $\tan(x) \approx 1 + 2\left(\frac{-\pi}{4} + x\right) + 2\left(\frac{-\pi}{4} + x\right)^2 + \frac{8}{3}\left(\frac{-\pi}{4} + x\right)^3 + \frac{10}{3}\left(\frac{-\pi}{4} + x\right)^4.$

(7) $\ln(x) \approx 1 + \frac{-e+x}{e} - \frac{(-e+x)^2}{2\,e^2} + \frac{(-e+x)^3}{3\,e^3} - \frac{(-e+x)^4}{4\,e^4}.$

(8) $x^{19} \approx 1 + 19\,(-1 + x) + 171\,(-1 + x)^2 + 969\,(-1 + x)^3 + 3876\,(-1 + x)^4.$

(9) $x^{-\frac{2}{3}} \approx \frac{1}{9} - \frac{2}{729}\,(-27 + x) + \frac{5}{59049}\,(-27 + x)^2 - \frac{40}{14348907}\,(-27 + x)^3 + \frac{110}{1162261467}\,(-27 + x)^4.$

(10) $\ln(\sin(x)) \approx -\frac{1}{2}\left(\frac{-\pi}{2} + x\right)^2 - \frac{1}{12}\left(\frac{-\pi}{2} + x\right)^4.$

Solutions for §19.7

(1) $(3 + 2i) + (1 - 5i) = 4 - 3i.$

(2) $(3 + 2i) \cdot (1 - 5i) = (3 + 2i) \cdot 1 - 5i(3 + 2i) = 13 - 13i.$

(3) $\frac{3+2i}{1-5i} = \frac{3+2i}{1-51} \cdot \frac{1+5i}{1+5i} = \frac{3(1+5i)+2i(1+5i)}{26} = \frac{-7+17i}{26}.$

(4) $(2 + 3i) \cdot (2 + 3i) = 2(2 + 3i) + 3i(2 + 3i) = -5 + 12i.$

(5) $(2 + 3i)^{-2} = \frac{1}{(2+3i)^2} \cdot \frac{(2-3i)^2}{(2-3i)^2} = \frac{(2-3i)^2}{13^2} = -\frac{5+12i}{169}.$

(6) $-2 + 2i.$

(7) Since

$$\begin{aligned} e^{(a+b)}i &= \cos(a + b) + i\sin(a + b) \\ &= e^{ai} \cdot e^{bi} = (\cos(a) + i\sin(a)) \cdot (\cos(b) + i\sin(b)) \\ &= (\cos(a)\cos(b) - \sin(a)\sin(b)) + (\cos(a)\sin(b) + \sin(a)\cos(b))i. \end{aligned}$$

Equating real and imaginary parts yields the desired formulae:

$$\begin{aligned} \cos(a + b) &= \cos(a)\cos(b) - \sin(a)\sin(b), \\ \sin(a + b) &= \sin(a)\cos(b) + \cos(a)\sin(b) \end{aligned}$$

(8) $\cos(3\theta) = \cos^3(\theta) - 3\sin^2(\theta)\cos(\theta)$.

(9) $\cos(N\theta) = \cos^N\theta - (cN2)\cos^{N-2}(\theta)\sin^2(\theta) + \cdots + +(-1)^k\,(cN2k)\cos^{N-2k}(\theta)\sin^{2k}(\theta) + \cdots = \mathrm{Re}\,(\cos(\theta) + i\sin(\theta))^N$.

(10) $\cos(\frac{\theta}{N}) = \mathrm{Re}\,(\cos(\theta) + i\sin(\theta))^{\frac{1}{N}}$.

Solutions for §19.8

(1) $\lim\limits_{x\to 3}\frac{x-3}{x^2-9} = \lim\limits_{x\to 3}\frac{1}{2x} = \frac{1}{6}$.

(2) $\lim\limits_{x\to 0}\frac{x}{\sin(x)} = \lim\limits_{x\to 0}\frac{1}{\cos(x)} = 1$.

(3) $\lim\limits_{x\to 3}\frac{x-3}{\cos\left(\frac{\pi x}{6}\right)} = \lim\limits_{x\to 3}\frac{1}{-\frac{\pi}{6}\sin\left(\frac{\pi x}{6}\right)} = -\frac{6}{\pi}$.

(4) $\lim\limits_{x\to 1}\frac{\ln(x)}{x^2-2x+1} = \lim\limits_{x\to 1}\frac{x^{-1}}{2x-2} = \infty$.

(5) $\lim\limits_{x\to 0}\frac{e^x-1-x}{\sin^2(x)} = \lim\limits_{x\to 0}\frac{e^x-1}{2\sin(x)\cos(x)} = \lim\limits_{x\to 0}\frac{e^x}{2(\cos^2(x)-\sin^2(x))} = \frac{1}{2}$.

(6) $\lim\limits_{x\to 0}\frac{\tan(x)-x}{x^3} = \lim\limits_{x\to 0}\frac{\sec^2(x)-1}{3x^2} = \lim\limits_{x\to 0}\frac{2\sec^2(x)\cdot\tan(x)}{6x}$
$= \lim\limits_{x\to 0}\frac{2\sec^2(x)\tan^2(x)+2\sec(x)(\sec^3(x)+\sec(x)\tan^2(x))}{6} = \frac{1}{3}$.

(7) $\lim\limits_{x\to 0}\frac{\sin(x)-x}{x^3} = \lim\limits_{x\to 0}\frac{\cos(x)-1}{3x^2} = \lim\limits_{x\to 0}\frac{-\sin(x)}{6x} = \lim\limits_{x\to 0}\frac{-\cos(x)}{6} = -\frac{1}{6}$.

(8) $\lim\limits_{x\to 0}\frac{\sinh(x)}{\sin(x)} = \lim\limits_{x\to 0}\frac{\cosh(x)}{\cos(x)} = 1$.

(9) $\lim\limits_{x\to 2}\frac{x^2-4}{\sin(\pi x)} = \infty$.

(10) $\lim\limits_{x\to 1}\left(\frac{1}{\ln(x)} - \frac{1}{x-1}\right) = \lim\limits_{x\to 1}\frac{x-1\ln(x)}{(x-1)\cdot\ln(x)} = \lim\limits_{x\to 1}\frac{1-x^{-1}}{\ln(x)+1-x^{-1}} = \lim\limits_{x\to 1}\frac{x^{-2}}{x^{-1}+x^{-2}} = \frac{1}{2}$.

(11) $\lim\limits_{x\to 0}\frac{a^x-b^x}{x} = \lim\limits_{x\to 0}\frac{\ln(a)\cdot a^x-\ln(b)\cdot b^x}{1} = \ln\left(\frac{a}{b}\right)$.

(12) $\lim\limits_{x\to 0}\frac{x}{\arcsin(4x)} = \lim\limits_{x\to 0}\frac{1}{4\cdot(1-16x^2)^{-\frac{1}{2}}} = \frac{1}{4}$.

(13) $\lim\limits_{x\to 0}\frac{\int_0^x\sin(\pi t^2)\,dt}{x^3} = \lim\limits_{x\to 0}\frac{\sin(\pi x^2}{3x^2} = \lim\limits_{x\to 0}\frac{2\pi x\cos(\pi x^2)}{6x} = \lim\limits_{x\to 0}\frac{2\pi\cos(\pi x^2)-(2\pi x)^2\sin(\pi x^2)}{6} = \frac{\pi}{3}$.

(14) Use the relation $\frac{d}{dx}f(x) = -\frac{1}{y^2}\frac{d}{dy}f\left(\frac{1}{y}\right)$ for both f and g, then apply the L'Hospital's rule for the limit of the ratio as y tends to zero.

Solutions for §19.9

(1) Let $y(x) = \sum\limits_{n=0}^{\infty} a_n x^n$. The differential equation $y''(x) = 1 \implies \sum\limits_{n=2}^{\infty} a_n n(n-1)x^{n-2} = 1$. Thus $2a_2 = 1$ and $a_n = 0$ for $n > 2$. We obtain $y(x) = a_0 + a_1 x + \frac{1}{2}x^2$.

(2) Let $y(x) = \sum\limits_{n=0}^{\infty} a_n x^n$. The differential equation $y'''(x) = x^2 \implies \sum\limits_{n=3}^{\infty} a_n n(n-1)(n-2)x^{n-3} = x^2$. Thus $a_5 = \frac{1}{60}$ and $a_n = 0$ for $n \geq 3$, $n \neq 5$. Hence $y = a_0 + a_1 x + a_2 x^2 + \frac{1}{60}x^5$. The initial conditions $y'(0) = 1$, $y''(0) = 1 \implies y(x) = 1 + x + \frac{x^2}{2} + \frac{x^5}{60}$.

(3) Let $y(x) = \sum\limits_{n=0}^{\infty} a_n x^n$. The differential equation $y' + 2y = 0 \implies (a_1 + 2a_2 x + 3a_3 x^2 + \cdots) + (2a_0 + 2a_1 x + 2a_2 x^2 + \cdots) = 0$. Thus $a_1 = -2a_0$, $a_2 = \frac{-2a_1}{2}$, $a_3 = \frac{-2a_2}{3}$, $a_4 = \frac{-2a_3}{4}, \ldots \implies a_1 = -2a_0$, $a_2 = \frac{-2\cdot 2}{2}a_0$, $a_3 = \frac{2\cdot2\cdot2}{2\cdot3}a_0$, $a_4 = -\frac{2\cdot2\cdot2\cdot2}{2\cdot3\cdot4}a_0$, $\implies y(x) = a_0 \sum\limits_{n=0}^{\infty} \frac{(-2x)^n}{n!} = a_0 e^{-2x}$.

(4) $y(x) = \sum\limits_{n=0}^{\infty} (-1)^n \frac{x^{2n}}{(2n)!} = \cos(x)$.

(5) $y(x) = \sum\limits_{n=0}^{\infty} a_n \frac{x^n}{n!}$, where $a_n = (n-2)a_{n-3}$, $a_2 = a_5 = \ldots \equiv 0$, while a_0 and a_1 should be defined by boundary conditions. Explicitly we have

$$y(x) = a_0 + a_1 x + a_0 \sum_{n=1}^{\infty} \frac{1.4\ldots(3n-2)}{(3n)!}x^{3n} + a_1 \sum_{n=1}^{\infty} \frac{2.5\ldots(3n-1)}{(3n+1)!}x^{3n+1}.$$

(6) $y(x) = \sum\limits_{n=0}^{\infty} a_n \frac{x^n}{n!}$, where $a_n = (n-4)a_{n-2}$, a_0 and a_1 should be defined by boundary conditions. Explicitly

$$y(x) = a_0 + a_1 x - a_0 x^2 - \frac{a_1}{3!}x^3 - a_1 \sum_{n=2}^{\infty} \frac{1.3\ldots(2n-3)}{(2n+1)!}x^{2n+1}.$$

Solutions for Additional Exercises for Chapter 19

(1a) $3x^2 - 6x + 4 = 3(x-2)^2 + 6(x-2) + 4$.

(1b) $x^4 - 5x^3 + 9x^2 = (x-2)^4 + 3(x-2)^3 + 3(x-2)^2 + 8(x-2) + 12$.

(2) Taylor polynomial of $\cos(x)$ at π of degree $2n$ is $-\frac{1}{0!} + \frac{1}{2!}(x-\pi)^2 - \frac{1}{4!}(x-\pi)^4 + \ldots + \frac{(-1)^{n+1}}{(2n)!}(x-\pi)^{2n}$.

(3) Taylor polynomial of $\exp(\cos(x))$ at 0 of degree 3 is $e - \frac{e}{2}x^2$.

(4) Taylor polynomial of $\frac{1}{1+x}$ at 0 of degree n is $1 - x + x^2 - x^3 + \ldots + (-1)^n x^n$.

(5) Taylor polynomial of $\sin(x)$ at $\frac{\pi}{2}$ of degree $2n$ is $\frac{1}{0!} - \frac{1}{2!}(x - \frac{\pi}{2})^2 + \frac{1}{4!}(x - \frac{\pi}{2})^4 + \ldots + \frac{(-1)^n}{(2n)!}(x - \frac{\pi}{2})^{2n}$.

(6) $\sin(5) \approx 5 - \frac{5^3}{3!} + \frac{5^5}{5!} - \cdots \approx -0.9589.$

(7) $\arctan(1) \approx 0.7854.$

(8) The ith coefficients of Taylor polynomials of $f+g$ and fg at a are $\frac{(f+g)^{(i)}(a)}{i!}$ and $\frac{(fg)^{(i)}(a)}{i!}$.

(9) Taylor polynomial of $\sum\limits_{-\infty}^{x} \exp(-\frac{1}{2}t^2)\, dt$ at 0 of degree 4 is $\sqrt{\pi} + x - \frac{1}{6}x^3$.

(10) $\exp(-\frac{1}{2}) \approx 0.6065.$

(11a) $y = \frac{1}{3}t^6 - 3\ln t.$

(11b) $y = \exp(t) + 2t^2 - 5\ln(t).$

(12a) The number of fish in the pond does not change.

(12b) If $0 \leqslant P < 1000$ then $\frac{dP}{dt} > 0$. If $P > 1000$ then $\frac{dP}{dt} < 0$. This means that if the number of fish in the pond is less than 1000, then the number of fish will increase; if it is greater than 1000, then the number of fish will decrease.

(12c) If fish were being caught at a rate of x per year, then the differential equation controlling the fish population would be $\frac{dP}{dt} = .001(1000 - P)P - xt$. The equilibrium level is obtained when $\frac{dP}{dt} = 0$. This occurs when $P = 500 \pm 10\sqrt{2500 - 10x}$.

(13a) The point $(3, 5)$ lies on the line $y = g'(3)x + 10$. Therefore $g'(3) = -\frac{5}{3}$.

(13b) Using a linear approximation, $g(2.95) \approx g(3) + g'(3)(2.95 - 3) \approx 5.0833$.

(14) Using a linear approximation, $f(.5) \approx f(1) + f'(1)(.5-1) = -\frac{1}{2}\sin(1) \approx -0.4207$. Using a quadratic approximation, $f(.5) \approx f(1) + f'(1)(.5-1) + \frac{1}{2}f'(1)(.5-1)^2 = -\frac{1}{2}\sin(1) + \frac{1}{2}\cos(1) \approx -0.1506$.

(15a) Error is $\leqslant 0.0009$.

(15b) Error is $\leqslant 0.000001$.

(15c) Error is $\leqslant 0.0002$.

(16) We are given $h(2) = 3$ and $h'(2) = -2$, and $-2 \leqslant h''(x) \leqslant 1$, if $0 \leqslant x \leqslant 4$. Using a linear approximation we can write $h(x) = h(2) + h'(2)(x-2) +$ (error term). The error term in absolute value is less than $\frac{|h''(c)|}{2}(x-2)^2$, where c is between x and 2. It follows that $h(3) = 3 + (-2)(3-2) +$ (error term) $=$ $1 +$ (error term), and the error term in absolute value is less than $\frac{2}{2}(3-2)^2 = 1$. So $-1 \leqslant h(3) - 1 \leqslant 1$, and therefore $0 \leqslant h(3) \leqslant 2$.

Solutions for §20.1

(1) $|\overrightarrow{(2,5)}| = \sqrt{2^2+5^2} = \sqrt{29}.$

(2) $|\overrightarrow{(0,-1,1)}| = \sqrt{0+1+1} = \sqrt{2}.$

(3) $|\overrightarrow{(-1,5,2)}| = \sqrt{1+25+4} = \sqrt{30}.$

(4) $|\overrightarrow{(\frac{1}{2},-\frac{1}{4},\frac{1}{2})}| = \sqrt{\left(\frac{1}{2}\right)^2+\left(\frac{1}{4}\right)^2+\left(\frac{1}{2}\right)^2} = 3/4.$

(5) $|\overrightarrow{(\sqrt{2},-\sqrt{3})}| = \sqrt{2+3} = \sqrt{5}.$

(6) $|\overrightarrow{(0,\sqrt{2},1)}| = \sqrt{0+2+1} = \sqrt{3}.$

(7) $|\overrightarrow{(\pi,1)}| = \sqrt{\pi^2+1} = \sqrt{\pi^2+1}.$

(8) $|\overrightarrow{(-3,1,0)}| = \sqrt{9+1} = \sqrt{10}.$

Solutions for §20.2

(1) $\overrightarrow{(-1,1)}.$

(2) $\overrightarrow{(1,5)}.$

(3) $\overrightarrow{(40,-10)}.$

(4) $\overrightarrow{(9,-1,7)}.$

(5) $\overrightarrow{(5,-14,-2)}.$

(6) $\overrightarrow{(1,0,-1/4)}.$

(7) $\overrightarrow{PQ} = \overrightarrow{OQ} - \overrightarrow{OP} = \overrightarrow{(1,-3)} - \overrightarrow{(2,6)} = \overrightarrow{(-1,-9)}$ which has length $\sqrt{82}$.

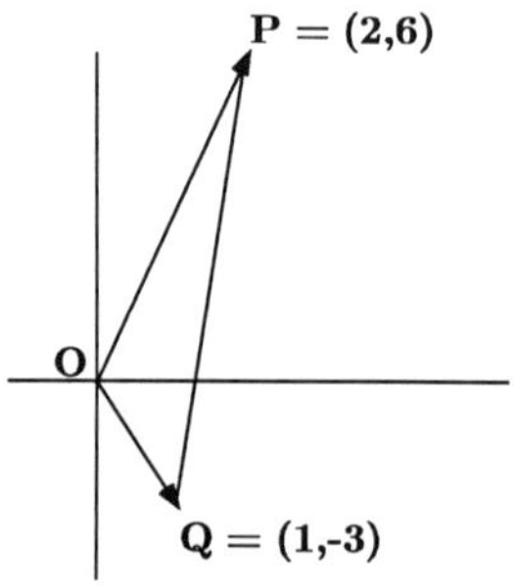

(8) $\overrightarrow{PQ} = \overrightarrow{(1,4,-3)} - \overrightarrow{(-2,3,0)} = \overrightarrow{(3,1,-3)}$ which has length $\sqrt{19}$.

Solutions for §20.4

(1) $\overrightarrow{(2,-4)} \cdot \overrightarrow{(1,3)} = 2 \cdot 1 + (-4) \cdot 3 = -10.$

(2) $2\vec{\imath} \cdot \vec{k} = \overrightarrow{(2,0,0)} \cdot \overrightarrow{(0,0,1)} = 0.$

(3) $(2\vec{\imath} - 3\vec{k}) \cdot (2\vec{\jmath} + 4\vec{k}) = \overrightarrow{(2,0,-3)} \cdot \overrightarrow{(0,2,4)} = -12.$

(4) $\overrightarrow{(1,-3,7)} \cdot \overrightarrow{(2,0,-2)} = 1 \cdot 2 + (-3) \cdot 0 + 7 \cdot (-2) = -12.$

(5) $(\vec{\imath} + 2\vec{\jmath}) \cdot (\vec{\imath} + 2\vec{\jmath}) = \overrightarrow{(1,2)} \cdot \overrightarrow{(1,2)} = 1 \cdot 1 + 2 \cdot 2 = 5.$

(6) $\overrightarrow{(1,-1,3)} \cdot \overrightarrow{(2,-3,4)} = 1 \cdot 2 + (-1)(-3) + 3 \cdot 4 = 17.$

(7) $\overrightarrow{(-2,9)} \cdot \overrightarrow{(1,-\frac{1}{2})} = -\frac{13}{2} \implies$ not perpendicular.

(8) $\overrightarrow{(-2,9)} \cdot \overrightarrow{(\frac{3}{2},\frac{1}{3})} = 0 \implies$ perpendicular.

(9) $\overrightarrow{(2,4)} \cdot \overrightarrow{(2,-1)} = 0 \implies$ perpendicular.

(10) $\overrightarrow{(1,3,-2)} \cdot \overrightarrow{(2,4,1)} = 12 \implies$ not perpendicular.

(11) $\overrightarrow{(3,-1,2)} \cdot \overrightarrow{(2,4,-1)} = 0 \implies$ perpendicular.

(12) $\overrightarrow{(2,1,-4)} \cdot \overrightarrow{(1,2,0)} = 4 \implies$ not perpendicular.

In the following four questions the perpendicular vectors obtained are not unique.

(13) Let $\overrightarrow{(x,y)}$ be a vector perpendicular to $\overrightarrow{(2,6)}$. Then $\overrightarrow{(2,6)} \cdot \overrightarrow{(x,y)} = 2x + 6y = 0$. A possible solution is $x = -3, y = 1$. The vector $\overrightarrow{(-3,1)}$ has length $\sqrt{10}$. Thus $\overrightarrow{\left(\frac{-3}{\sqrt{10}}, \frac{1}{\sqrt{10}}\right)}$ has length 1. Similarly, $\overrightarrow{\left(\frac{-3\sqrt{5}}{\sqrt{10}}, \frac{\sqrt{5}}{\sqrt{10}}\right)} = \overrightarrow{\left(\frac{-3}{\sqrt{2}}, \frac{1}{\sqrt{2}}\right)}$ has length 5 and $\overrightarrow{(-3\sqrt{10}, \sqrt{10})}$ has length 10. These latter 3 vectors are all perpendicular to $\overrightarrow{(2,6)}$.

(14) $\overrightarrow{(\frac{4}{5},\frac{3}{5})}, \overrightarrow{(4,3)}, \overrightarrow{(8,6)}.$

(15) $\overrightarrow{(\frac{5}{\sqrt{27}}, \frac{-1}{\sqrt{27}}, \frac{1}{\sqrt{27}})}, \overrightarrow{5(\frac{5}{\sqrt{27}}, \frac{-1}{\sqrt{27}}, \frac{1}{\sqrt{27}})}, \overrightarrow{10(\frac{5}{\sqrt{27}}, \frac{-1}{\sqrt{27}}, \frac{1}{\sqrt{27}})}.$

(16) $\overrightarrow{(\frac{3}{\sqrt{14}}, \frac{1}{\sqrt{14}}, \frac{2}{\sqrt{14}})}, \overrightarrow{5(\frac{3}{\sqrt{14}}, \frac{1}{\sqrt{14}}, \frac{2}{\sqrt{14}})}, \overrightarrow{10(\frac{3}{\sqrt{14}}, \frac{1}{\sqrt{14}}, \frac{2}{\sqrt{14}})}.$

(17) $\overrightarrow{(3,-1)} \cdot \overrightarrow{(2,1)} = 5 = |\overrightarrow{(3,-1)}| \cdot |\overrightarrow{(2,1)}| \cos(\theta) = \sqrt{50}\cos(\theta)$. Thus $\cos(\theta) = \frac{1}{\sqrt{2}} \implies \theta = \frac{\pi}{4}$ is the angle between the vectors.

(18) $\overrightarrow{(1,1,0)} \cdot \overrightarrow{(-1,-2,1)} = -3 = \sqrt{2} \cdot \sqrt{6}\cos(\theta)$. Thus $\cos(\theta) = -\frac{\sqrt{3}}{2} \implies \theta = \frac{2\pi}{3}$ is the angle between the vectors.

(19) $\overrightarrow{(3,1)} \cdot \overrightarrow{(2,4)} = 10 = \sqrt{10} \cdot \sqrt{20}\cos(\theta)$. Thus $\cos(\theta) = -\frac{1}{\sqrt{2}} \implies \theta = \frac{\pi}{4}$ is the angle between the vectors.

(20) $\overrightarrow{(-6,0,2)} \cdot \overrightarrow{(-5,3,-2)} = 26 = \sqrt{40} \cdot \sqrt{38}\cos(\theta)$. Thus $\theta = \cos^{-1}\left(\frac{13}{2\sqrt{95}}\right)$ is the angle between the vectors.

(21) Angle between the vectors $= 0$.

(22) Since $\overrightarrow{(\frac{1}{7},-3)} \cdot \overrightarrow{(28,\frac{4}{3})} = 0$, the angle between the vectors is $\frac{\pi}{2}$.

(23) $\vec{v}$ projected onto $\vec{w}$ is $\left(\frac{\vec{v}\cdot\vec{w}}{|\vec{w}|^2}\right)\vec{w}$ in general, yielding here $\frac{-7}{34}\vec{w}$.

(24) Using (23), $\frac{7}{26}\vec{w}$.

(25) Using (23), $\frac{8}{7}\vec{w}$.

(26) Using (23), $\frac{11}{26}\vec{w}$.

(27) The angles are respectively $\cos^{-1}\left(\frac{4}{\sqrt{17}}\right) \approx 0.24$ radians and $\cos^{-1}\left(\frac{-3}{\sqrt{29}}\right) \approx 2.16$ radians.

(28) We know $|\cos(\theta)| \leq 1 \Rightarrow \left|\frac{\vec{v}\cdot\vec{w}}{|\vec{v}||\vec{w}|}\right| \leq 1 \Rightarrow |\vec{v}\cdot\vec{w}| \leq |\vec{v}||\vec{w}|$.

(29) $|\vec{v}+\vec{w}|^2 = (\vec{v}+\vec{w})\cdot(\vec{v}+\vec{w}) = |\vec{v}|^2 + |\vec{w}|^2 + 2\vec{v}\cdot\vec{w} \leq |\vec{v}|^2 + |\vec{w}|^2 + 2|\vec{v}\cdot\vec{w}| \leq$ (by question 28) $|\vec{v}|^2 + |\vec{w}|^2 + 2|\vec{v}||\vec{w}| = (|\vec{v}| + |\vec{w}|)^2$.

(30) $(\vec{v}+\vec{w})\cdot(\vec{v}-\vec{w}) = |\vec{v}|^2 - |\vec{w}|^2 = 0$.

(31) $(\vec{v} - (\vec{v}\cdot\vec{u})\vec{u})\cdot\vec{u} = \vec{v}\cdot\vec{u} - (\vec{v}\cdot\vec{u})|\vec{u}|^2 = 0$.

Solutions for §20.5

(1) $\left(2\vec{i} - 3\vec{j}\right) \times \vec{i} = 2\vec{i}\times\vec{i} - 3\vec{j}\times\vec{i} = 0 + 3\vec{k} = 3\vec{k}$.

(2)
$$\begin{aligned}\left(-2\vec{i}+\vec{j}-4\vec{k}\right)\times\left(2\vec{i}-\vec{j}\right) &= \left(-2\vec{i}+\vec{j}-4\vec{k}\right)\times(2\vec{i}) - \left(-2\vec{i}+\vec{j}-4\vec{k}\right)\times\vec{j}\\ &= -4\vec{i}\times\vec{i} + 2\vec{j}\times\vec{i} - 8\vec{k}\times\vec{i} + 2\vec{i}\times\vec{j} - \vec{j}\times\vec{j} + 4\vec{k}\times\vec{j}\\ &= -2\vec{k} - 8\vec{j} + 2\vec{k} - 4\vec{i} = -4\vec{i} - 8\vec{j}.\end{aligned}$$

(3) $\overrightarrow{(1,4,-1)} \cdot \overrightarrow{(2,1,0)} = \begin{vmatrix}\vec{i} & \vec{j} & \vec{k}\\ 1 & 4 & -1\\ 2 & 1 & 0\end{vmatrix} = \vec{i}\cdot\begin{vmatrix}4 & -1\\ 1 & 0\end{vmatrix} - \vec{j}\cdot\begin{vmatrix}1 & -1\\ 2 & 0\end{vmatrix} + \vec{k}\cdot\begin{vmatrix}1 & 4\\ 2 & 1\end{vmatrix} = \overrightarrow{(1,-2,-7)}$.

(4) $\overrightarrow{(-3,2,5)} \cdot \overrightarrow{(0,-1,-3)} = \begin{vmatrix} \vec{i} & \vec{j} & \vec{k} \\ -3 & 2 & 5 \\ 0 & -1 & -3 \end{vmatrix} = \vec{i} \cdot \begin{vmatrix} 2 & 5 \\ -1 & -3 \end{vmatrix} - \vec{j} \cdot \begin{vmatrix} -3 & 5 \\ 0 & -3 \end{vmatrix} + \vec{k} \cdot \begin{vmatrix} -3 & 2 \\ 0 & -1 \end{vmatrix}$

hence $\overrightarrow{(-3,2,5)} \cdot \overrightarrow{(0,-1,-3)} = \overrightarrow{(-1,-9,3)}$.

(5) $\overrightarrow{(1,-1,1)} \cdot \overrightarrow{(2,-3,-5)} = \begin{vmatrix} \vec{i} & \vec{j} & \vec{k} \\ 1 & -1 & 1 \\ 2 & -3 & -5 \end{vmatrix} = \vec{i} \cdot \begin{vmatrix} -1 & 1 \\ -3 & -5 \end{vmatrix} - \vec{j} \cdot \begin{vmatrix} 1 & 1 \\ 2 & -5 \end{vmatrix} + \vec{k} \cdot \begin{vmatrix} 1 & -1 \\ 2 & -3 \end{vmatrix} = \overrightarrow{(8,7,-1)}.$

(6) $\overrightarrow{(2,-3,2)} \cdot \overrightarrow{(1,2,2)} = \begin{vmatrix} \vec{i} & \vec{j} & \vec{k} \\ 2 & -3 & 2 \\ 1 & 2 & 2 \end{vmatrix} = \vec{i} \cdot \begin{vmatrix} -3 & 2 \\ 2 & 2 \end{vmatrix} - \vec{j} \cdot \begin{vmatrix} 2 & 2 \\ 1 & 2 \end{vmatrix} + \vec{k} \cdot \begin{vmatrix} 2 & -3 \\ 1 & 2 \end{vmatrix} = \overrightarrow{(-10,-2,7)}.$

(7) $\begin{vmatrix} 1 & 2 \\ 6 & -3 \end{vmatrix} = 1 \cdot (-3) - 2 \cdot 6 = -15.$

(8) $\begin{vmatrix} 1 & 6 & 2 \\ -1 & 1 & 1 \\ 2 & -3 & 1 \end{vmatrix} = 1 \cdot \begin{vmatrix} 1 & 1 \\ -3 & 1 \end{vmatrix} - 6 \begin{vmatrix} -1 & 1 \\ 2 & 1 \end{vmatrix} + 2 \begin{vmatrix} -1 & 1 \\ 2 & -3 \end{vmatrix} = 4 - 6 \cdot (-3) + 2 \cdot 1 = 24.$

(9) $\begin{vmatrix} \vec{i} & \vec{j} & \vec{k} \\ 4 & -1 & 0 \\ 1 & 2 & 1 \end{vmatrix} = \vec{i} \cdot \begin{vmatrix} -1 & 0 \\ 2 & 1 \end{vmatrix} - \vec{j} \begin{vmatrix} 4 & 0 \\ 1 & 1 \end{vmatrix} + \vec{k} \begin{vmatrix} 4 & -1 \\ 1 & 2 \end{vmatrix} = -\vec{i} - 4\vec{j} + 9\vec{k}.$

Solutions for §20.7

(1) $\overrightarrow{PQ} = \overrightarrow{(3,5,-2)} - \overrightarrow{(1,-1,2)} = \overrightarrow{(2,6,-4)}$, $\qquad \overrightarrow{PR} = \overrightarrow{(-2,6,0)} - \overrightarrow{(1,-1,2)} = \overrightarrow{(-3,7,-2)}$.

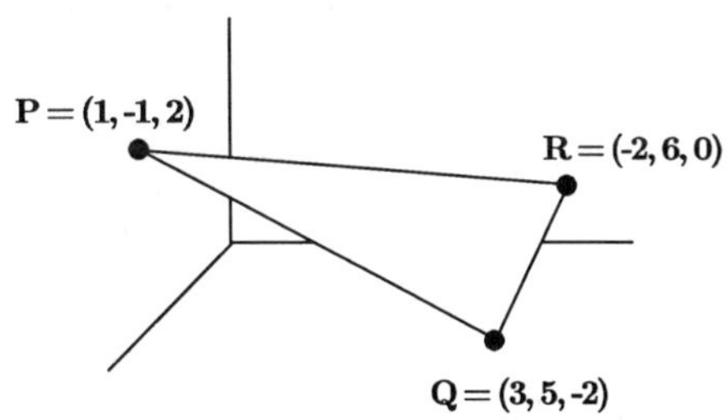

$$\text{Area} = \left|\tfrac{1}{2}\overrightarrow{PQ} \times \overrightarrow{PR}\right| = \tfrac{1}{2} \begin{vmatrix} \vec{i} & \vec{j} & \vec{k} \\ 2 & 6 & -4 \\ -3 & 7 & -2 \end{vmatrix} = \tfrac{1}{2}\left|16\vec{i} - 16\vec{j} + 32\vec{k}\right| = \tfrac{1}{2}\sqrt{16^2 + 16^2 + 32^2} = 8\sqrt{6}.$$

(2) $\text{Area} = |\vec{v} \times \vec{w}| = \left\| \begin{matrix} \vec{i} & \vec{j} & \vec{k} \\ 3 & -4 & 1 \\ 1 & 2 & -2 \end{matrix} \right\| = \left|6\vec{i} + 8\vec{j} + 10\vec{k}\right| = \sqrt{6^2 + 8^2 + 10^2} = 10\sqrt{2}.$

(3) $\vec{a} = \overrightarrow{(1,1,1)},\ \vec{b} = \overrightarrow{(-2,3,2)},\ \vec{c} = \overrightarrow{(0,0,1)}$. The volume of the parallelepiped determined by $\vec{a}, \vec{b}, \vec{c}$ is

$$\text{Volume} = \left\| \begin{matrix} 1 & 1 & 1 \\ -2 & 3 & 2 \\ 0 & 0 & 1 \end{matrix} \right\| = 5.$$

(4) $$\text{Volume} = \left\| \begin{matrix} 2 & -1 & 3 \\ 1 & -2 & -1 \\ 1 & -1 & 9 \end{matrix} \right\| = \left| 2 \cdot \begin{vmatrix} -2 & -1 \\ -1 & 9 \end{vmatrix} + 1 \cdot \begin{vmatrix} 1 & -1 \\ 1 & 9 \end{vmatrix} + 3 \cdot \begin{vmatrix} 1 & -2 \\ 1 & -1 \end{vmatrix} \right| = 25.$$

(5) $$x = \frac{\overrightarrow{(0,6,6)} \cdot (\overrightarrow{(4,1,-1)} \times \overrightarrow{(1,3,1)})}{\overrightarrow{(2,1,3)} \cdot (\overrightarrow{(4,1,-1)} \times \overrightarrow{(1,3,1)})} = \frac{\overrightarrow{(0,6,6)} \cdot \overrightarrow{(4,-5,11)}}{\overrightarrow{(2,1,3)} \cdot \overrightarrow{(4,-5,11)}} = \frac{36}{36} = 1.$$

$$y = \frac{\overrightarrow{(0,6,6)} \cdot (\overrightarrow{(2,1,3)} \times \overrightarrow{(1,3,1)})}{\overrightarrow{(4,1,-1)} \cdot (\overrightarrow{(2,1,3)} \times \overrightarrow{(1,3,1)})} = \frac{\overrightarrow{(0,6,6)} \cdot \overrightarrow{(-8,1,5)}}{\overrightarrow{(4,1,-1)} \cdot \overrightarrow{(-8,1,5)}} = \frac{36}{-36} = -1.$$

$$z = \frac{\overrightarrow{(0,6,6)} \cdot (\overrightarrow{(2,1,3)} \times \overrightarrow{(4,1,-1)})}{\overrightarrow{(1,3,1)} \cdot (\overrightarrow{(2,1,3)} \times \overrightarrow{(4,1,-1)})} = \frac{\overrightarrow{(0,6,6)} \cdot \overrightarrow{(-4,14,-2)}}{\overrightarrow{(1,3,1)} \cdot \overrightarrow{(-4,14,-2)}} = \frac{72}{36} = 2.$$

(6) $$x = \frac{\overrightarrow{(3,2,-7)} \cdot (\overrightarrow{(-1,1,3)} \times \overrightarrow{(1,4,-1)})}{\overrightarrow{(3,2,-2)} \cdot (\overrightarrow{(-1,1,3)} \times \overrightarrow{(1,4,-1)})} = \frac{\overrightarrow{(3,2,-7)} \cdot \overrightarrow{(-13,2,-5)}}{\overrightarrow{(3,2,-2)} \cdot \overrightarrow{(-13,2,-5)}} = 0.$$

$$y = \frac{\overrightarrow{(3,2,-7)} \cdot (\overrightarrow{(3,2,-2)} \times \overrightarrow{(1,4,-1)})}{\overrightarrow{(-1,1,3)} \cdot (\overrightarrow{(3,2,-2)} \times \overrightarrow{(1,4,-1)})} = \frac{\overrightarrow{(3,2,-7)} \cdot \overrightarrow{(6,1,10)}}{\overrightarrow{(-1,1,3)} \cdot \overrightarrow{(6,1,10)}} = \frac{-50}{25} = -2.$$

$$z = \frac{\overrightarrow{(3,2,-7)} \cdot (\overrightarrow{(3,2,-2)} \times \overrightarrow{(-1,1,3)})}{\overrightarrow{(1,4,-1)} \cdot (\overrightarrow{(3,2,-2)} \times \overrightarrow{(-1,1,3)})} = \frac{\overrightarrow{(3,2,-7)} \cdot \overrightarrow{(8,-7,5)}}{\overrightarrow{(1,4,-1)} \cdot \overrightarrow{(8,-7,5)}} = \frac{-25}{-25} = 1.$$

(7) $\vec{v} \cdot \vec{w} = |\vec{v}| \cdot |\vec{w}| \cos(\theta)$, $|\vec{v} \times \vec{w}| = |\vec{v}| \cdot |\vec{w}| \sin(\theta) \implies$

$$|\vec{v} \cdot \vec{w}|^2 + |\vec{v} \times \vec{w}|^2 = (|\vec{v}| \cdot |\vec{w}|)^2 \left(\cos^2(\theta) + \sin^2(\theta)\right) = |\vec{v}|^2 |\vec{w}|^2.$$

(8) Note that, by definition, $\vec{a} + \vec{b} + \vec{c} = 0 \implies \vec{a} \times (\vec{a} + \vec{b} + \vec{c}) = 0$. But this implies $\vec{a} \times \vec{b} = -\vec{a} \times c \implies |\vec{a}||\vec{b}| \sin(\gamma) = |\vec{a}||\vec{c}| \sin(\beta)$, i.e., $\frac{\sin(\gamma)}{|\vec{c}|} = \frac{\sin(\beta)}{|\vec{b}|}$. The remaining identities are similar.

(9) By choosing our basis vectors suitably we can simplify this calculation as follows: let $\vec{a} = (a_1, 0, 0)$, $\vec{b} = (b_1, b_2, 0)$ and $\vec{c} = (c_1, c_2, c_3)$. Then $\vec{a} \times (\vec{b} \times \vec{c}) = (0, -a_1 b_1 c_2 + a_1 b_2 c_1, -a_1 b_1 c_3)$ which equals $(\vec{a} \cdot \vec{c})\vec{b} - (\vec{a} \cdot \vec{b})\vec{c}$.

(11) This is easily seen since both $\vec{v}$ and $\vec{w}$ are perpendicular to $\vec{v} \times \vec{w}$.

Solutions To Additional Exercises for Chapter 20

(1) Let $\vec{u}$ denote the vector from A to B and let $\vec{v}$ denote the vector from A to D. Then, by definition, $AM = \{s(\vec{u} + \frac{1}{2}\vec{v})\colon |\colon 0 \leqslant s \leqslant 1\}$ and $BD = \{\vec{u} + t(\vec{v} - \vec{u})\colon |\colon 0 \leqslant t \leqslant 1\}$. So AM will intersect BD at the values of s and t between 0 and 1 that satisfy $s(\vec{u} + \frac{1}{2}\vec{v}) = \vec{u} + t(\vec{v} - \vec{u})$. This equation implies that $(1-t)\vec{u} = s\vec{u}$ and $t\vec{v} = \frac{1}{2}s\vec{v}$. It follows that $1 - t = s$ and $t = \frac{1}{2}s$. So $t = \frac{1}{3}$ and $s = \frac{2}{3}$. This shows that both AM trisects the diagonal BD and that the diagonal BD also trisects AM.

(2) Computing the lengths of the sides of triangle ABC, we have $\overrightarrow{AB} = \overrightarrow{OB} - \overrightarrow{OA} = (3\vec{\imath} + \vec{k}) - (\vec{\imath} + \vec{\jmath}) = 2\vec{\imath} - \vec{\jmath} + \vec{k}$, so that $|\overrightarrow{AB}| = \sqrt{(2)^2 + (-1)^2 + (1)^2} = \sqrt{6}$. $\overrightarrow{BC} = \overrightarrow{OC} - \overrightarrow{OB} = (4\vec{\imath} - 3\vec{\jmath} - 4\vec{k}) - (3\vec{\imath} + \vec{k}) = \vec{\imath} - 3\vec{\jmath} - 5\vec{k}$, so that $|\overrightarrow{BC}| = \sqrt{(1)^2 + (-3)^2 + (-5)^2} = \sqrt{35}$. $\overrightarrow{AC} = \overrightarrow{OC} - \overrightarrow{OA} = (4\vec{\imath} - 3\vec{\jmath} - 4\vec{k}) - (\vec{\imath} + \vec{\jmath}) = 3\vec{\imath} - 4\vec{\jmath} - 4\vec{k}$, so that $|\overrightarrow{AC}| = \sqrt{(3)^2 + (-4)^2 + (-4)^2} = \sqrt{41}$. Since $|\overrightarrow{AB}|^2 + |\overrightarrow{BC}|^2 = |\overrightarrow{AC}|^2$, we see that triangle ABC satisfies the Pythagorean Theorem and is therefore a right triangle with hypotenuse AC. Its area is given by $\frac{1}{2}|\overrightarrow{AB}|\,|\overrightarrow{BC}| = \frac{1}{2}\sqrt{210}$. Alternatively, $\overrightarrow{AB} \cdot \overrightarrow{BC} = (2)(1) + (-1)(-3) + (1)(-5) = 0$, so we see immediately that $\overrightarrow{AB}$ and $\overrightarrow{BC}$ are perpendicular. And $\overrightarrow{AB} \times \overrightarrow{BC} = [(-1)(-5) - (1)(-3)]\vec{\imath} - [(2)(-5) - (1)(1)]\vec{\jmath} + [(2)(-3) - (1)(-1)]\vec{k} = 8\vec{\imath} - 11\vec{\jmath} - 5\vec{k}$, so that $AREA = \frac{1}{2}|\overrightarrow{AB}|\,|\overrightarrow{BC}| = \frac{1}{2}|\overrightarrow{AB}|\,|\overrightarrow{BC}|\sin\frac{\pi}{2} = \frac{1}{2}|\overrightarrow{AB} \times \overrightarrow{BC}| = \frac{1}{2}\sqrt{(8)^2 + (-11)^2 + (-5)^2} = \frac{1}{2}\sqrt{210}$.

(3) It means that $\vec{v}$ and $\vec{w}$ are parallel or $\vec{v}$ and $-\vec{w}$ are parallel. (The sine of the angle between $\vec{v}$ and $\vec{w}$ is zero, so the angle between them is either 0 or π.)

(4) If $\vec{a} = \vec{\imath} + \vec{\jmath} + \vec{k}$, $\vec{b} = \vec{\imath} - \vec{\jmath} - 2\vec{k}$, and $\vec{c} = 2\vec{\jmath} - \vec{k}$, then $\vec{b} \times \vec{c} = 5\vec{\imath} + \vec{\jmath} + 2\vec{k}$. The volume of the parallelepiped determined by $\vec{a}$, $\vec{b}$, and $\vec{c}$ is then given by $\vec{a} \cdot (\vec{b} \times \vec{c}) = 8$.

(5) Yes.

(6) No.

(7) In parallelogram $ABCD$, let $\vec{u}$ denote the vector from A to B and let $\vec{v}$ denote the vector from A to D. Then the diagonals are $AC = \{s(\vec{u} + \vec{v})\colon |\colon 0 \leqslant s \leqslant 1\}$ and $BD = \{\vec{u} + t(\vec{v} - \vec{u})\colon |\colon 0 \leqslant t \leqslant 1\}$. So AC will intersect BD at the values of s and t between 0 and 1 that satisfy $s(\vec{u} + \vec{v}) = \vec{u} + t(\vec{v} - \vec{u})$. This equation implies that $(1 - t)\vec{u} = s\vec{u}$ and $t\vec{v} = s\vec{v}$. It follows that $1 - t = s$ and $t = s$. So $t = \frac{1}{2}$ and $s = \frac{1}{2}$. This shows that both diagonals AC and BD bisect the other. The area of the triangle ABC is $\frac{1}{2}|\vec{u} \times \vec{v}|$.

(8a) They lie in the same plane.

(8b) If $\vec{b} \times \vec{c} = \vec{0}$, then $\vec{a} \times (\vec{b} \times \vec{c}) = \vec{0} = 0\vec{b} + 0\vec{c}$. If $\vec{b} \times \vec{c} \neq 0$, then $\vec{b}$ and $\vec{c}$ span a plane containing $\vec{a} \times (\vec{b} \times \vec{c})$. So $\vec{a} \times (\vec{b} \times \vec{c}) = s\vec{b} + t\vec{c}$.

(8c) $\vec{a} \cdot (s\vec{b} + t\vec{c}) = 0$, so $s(\vec{a} \cdot \vec{b}) + t(\vec{a} \cdot \vec{c}) = 0$.

(8d) Since $s(\vec{a} \cdot \vec{b}) + t(\vec{a} \cdot \vec{c}) = 0$, we take $s = (\vec{a} \cdot \vec{c})$ and $t = -(\vec{a} \cdot \vec{b})$. So $\vec{a} \times (\vec{b} \times \vec{c}) = (\vec{a} \cdot \vec{c})\vec{b} - (\vec{a} \cdot \vec{b})\vec{c}$.

(8e) No.

(9) Combine $\vec{OP}$ with a vector of magnitude $2a$ pointing to the left in direction $60°$ above the horizontal.

(10) $\vec{u} \times \vec{v}$ is of magnitude 15 and points towards you.

(11) Combine the 1 lb force with a force of magnitude $\sqrt{5 - 2\sqrt{3}}$ lb pulling in a direction θ north of east, where $\tan(\theta) = \frac{1}{\sqrt{3}-1}$.

(12) $\vec{\imath} \cdot (\vec{\jmath} \times \vec{k}) = \vec{\imath} \cdot \vec{\imath} = 1$. $\quad \vec{\jmath} \cdot (\vec{k} \times \vec{\imath}) = \vec{\jmath} \cdot \vec{\jmath} = 1$. $\quad \vec{k} \cdot (\vec{\imath} \times \vec{\jmath}) = \vec{k} \cdot \vec{k} = 1$.

(13) Algebraically, let $\vec{u} = \vec{\imath}$, $\vec{v} = \vec{\imath} + \vec{\jmath}$, and $\vec{w} = \vec{\imath} + \vec{\jmath} + \vec{k}$. Then $\vec{u} \times \vec{v} = \vec{k}$, and so $(\vec{u} \times \vec{v}) \cdot \vec{w} = \vec{k} \cdot (\vec{\imath} + \vec{\jmath} + \vec{k}) = 1$. Geometrically, the vectors $\vec{u}$, $\vec{v}$, and $\vec{w}$ all point from the origin to corners of the unit cube spanned by $\vec{\imath}$, $\vec{\jmath}$, and $\vec{k}$. If ϕ is the angle between $\vec{u}$ and $\vec{v}$, and if θ is the angle between $\vec{u} \times \vec{v}$ and $\vec{w}$, then the sides of the cube have lengths $|\vec{u}|$, $|\vec{v}|\sin(\phi)$, and $|\vec{w}|\cos(\theta)$. The volume of the cube is therefore equal to $(|\vec{u}|\,|\vec{v}|\sin(\phi))(|\vec{w}|\cos(\theta)) = (\vec{u} \times \vec{v}) \cdot \vec{w}$. So $(\vec{u} \times \vec{v}) \cdot \vec{w} = 1$.

(14) $\arcsin\left(\frac{1}{\sqrt{7}}\right)$

(15) Let $\vec{a} = (\cos(\alpha))\vec{\imath} + (\sin(\alpha))\vec{\jmath}$ and let $\vec{b} = (\cos(\beta))\vec{\imath} + (\sin(\beta))\vec{\jmath}$. The angle between $\vec{a}$ and $\vec{b}$ is $(\beta - \alpha)$. Now $\vec{a} \times \vec{b} = (\sin(\beta)\cos(\alpha) - \cos(\beta)\sin(\alpha))\vec{k}$. So $\sin(\beta - \alpha) = |\vec{a}|\,|\vec{b}|\sin(\beta - \alpha) = |\vec{a} \times \vec{b}| = (\sin(\beta)\cos(\alpha) - \cos(\beta)\sin(\alpha))$.

(16) $x = 1,\ y = \frac{7}{2},\ z = \frac{3}{2}$

(17) $x = 0,\ y = -1,\ z = 3$

(18) $x = \frac{1}{2},\ y = -2,\ z = \frac{1}{4}$

(19) $x = \frac{1}{8},\ y = 30,\ z = -10$

(20) $x = \frac{1}{3},\ y = \frac{1}{4},\ z = -\frac{1}{12}$.

Solutions for §21.1

(1) Assume the point (x, y, z) lies on the line which passes through the points $(2, -4, 0)$ and $(1, 3, 0)$. The vector $\overrightarrow{(x - 1, y - 3, z)}$ must be a multiple of the vector $\overrightarrow{(2, -4, 0)} - \overrightarrow{(1, 3, 0)} = \overrightarrow{(1, -7, 0)}$.

Thus $\overrightarrow{(x - 1, y - 3, z)} = t\overrightarrow{(1, -7, 0)}$ and we obtain the parametric equation of the line: $x = 1 + t$, $y = 3 - 7t$, $z = 0$.

(2) $\overrightarrow{(x, y, z)} - \overrightarrow{(-1, 0, 3)} = t\left(\overrightarrow{(2, 0, 8)} - \overrightarrow{(-1, 0, 3)}\right) \implies \overrightarrow{(x + 1, y, z - 3)} = t\overrightarrow{(3, 0, 5)} \implies x = 1 + 3t,\ y = 0,\ z = 3 + 5t.$

(3) $\overrightarrow{(x, y, z)} - \overrightarrow{(1, 1, 2)} = t\left(\overrightarrow{(2, -4, 0)} - \overrightarrow{(1, 1, 2)}\right) = t\overrightarrow{(1, -5, -2)} \implies x = 1 + t,\ y = 1 - 5t,\ z = 2 - 2t.$

(4) $\overrightarrow{(x, y, z)} - \overrightarrow{(0, -1, 3)} = t\left(\overrightarrow{(4, -1, 3)} - \overrightarrow{(0, -1, 3)}\right) = t\overrightarrow{(4, 0, 0)} \implies x = 4t,\ y = -1,\ z = 3.$

(5) $\overrightarrow{(x, y, z)} - \overrightarrow{(-2, 9, -3)} = t\left(\overrightarrow{(4, 0, 8)} - \overrightarrow{(-2, 9, -3)}\right) = t\overrightarrow{(6, -9, 11)} \implies x = -2 + 6t,\ y = 9 - 9t,\ z = -3 + 11t.$

(6) $\overrightarrow{(x, y, z)} - \overrightarrow{(-2, -1, -3)} = t\left(\overrightarrow{(-2, -1, -3)} - \overrightarrow{(1, 2, 3)}\right) = t\overrightarrow{(-1, -3, -6)} \implies x = -2 - t,\ y = -1 - 3t,\ z = -3 - 6t.$

(7) Let QC and PB be two altitudes. Draw RA passing through D. We must show $\overrightarrow{RD} \cdot \overrightarrow{PQ} = 0$ given that $\overrightarrow{QD} \cdot \overrightarrow{PR} = 0$ and $\overrightarrow{PD} \cdot \overrightarrow{QR} = 0$.

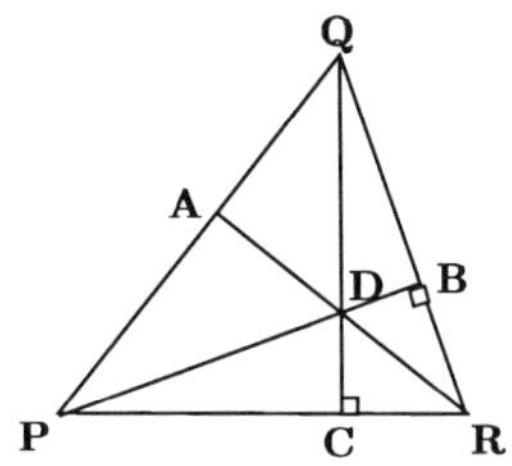

From the figure above we see that

$$\begin{aligned}\overrightarrow{RD} \cdot \overrightarrow{PQ} &= \overrightarrow{RD} \cdot \left(\overrightarrow{PR} + \overrightarrow{RQ}\right) = \overrightarrow{RD} \cdot \overrightarrow{PR} + \overrightarrow{RD} \cdot \overrightarrow{RQ} \\ &= \left(\overrightarrow{RQ} + \overrightarrow{QD}\right) \cdot \overrightarrow{PR} + \left(\overrightarrow{RP} + \overrightarrow{PD}\right) \cdot \overrightarrow{RQ} \\ &= \overrightarrow{RQ} \cdot \overrightarrow{PR} + \overrightarrow{RQ} \cdot \overrightarrow{RP} = 0,\end{aligned}$$

since $\overrightarrow{RP} = -\overrightarrow{PR}$.

(8) The parametric form can be derived by setting $\frac{x-x_0}{a} = \frac{y-y_0}{b} = \frac{z-z_0}{c} = t$ in th form above (which is termed **the symmetric form of the line**), unless all points (x, y, z) on the line have the same x (respectively, y, z) coordinate.

Solutions for §21.2

(1) $\left(\overrightarrow{(x,y,z)} - \overrightarrow{(2,-3,1)}\right) \cdot \overrightarrow{(3,-2,4)} = 0 \implies 3x - 2y + 4z = -16.$

(2) $\overrightarrow{(x-2, y+3, z-1)} \cdot \overrightarrow{(-1,4,2)} = 0 \implies -x + 4y + 2z = -14.$

(3) $\overrightarrow{(1,-2,4)} - \overrightarrow{(-1,3,5)} = \overrightarrow{(2,-5,-1)}$ and $\overrightarrow{(1,-2,4)} - \overrightarrow{(-2,1,-8)} = \overrightarrow{(3,-3,12)}$ lie on the plane. Thus

$$\overrightarrow{(2,-5,-1)} \times \overrightarrow{(3,-3,12)} = \begin{vmatrix} \vec{i} & \vec{j} & \vec{k} \\ 2 & -5 & -1 \\ 3 & -3 & 12 \end{vmatrix} = \overrightarrow{(-63,-27,9)}.$$

is a normal vector to the plane. The equation of the plane is given by

$$\overrightarrow{(x-1, y+2, z-4)} \cdot \overrightarrow{(-63,-27,9)} = 0 \implies 7x + 3y - z = -3.$$

(4) $\overrightarrow{(1,1,8)} - \overrightarrow{(2,-2,3)} = \overrightarrow{(-1,3,5)}$ and $\overrightarrow{(1,1,8)} - \overrightarrow{(0,-1,-3)} = \overrightarrow{(1,2,11)}$ lie on the plane then

$$\overrightarrow{(-1,3,5)} \times \overrightarrow{(1,2,11)} = \begin{vmatrix} \vec{i} & \vec{j} & \vec{k} \\ -1 & 3 & 5 \\ 1 & 2 & 11 \end{vmatrix} = \overrightarrow{(23,16,-5)}$$

is a normal vector. The equation of the plane is given by

$$\overrightarrow{(x-1, y-1, z-8)} \cdot \overrightarrow{(23,16,-5)} = 23x + 16y - 5z + 1 = 0.$$

(5) The line $x = 1 + t$, $y = 2 - t$, $z = 3t$ contains the two points $(1, 2, 0)$, $(2, 1, 3)$ corresponding to $t = 0, 1$. With these two points plus the given point $(1, -1, 5)$ we have found 3 points on the plane. The problem can now be solved as in the previous 2 problems. We obtain: $-x + 5y + 3z = 6$.

(6) $5x + 2y + 4z = 17$.

(7) The general equation of a plane is $ax + by + cz = d$. If the plane contains $(0, 0, 1)$ and $(0, -1, 0) \implies c = d$ and $b = -d$. Thus, we may write the equation of the plane in the form $ax + by - bz = -b$.

Recall that the vector $\overrightarrow{(A, B, C)}$ is always a normal vector to the plane $Ax + By + Cz = D$. Two planes are perpendicular if and only if their normal vectors are perpendicular to the plane $3x + y - 2z = 4$ if and only if $\overrightarrow{(3, 1, -2)} \cdot \overrightarrow{(a, b, -b)} = 0 \implies 3a + 3b = 0 \implies a = -b$. So one possible answer is the plane $x - y + z = 1$.

(8) $\overrightarrow{(2, -3, 1)} \cdot \overrightarrow{(3, -3, -15)} = 0 \implies$ normal vectors are perpendicular $\implies$ the planes are perpendicular.

(9) $\overrightarrow{(2, -1, 1)} = \frac{1}{3}\overrightarrow{(6, -3, 3)} \implies$ normal vectors are parallel $\implies$ planes are parallel.

(10) Planes are skew.

(11) $\overrightarrow{(7, -1, -1)} \cdot \overrightarrow{(2, 3, 11)} = 0 \implies$ normal vectors are perpendicular $\implies$ planes are perpendicular.

(12) $\overrightarrow{(3, -7, 2)} \cdot \overrightarrow{(-5, 1, 4)} = -14 = \sqrt{3^2 + 7^2 + 2^2}\sqrt{5^2 + 1 + 4^2}\cos(\theta) = 2\sqrt{7 \cdot 93}\cos(\theta)$ hence we see that $\theta = \arccos\left(-\sqrt{\frac{7}{93}}\right)$.

(13) $\overrightarrow{(8, 1, -2)} \cdot \overrightarrow{(2, -3, 2)} = 11 = \sqrt{66} \cdot \sqrt{17}\cos(\theta) \implies \arccos\left(\sqrt{\frac{11}{102}}\right)$.

(14) $\arccos\left(\frac{8}{5\sqrt{3}}\right)$.

(15) $\arccos\left(\frac{1}{\sqrt{11}}\right)$.

Recall that the distance from a point P to a plane is given by $\frac{\overrightarrow{PQ} \cdot \vec{n}}{|\vec{n}|}$ where $\vec{n}$ is a normal vector and Q is another point on the plane.

(16) Choose the points $Q = (1, 0, 0)$ which lies on our plane $3x - y + 4z = 3$. Then $\vec{n} = \overrightarrow{(3, -1, 4)}$. The distance from $P = (1, -1, 5)$ to the plane is given by

$$\left| \frac{\left(\overrightarrow{(1, 0, 0)} - \overrightarrow{(1, -1, 5)}\right) \cdot \overrightarrow{(3, -1, 4)}}{\sqrt{3^2 + 1 + 4^2}} \right| = \frac{21}{\sqrt{26}}.$$

(17) $Q = (2, 1, 0)$ lies on the plane $2x + y - z = 5$ and $\vec{n} = \overrightarrow{(2, 1, -1)}$. Then the distance from $P = (0, 1, 2)$ to the plane is given by

$$\left| \frac{\left(\overrightarrow{(2, 1, 0)} - \overrightarrow{(0, 1, 2)}\right) \cdot \overrightarrow{(2, 1, -1)}}{\sqrt{2^2 + 1 + 1}} \right| = \sqrt{6}.$$

(18) The point is located on the plane, so the distance is 0.

(19) $32\sqrt{3}$.

(20) Note that the distance is given by the projection of a vector from P to any point of the plane onto the direction normal to the plane, that is, $\frac{\overrightarrow{(p-x,q-y,r-z)}\cdot\overrightarrow{(a,b,c)}}{|\overrightarrow{(a,b,c)}|}$, where (x, y, z) is any point on the plane, that is, $ax + by + cz = d$, and the answer follows.

(21) The points (x, y, z) which simultaneously satisfy the two equations $2x-4y+4z = 8$ and $8x+7y+z = 2$ lie on the line of intersection. Multiplying the first equation by 4 and subtracting from the second (eliminates x) and we have

$$23y - 15z = -30 \implies y = \frac{15}{23}(z-2).$$

Similarly,eliminating z yields

$$-30x - 32y = 0 \implies x = -\frac{16}{15}y.$$

So the equation of the line can be written in the form $-\frac{x}{16} = \frac{y}{15} = \frac{z-2}{23}$.

(22) Assume that (p, q, r) is a point on the second plane; what we need is a distance from this point to the plane $ax + by + cz = d_1$. This distance is then given by the formula of **(20)**, where one substitutes $ap + bq + cr = d_2$.

(23) Check that for any point (p, q, r), which is not on the line of intersection of the two original planes, there exists one and only one choice of m, n, such that the corresponding equation defines a plane passing through the point and the line.

Solutions for §21.3

(1) $\overrightarrow{f(t)} = 7\vec{i} + 4\cos(t)\vec{j} + 4\sin(t)\vec{k}$.

(2) $\overrightarrow{f(t)} = 2\cos(t)\vec{i} - 5\vec{j} + 2\sin(t)\vec{k}$.

(3) Graph is a line in the plane $y = 0$.

(4)

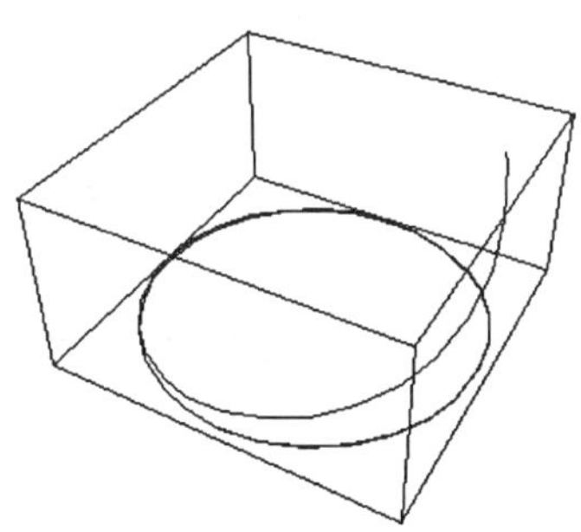

$\overrightarrow{f(t)} = \cos(t)\vec{i} + \sin(t)\vec{j} + e^t\vec{k}$

(5) Graph is a parabola in the plane $x = 1$.

(6) Graph is a hyperbola in the plane $z = 10$.

Solutions for §21.4

In rectangular coordinates, the corresponding points are:

(1) $(2, 2\sqrt{3})$ and $(-2, -2\sqrt{3})$.

(2) $(-\sqrt{2}, \sqrt{2})$ and $(\sqrt{2}, -\sqrt{2})$.

(3) $(3, 0)$ and $(-3, 0)$.

(4) $(\sqrt{3}, -1)$ and $(-\sqrt{3}, 1)$.

(5) $(-\sqrt{3}, -1)$ and $(\sqrt{3}, 1)$.

(6) $(\frac{5}{2}, \frac{5}{2}\sqrt{3}, -1)$ and $(-\frac{5}{2}, -\frac{5}{2}\sqrt{3}, -1)$.

(7) Use the relations $x = r\cos(\theta)$, $y = r\sin(\theta)$, and $r = \sqrt{x^2 + y^2}$.

(8) Apply the Cosine Theorem for the triangle with vertices at two given points and at the origin.

(9) We have $x = r\cos(\theta) = 3(1 - 2\sin(\theta)) \cdot \cos(\theta)$ and $y = r\sin(\theta) = 3(1 - 2\sin(\theta)) \cdot \sin(\theta)$. It follows that $\sqrt{x^2 + y^2} = 3(1 - 2\sin(\theta)) = 3\left(1 - 2\frac{y}{\sqrt{x^2+y^2}}\right)$.

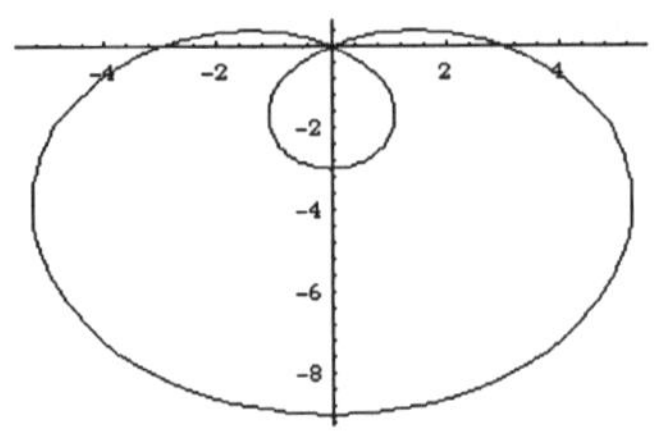

$r = 3(1 - 2\sin(\theta))$

(10) The Cartesian equation is $(x^2 + y^2)^2 = 2y(3x^2 - y^2)$.

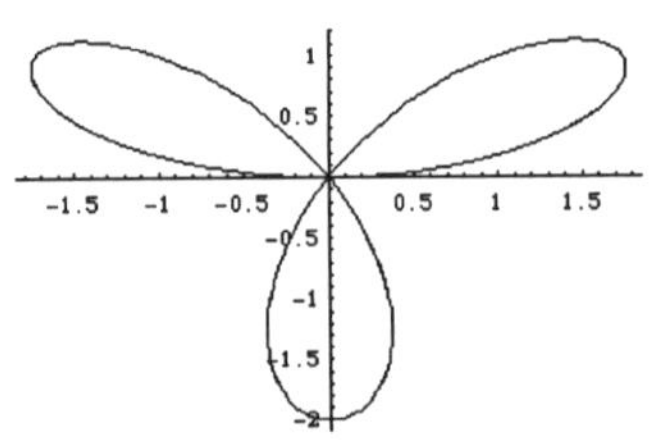

$r = 2\sin(3\theta)$

(11) The polar equation for the line $y = mx + b$ is given by $r = \frac{b}{\sin\theta - m\cos\theta}$.

(12) The polar equation $r = \frac{5}{\sin(\theta)}$ is just the Cartesian equation of the line $y = 5$.

(13) Intersection occurs at points $\theta = 0$ and $\theta = \pi/6$.

(14) $r = \sqrt{3 + \sin^2\theta}$.

(15) In rectangular coordinates, the equation takes the form $x^2 + y^2 = Ay + Bx$, which is the same as the equation of a sphere $\left(x - \frac{B}{2}\right)^2 + \left(y - \frac{A}{2}\right)^2 = \frac{A^2+B^2}{4}$.

(16) $(-1, \sqrt{2}, 5)$.

(17) $(\sqrt{2}, -\sqrt{2}, -2)$.

(18) $(5, 0, 1)$.

(19) $(0, -2, 1)$.

(20) $(0, 3, 2)$.

(21) $(-1, -\sqrt{2}, -6)$.

Solutions for §21.5

(1) $\overrightarrow{f(t)} = 4\sin(5t)\cos(t)\vec{i} + 4\sin(5t)\sin(t)\vec{j} - 3\vec{k}$.

(2) $\overrightarrow{f(t)} = 4\sin\left(5\left(t + \frac{\pi}{4}\right)\right)\cos\left(t + \frac{\pi}{4}\right)\vec{i} + 4\sin\left(5\left(t + \frac{\pi}{4}\right)\right)\sin\left(t + \frac{\pi}{4}\right)\vec{j} - 3\vec{k}$.

(3) $\overrightarrow{f(t)} = \vec{i} + (6 + \left(\frac{t}{3}\right)^2\cos(t))\vec{j} + (6 + \left(\frac{t}{3}\right)^2\sin t)\vec{k}$.

(4) $\overrightarrow{f(t)} = \vec{i} + \left(6 + \left(\frac{t+\frac{\pi}{3}}{3}\right)^2\right)\cos\left(t + \frac{\pi}{3}\right)\vec{j} + 6 + \left(\frac{t+\frac{\pi}{3}}{3}\right)^2 \cdot \sin\left(t + \frac{\pi}{3}\right)\vec{k}$.

(5) $\overrightarrow{f(t)} = \left(\frac{t}{3}\right)^2\cos(t)\vec{j} + (7 + \left(\frac{t}{3}\right)^2\sin(t))\vec{k}$.

(6) $\overrightarrow{f(t)} = R(t - \sin(t))\vec{i} + R(1 - \cos(t))\vec{j}$.

Solutions to Additional Exercises for Chapter 21

(1) $\vec{v} = \frac{4}{\sqrt{6}}(1, -2, 1)$

(2) Let $M = (x, y, z)$ and let $M_0 = (x_0, y_0, z_0)$ denote the closest point on the plane to M. If d is the distance between M and the plane (i.e. between M and M_0), then $\overrightarrow{(x_0, y_0, z_0)} - \overrightarrow{(x, y, z)} = \pm d\vec{n}$. So $d = |\vec{n} \cdot \overrightarrow{(x_0, y_0, z_0)} - \vec{n} \cdot \overrightarrow{(x, y, z)}| = |e - \vec{n} \cdot \overrightarrow{(x, y, z)}|$.

(3a) $\vec{n} = \frac{1}{\sqrt{3}}(1, 1, 1)$, $e = \frac{1}{\sqrt{3}}$, so distance $= |\frac{1}{\sqrt{3}| - \frac{1}{\sqrt{3}}\overrightarrow{(1,1,1)} \cdot \overrightarrow{(-1,0,1)}} = \frac{1}{\sqrt{3}}$.

(3b) $\vec{n} = \frac{1}{\sqrt{14}}(2,3,1)$, $e = \frac{1}{\sqrt{14}}$, so distance $= \left|\frac{1}{\sqrt{14}} - \frac{1}{\sqrt{14}}(2,3,1)\cdot(-1,0,1)\right| = \frac{2}{\sqrt{14}}$.

(3c) $\vec{n} = (\frac{3}{5},\frac{4}{5})$, $e = 0$, so distance $= \overrightarrow{(\frac{3}{5},\frac{4}{5})}\cdot\overrightarrow{(1,2)} = \frac{11}{5}$.

(4) The planes are identical.

(5) A vector pointing along the direction of the line is $(5,2,1)$ and a vector pointing perpendicular to the plane is $(1,2,-9)$. Since $(5,2,1)\cdot(1,2,-9) = 0$ the line is parallel to the plane.

(6) It is easy to check that $\vec{u} = \overrightarrow{(\frac{3}{2},\frac{3}{2},0)}$ is a vector that points from the origin to the intersection of the two planes. Normal vectors to the two planes are given by $\vec{n}_1 = \overrightarrow{(1,3,2)}$ and $\vec{n}_2 = \overrightarrow{(3,1,-1)}$. Thus $\vec{n}_1 \times \vec{n}_2 = \overrightarrow{(-5,7,-8)}$ is a vector pointing along the line of intersection. Parametric equations of this line of intersection are therefore given by $x = \frac{3}{2} - 5t$, $y = \frac{3}{2} + 7t$, and $z = -8t$. The angle θ between the two planes is given by the equation $\vec{n}_1 \cdot \vec{n}_2 = |\vec{n}_1|\,|\vec{n}_2|\cos\theta$. Thus $\cos(\theta) = \frac{2}{\sqrt{77}}$.

(7) $x^2 + y^2 = 16$ (The graph is a circle of radius 4 centered at the origin.)

(8) $4y - y^2 = x^3$ **(9)** $y = 1 + \sqrt{\frac{1}{x} - 1}$ **(10)** $y = \frac{x(1-x)}{2x-1}$

(11a) $(\frac{3\sqrt{3}}{2},\frac{3}{2},4)$ **(11b)** $(\frac{5}{2},-\frac{5\sqrt{3}}{2},-1)$ **(11c)** $(\frac{5}{2},-\frac{5\sqrt{3}}{2},-1)$ **(11d)** $(\frac{1}{2},-\frac{\sqrt{3}}{2},0)$

(12a) Half the interior of a cylinder of radius $\frac{1}{2}$ and height 5 whose base is parallel to the xy-plane and centered at $(\frac{1}{2},0,0)$, with also a small slab removed consisting of all points above the plane $y = \sqrt{3}x$.

(12b) The wall of a cylinder of radius 2 and height 5 whose base parallel to the xyplane and centered at $(0,2,0)$.

(12c) A planar strip of height 1 at an angle of $\frac{\pi}{3}$ from the xz-plane.

(13a) A circle of radius 1 parallel to the xy-plane at a height of 3.

(13b) A gradually expanding spiral starting at the origin and looping counterclockwise around the z-axis, inscribed in the cone $z = \sqrt{x^2 + y^2}$.

(14)

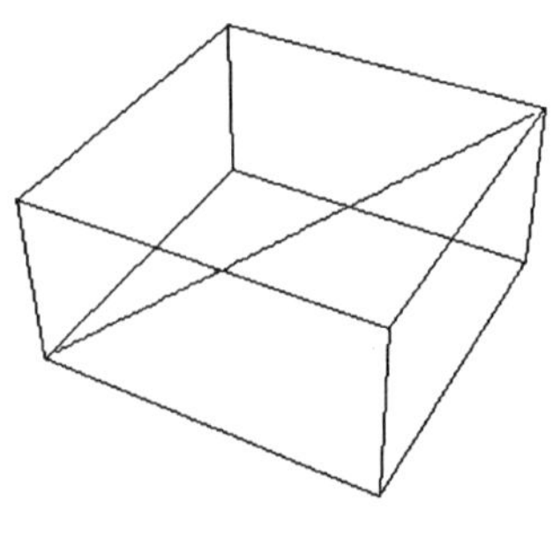

$\overrightarrow{f(t)} = 3t\vec{\imath} + t\vec{\jmath} + 4t\vec{k}$

(15)

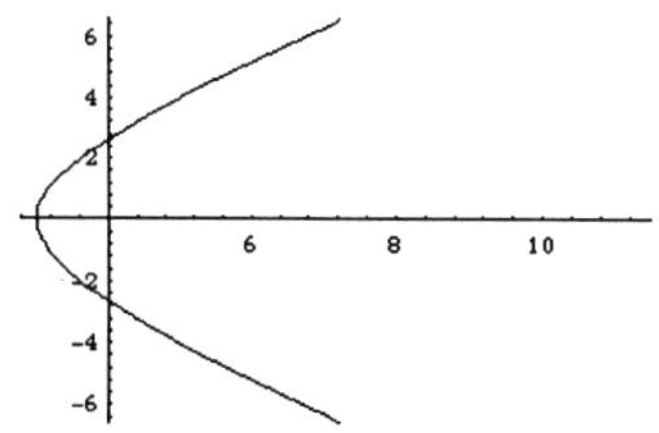

$$\overrightarrow{f(t)} = 3\cosh(t)\vec{i} + e\sinh(t)\vec{j}$$

(16)

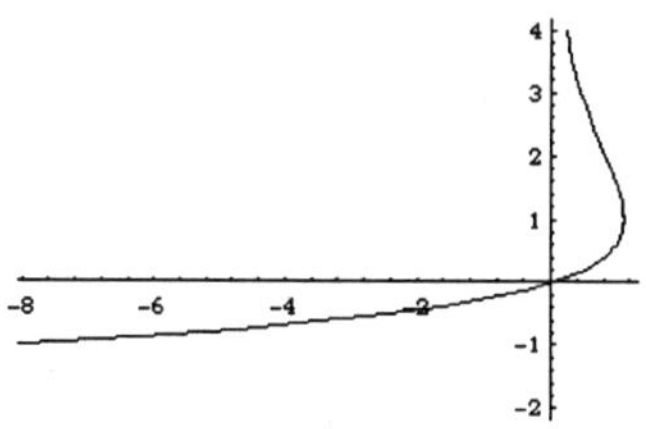

$$\overrightarrow{f(t)} = 3te^{-t}\vec{i} + t\vec{k}$$

(17)

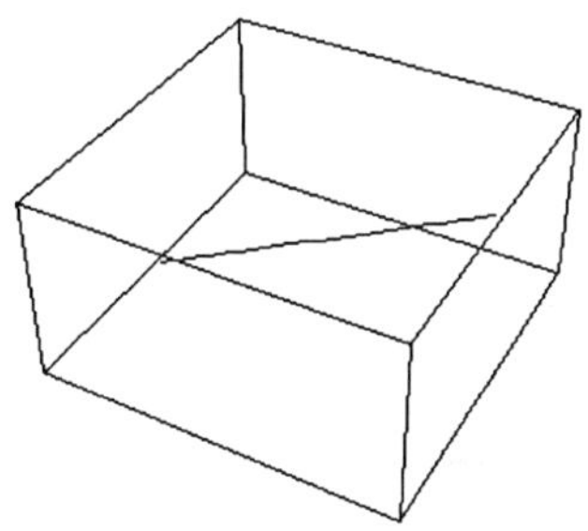

$$\overrightarrow{f(t)} = 3\cos(t)\vec{i} + 2\vec{j} + 3\cos(t)\vec{k}$$

(18)

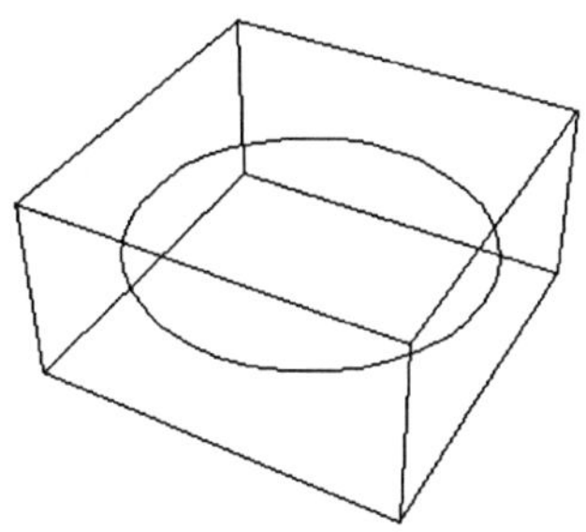

$$\overrightarrow{f(t)} = 2\cos(2t)\vec{i} + 2\sin(2t)\vec{j} + 3\vec{k}$$

Solutions for §22.1

If $\overrightarrow{f(t)}$ is a space curve, then a tangent vector at $t = t_0$ is given by $\overrightarrow{f'(t_0)}$. The space curve $\overrightarrow{f(t_0)} + t\overrightarrow{f'(t_0)}$ passes through the point $f(t_0)$ and is tangent to the curve. If $\vec{n}$ is perpendicular to $\overrightarrow{f'(t_0)}$ (i.e., $\vec{n} \cdot \overrightarrow{f'(t_0)} = 0$) then $\overrightarrow{f(t_0)} + t\vec{n}$ is a normal line. There are many possible normal lines.

(1) $\overrightarrow{f(t)} = \overrightarrow{(1+5t^2, 2-t, 1+7t)}$ and $\overrightarrow{f'(t)} = \overrightarrow{(10t, -1, 7)}$. Thus $f(0) = (1, 2, 1)$ and $\overrightarrow{f'(0)} = \overrightarrow{(0, -1, 7)}$.

Tangent Line: $\overrightarrow{(x(t), y(t), z(t))} = \overrightarrow{(1, 2, 1)} + t\overrightarrow{(0, -1, 7)} = \overrightarrow{(1, 2-t, 1+7t)}$.

Further: $\overrightarrow{(0, -1, 7)} \cdot \overrightarrow{(0, 7, 1)} = 0$, so we may take $\vec{n} = \overrightarrow{(0, 7, 1)}$.

Normal Line: $\overrightarrow{(x(t), y(t).z(t))} = \overrightarrow{(1, 2, 1)} + t\overrightarrow{(0, 7, 1)} = \overrightarrow{(1, 2+7t, 1+t)}$.

(2) $\overrightarrow{f(t)} = \overrightarrow{(t^3, -3+t^2, 1+t)}$ and $\overrightarrow{f'(t)} = \overrightarrow{(3t^2, 2t, 1)}$. Thus $f(2) = (8, 1, 3)$ and $\overrightarrow{f'(2)} = \overrightarrow{(12, 4, 1)}$.

Tangent Line: $\overrightarrow{(x(t), y(t), z(t))} = \overrightarrow{(8, 1, 3)} + t\overrightarrow{(12, 4, 1)} = \overrightarrow{(8+12t, 1+4t, 3+t)}$.

Further: $\overrightarrow{(1, -3, 0)} \cdot \overrightarrow{(12, 4, 1)} = 0$, so we may take $\vec{n} = \overrightarrow{(1, -3, 0)}$.

Normal Line: $\overrightarrow{(x(t), y(t).z(t))} = \overrightarrow{(8, 1, 3)} + t\overrightarrow{(1, -3, 0)} = \overrightarrow{(8+t, 1-3t, 3)}$.

(3) Tangent Line: $x = 3, y = \sqrt{2}(1-t), z = \sqrt{2}(1+t)$.

Normal Line: $x = 3, y = \sqrt{2}+t, z = \sqrt{2}+t$. (Here $\vec{n} = \overrightarrow{(0, 1, 1)}$).

(4) Tangent Line: $x = 3t, y = 7, z = -2$.

Normal Line: $x = 0, y = t, z = t$. (Here $\vec{n} = \overrightarrow{(0, 1, 1)}$).

(5) Tangent Line: $x = 3, y = 1, z = e^{\frac{4\pi}{5}}\left(1 + \frac{1}{5}t\right)$.

Normal Line: $x = -\frac{1}{5}e^{\frac{4\pi}{5}} \cdot t, y = 1, z = e^{\frac{4\pi}{5}}\left(1 + \frac{1}{5}t\right)$. (Here $\vec{n} = \overrightarrow{(-\frac{1}{5}e^{\frac{4\pi}{5}}, 0, 1)}$).

(6) Tangent Line: $x = 9\pi^2 + 6\pi t, y = 2t, z = e^{\frac{3\pi}{4}}(1 + \frac{t}{4})$.

Normal Line: $x = 9\pi^2 + t, y = -3\pi t, z = e^{\frac{3\pi}{4}}$. (Here $\vec{n} = \overrightarrow{(1, -3\pi, 0)}$).

Solutions for §22.2

(1) $(\frac{t^4}{4} - t)\vec{i} + t^2\vec{j} + (e^t - t)\vec{k} + \vec{C}$, where $\vec{C}$ is a vector-valued constant of integration.

(2) $\sin(t)\vec{i} + (t\ln(t) - t)\vec{j} + e^t\vec{k} + \vec{C}$.

(3) $(-e^{-t} - te^{-t})\vec{i} + \frac{2}{3}(t+1)^{\frac{3}{2}}\vec{j} + \frac{t^5}{5}\vec{k} + \vec{C}$.

(4) $L = \int_{-1}^{2} |\overrightarrow{f'(t)}|\, dt = \int_{-1}^{2} \sqrt{1+4+1}\, dt = 3\sqrt{6}$.

(5) Here we have

$$L = \int_1^3 \left[\left(\frac{t^2}{2} - \frac{1}{2t^2}\right)^2 + 1\right]^{\frac{1}{2}} dt$$

$$= \int_1^3 \left[\frac{t^4}{4} - \frac{1}{2} + \frac{1}{4}t^{-4} + 1\right]^{\frac{1}{2}} dt$$

$$= \int_1^3 \left[\frac{t^4}{4} + \frac{1}{2} + \frac{1}{4}t^{-4}\right]^{\frac{1}{2}} dt$$

$$= \int_1^3 \left(\frac{t^2}{2} + \frac{t^{-2}}{2}\right) dt$$

$$= \frac{t^3}{6} - \frac{t^{-1}}{2}\Big]_1^3 = \left(\frac{27}{6} - \frac{1}{6}\right) - \left(\frac{1}{6} - \frac{1}{2}\right) = \frac{14}{3}.$$

(6) $L = \int_0^{10} \sqrt{25\cos^2(t) + 49 + 25\sin^2(t)}\, dt = \sqrt{74}\int_0^{10} dt = 10\sqrt{74}.$

(7) $L = 21 + \ln\frac{5}{2}$.

(8) $L = \frac{4}{3}\sqrt{2}$.

(9) $L = \ln(3 + 2\sqrt{2})$.

(10) $y = (x-1)^{\frac{3}{2}}$. Set $x = t$. $y = (t-1)^{\frac{3}{2}}$. So $\overrightarrow{f(t)} = t\vec{i} + (t-1)^{\frac{3}{2}}\vec{j}$ is a space curve whose graphical display is identical to the graphical display of $y = (x-1)^{\frac{3}{2}}$. We compute the arclength:

$$L = \int_2^4 |\overrightarrow{f'(t)}|\, dt = \int_2^4 \sqrt{1 + \left(\frac{3}{2}(t-1)^{\frac{1}{2}}\right)^2}\, dt$$

$$= \int_2^4 \sqrt{1 + \frac{9}{4}(t-1)}\, dt$$

$$= \frac{4}{9}\int_{\frac{13}{4}}^{\frac{31}{4}} \sqrt{u}\, du = \frac{1}{27}(31^{\frac{3}{2}} - 13^{\frac{3}{2}})$$

after setting $u = 1 + \frac{9}{4}(t-1)$, $dt = \frac{4}{9}\, du$.

(11) $\overrightarrow{f(t)} = t\vec{i} + (\frac{t^3}{6} + \frac{1}{2t})\vec{j}, L = 4\frac{2}{3}$.

(12) $\overrightarrow{f(t)} = t\vec{i} + e^t\vec{j}, L = \sqrt{1+e^4} + 2 + \frac{1}{2}\ln(3 + 2\sqrt{2}) - \frac{1}{2}\ln(2 + e^4 + 2\sqrt{1+e^4}) - \sqrt{2}.$

Solutions for §22.3

(1) $A = \frac{1}{2}\int\limits_{-\frac{\pi}{8}}^{\frac{\pi}{8}} 16\cos^2(4\theta)d\theta = \pi$, $L = \int\limits_{-\frac{\pi}{8}}^{\frac{\pi}{8}} 4\sqrt{1+15\sin^2(4\theta)}d\theta$.

(2) $A = \frac{1}{2}\int\limits_{-\frac{\pi}{16}}^{\frac{\pi}{16}} 4\cos^2(8\theta)d\theta = \frac{\pi}{8}$, $L = \int\limits_{-\frac{\pi}{16}}^{\frac{\pi}{16}} 2\sqrt{1+63\sin^2(8\theta)}d\theta$.

(3) $A = \frac{1}{2}\int\limits_{-\frac{\pi}{10}}^{\frac{\pi}{10}} 49\cos^2(5\theta)d\theta = \frac{49}{20}\pi$, $L = \int\limits_{-\frac{\pi}{10}}^{\frac{\pi}{10}} 7\sqrt{1+24\sin^2(5\theta)}d\theta$.

(4) $A = \frac{1}{2}\int\limits_{-\frac{\pi}{3}}^{0} 4\cos^2(3\theta+\frac{\pi}{2})d\theta = \frac{\pi}{3}$, $L = \int\limits_{-\frac{\pi}{3}}^{0} 2\sqrt{1+8\sin^2(3\theta+\frac{\pi}{2})}d\theta$.

(5) $A = \frac{1}{2}\int\limits_{-\frac{11\pi}{36}}^{\frac{7\pi}{36}} 16\cos^2(2\theta+\frac{\pi}{9})d\theta = 2\pi$, $L = \int\limits_{-\frac{11\pi}{36}}^{\frac{7\pi}{36}} 4\sqrt{1+3\sin^2(2\theta+\frac{\pi}{9})}d\theta$.

(6) $A = \frac{1}{2}\int\limits_{-\frac{\pi}{14}}^{\frac{\pi}{14}} 9\cos^2(7\theta)d\theta = \frac{9}{28}\pi$, $L = \int\limits_{-\frac{\pi}{14}}^{\frac{\pi}{14}} 3\sqrt{1+48\sin^2(7\theta)}d\theta$.

(7) $L = \int\limits_{-10}^{10} \sqrt{e^{\frac{2}{3}\theta}+\frac{1}{9}e^{\frac{2}{3}\theta}}\,d\theta = \frac{\sqrt{10}}{3}\int\limits_{-10}^{10} e^{\frac{\theta}{3}}\,d\theta = \sqrt{10}(e^{\frac{10}{3}} - e^{-\frac{10}{3}})$.

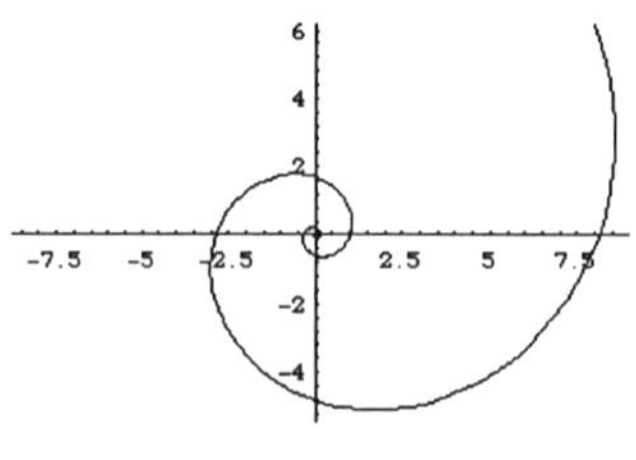

$r = e^{\frac{\theta}{3}}$ for $-10 \le \theta \le 10$

(8) Here $r = 5$ since $\sin\left(\frac{\pi}{2}\right) = 1 \implies L = \int\limits_{0}^{\pi} \sqrt{5^2}\,d\theta = 5\pi$. The graph is just a semicircle of radius 5 centered at the origin.

(9) Here we have

$$L = \int\limits_{-\frac{\pi}{3}}^{\frac{\pi}{3}} \sqrt{\tan^4(\theta)+(2\tan(\theta)\sec^2(\theta))^2}\,d\theta = \int\limits_{-\frac{\pi}{3}}^{\frac{\pi}{3}} \frac{\sin(\theta)}{\cos^2(\theta)}\sqrt{\sin^2(\theta)+4}\,d\theta.$$

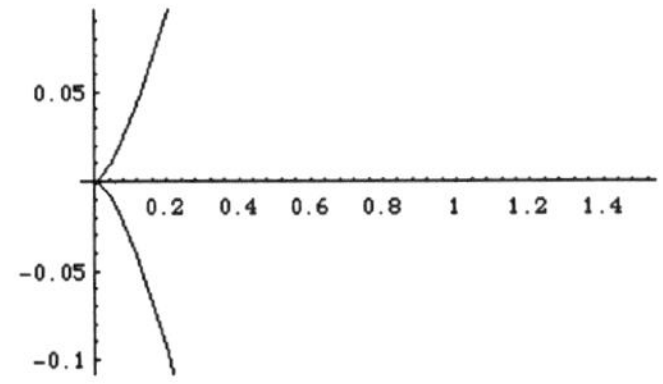

$$r = \tan^2(\theta) \qquad \text{for} - \frac{\pi}{3} \leq \theta \leq \frac{\pi}{3}$$

(10) $r = 3\sqrt{1 + 3\sin^2(\theta)} \implies A = \int_{-\pi}^{\pi} \frac{9}{2}(1 + 3\sin^2(\theta))\, d\theta = \frac{45\pi}{2}.$

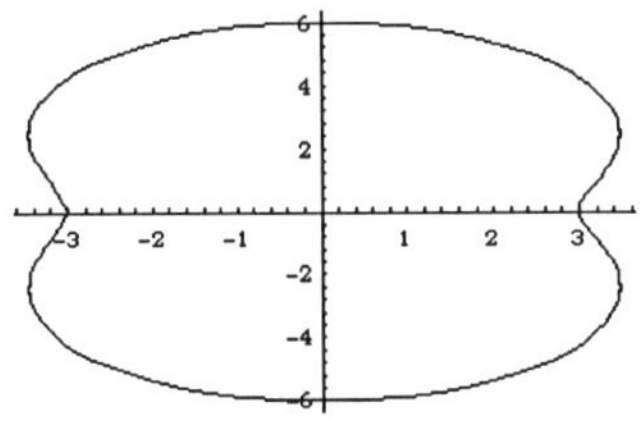

$$r^2 = 9(1 + 3\sin^2(\theta)) \qquad \text{for} - \pi \leq \theta \leq \pi$$

(11) The area bounded by the parabola $r = \frac{3}{1-\cos(\theta)}$ and the line $\theta = \frac{\pi}{3}$ is displayed below. The parabola and line intersect at

$$r = \frac{3}{1 - \cos\left(\frac{\pi}{3}\right)} = 6, \quad \theta = \frac{\pi}{3} \text{ and } r = \frac{3}{1 - \cos\left(\frac{4\pi}{3}\right)} = 2, \quad \theta = \frac{4\pi}{3}.$$

Thus, $A = \int_{\frac{\pi}{3}}^{\frac{4\pi}{4}} \frac{1}{2}\left(\frac{3}{1-\cos(\theta)}\right)^2 d\theta = \frac{16\sqrt{3}}{3}.$

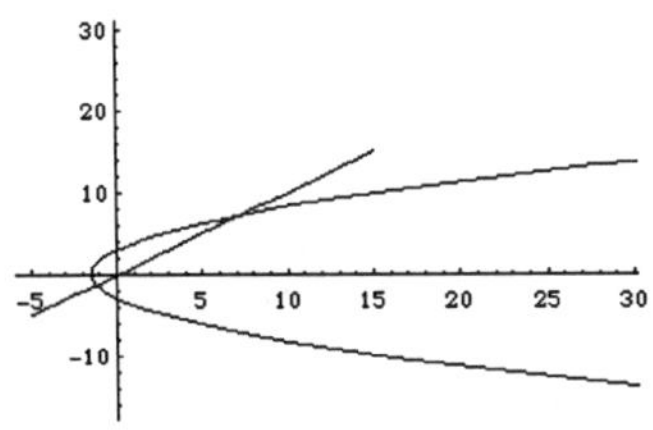

$$r = \frac{3}{1 - \cos(\theta)} \qquad \text{for} 0 \leq \theta \leq 2\pi \text{ and the line } \theta = \frac{\pi}{3}$$

(12) The limacon $r = \frac{1}{2} + \cos(\theta)$ is depicted below:

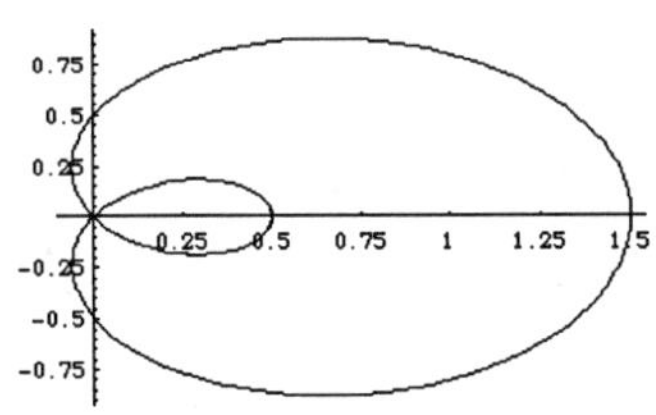

The smaller loop occurs for $\frac{2\pi}{3} \leq \theta \leq \frac{4\pi}{3}$. Thus

$$A = \int_{\frac{2\pi}{3}}^{\frac{4\pi}{3}} \frac{1}{2}\left(\frac{1}{2} + \cos(\theta)\right)^2 d\theta = \frac{2\pi - 3\sqrt{3}}{8}.$$

(13) The intersections of the curves are at the points $\frac{\pi}{3}$ and $\frac{2\pi}{3}$, and one switches from one curve to the other at these points when evaluating the area integral:

$$A = \frac{1}{2}\int_0^{\frac{\pi}{3}} \sin^2(\theta)d\theta + \int_{\frac{\pi}{3}}^{\frac{\pi}{2}} \sin^2(2\theta)d\theta + \int_{\frac{\pi}{2}}^{\frac{2\pi}{3}} \sin^2(2\theta)d\theta + \int_{\frac{2\pi}{3}}^{\pi} \sin^2(\theta)d\theta = \frac{\pi}{4} - \frac{3\sqrt{3}}{16}.$$

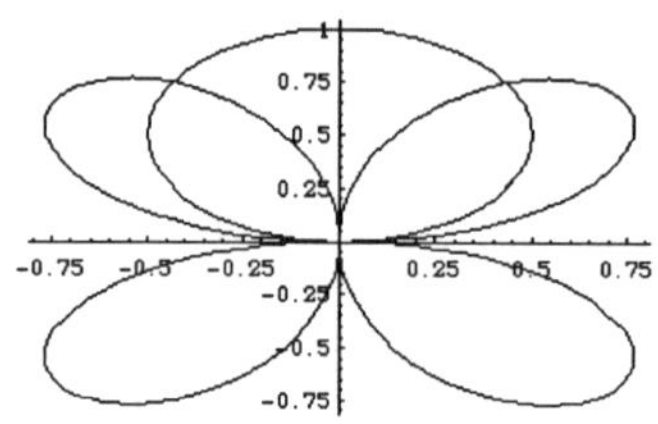

$r = \sin(2\theta)$ and $r = \sin(\theta)$

$r = \sin(\theta)$ and $r = \sin(\theta)$

(14) $A = \frac{1}{2}\int_{-\pi}^{\pi} 9(1+\cos(\theta))^2\,d\theta - \frac{1}{2}\int_{-\frac{\pi}{2}}^{\frac{\pi}{2}} 4(1+\cos(\theta))\,d\theta = \frac{23\pi-8}{2}$.

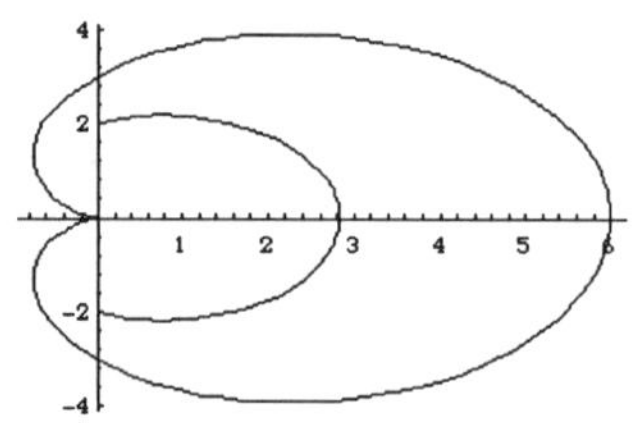

$r = 3(1+\cos(\theta)) \qquad -\pi \leq \theta \leq \pi$

$r = 2\sqrt{1+\cos(\theta)} \qquad \frac{-\pi}{2} \leq \theta \leq \frac{\pi}{2}$

Solutions for §22.4

(1) $\overrightarrow{f'(t)} = 2\cos(2t)\vec{i} + 5\vec{j} - 2\sin(2t)\vec{k}$ and $|\overrightarrow{f'(t)}| = \sqrt{4\cos^2(2t) + 25 + 4\sin^2(2t)} = \sqrt{29}$. Thus,

$$\begin{aligned}
\overrightarrow{T(t)} &= \frac{\overrightarrow{f'(t)}}{|\overrightarrow{f'(t)}|} = \frac{1}{\sqrt{29}}\left(2\cos(2t)\vec{i} + 5\vec{j} - 2\sin(2t)\vec{k}\right) \\
\overrightarrow{T'(t)} &= \frac{1}{\sqrt{29}}\left(-4\sin(2t)\vec{i} - 4\cos(2t)\vec{k}\right) \\
\overrightarrow{N(t)} &= \frac{\overrightarrow{T'(t)}}{|\overrightarrow{T'(t)}|} = -\sin(2t)\vec{i} - \cos(2t)\vec{k} \\
\kappa(t) &= \frac{|\overrightarrow{T'(t)}|}{|\overrightarrow{f'(t)}|} = \frac{4}{29}.
\end{aligned}$$

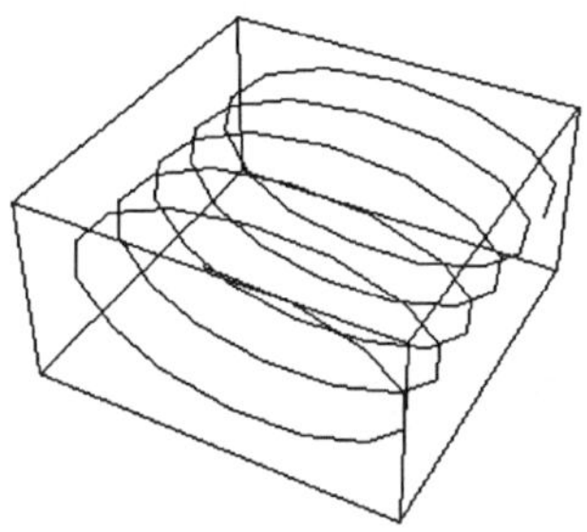

$$\overrightarrow{f(t)} = \sin(2t)\vec{i} + 5t\vec{j} + \cos(2t)\vec{k}$$

(2) $\overrightarrow{f'(t)} = e^t\vec{i} - e^{-t}\vec{j} + \sqrt{2}\vec{k}$ and $|\overrightarrow{f'(t)}| = \sqrt{e^{2t} + e^{-2t} + 2} = e^t + e^{-t}$. Thus,

$$\begin{aligned}
\overrightarrow{T(t)} &= \frac{1}{e^t + e^{-t}}\left(e^t\vec{i} - e^{-t}\vec{j} + \sqrt{2}\vec{k}\right) \\
\overrightarrow{N(t)} &= \frac{\sqrt{2}e^t}{1 + e^{2t}}\vec{i} + \frac{\sqrt{2}e^t}{1 + e^{2t}}\vec{j} + \frac{1 - e^{2t}}{1 + e^{2t}}\vec{k} \\
\kappa(t) &= \frac{|\overrightarrow{T'(t)}|}{|\overrightarrow{f'(t)}|} = \frac{\sqrt{2}e^t}{(e^t + e^{-t})^2}.
\end{aligned}$$

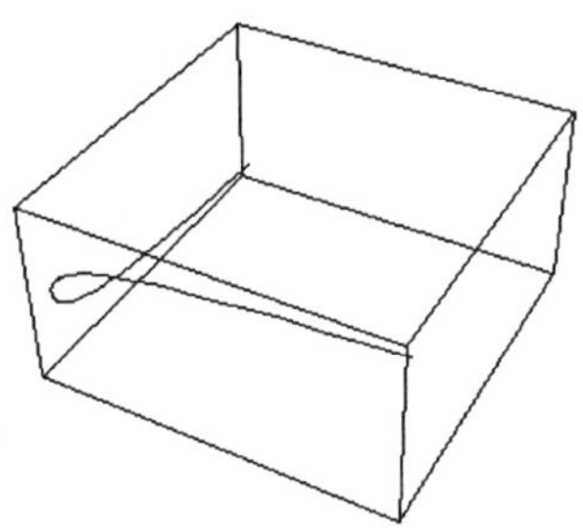

$$\overrightarrow{f(t)} = e^t\vec{i} + e^{-t}\vec{j} + \sqrt{2}t\vec{k}$$

(3) $\overrightarrow{T(t)} = -\sin(t)\vec{i} - \sin(t)\cos(t)\vec{j} + \cos^2(t)\vec{k}$, $\quad \overrightarrow{N(t)} = \frac{1}{1+\cos^2(t)}(\cos(t)\vec{i} - \cos(2t))\vec{j} - \sin(2t)\vec{k})$, $\kappa(t) = (1+\cos^2(t))\cos(t)$.

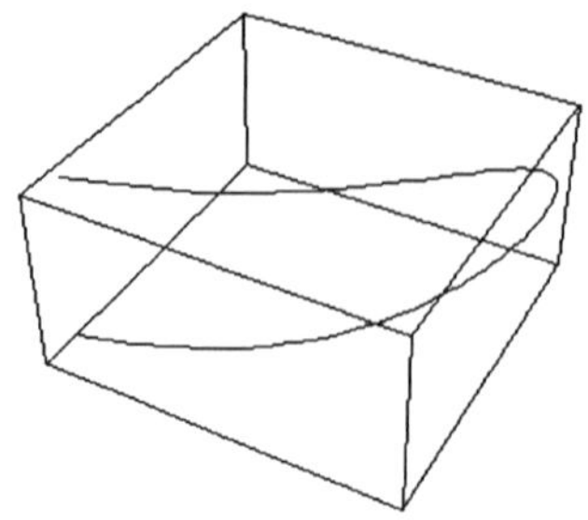

$$\overrightarrow{f(t)} = \ln(\cos(t))\vec{i} + \cos(t)\vec{j} + \sin(t)\vec{k}$$

(4) $\overrightarrow{T(t)} = \frac{1}{\sqrt{2}(1+t^2)}\left((1-t^2)\vec{i} + 2t\vec{j} + (1+t^2)\vec{k}\right)$, $\quad \overrightarrow{N(t)} = \frac{1}{1+t^2}(-2t\vec{i} + (1-t^2)\vec{j})$, $\kappa(t) = \frac{1}{(1+t^2)^2}$.

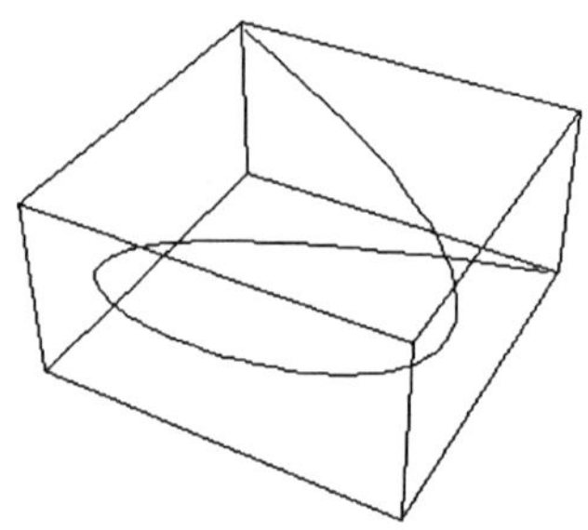

$$\overrightarrow{f(t)} = \left(t - \frac{t^3}{3}\right)\vec{i} + t^2\vec{j} + \left(t + \frac{t^3}{3}\right)\vec{k}$$

(5) We have $\kappa(0) = 1$. So we require a circle of radius one which passes through the points $A = \left(\frac{11}{24}, \frac{1}{4}, \frac{13}{24}\right)$ and $B = \left(-\frac{11}{24}, \frac{1}{4}, -\frac{13}{24}\right)$.

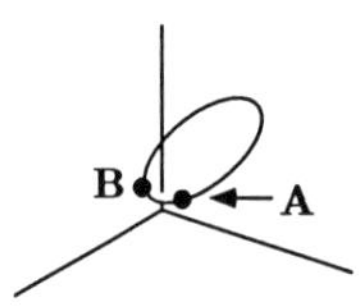

Let $C = \left(c_1, \frac{1}{4}, c_3\right)$ be the center of the circle. Remark that $|CA| = |CB| = 1$. This implies

$$\left(\frac{11}{24} - c_1\right)^2 + \left(\frac{13}{24} - c_3\right)^2 = \left(\frac{-11}{24} - c_1\right)^2 + \left(\frac{-13}{24} - c_3\right)^2 = 1$$

$$\Longrightarrow c_1^2 - \frac{11}{12}c_1 + \left(\frac{11}{24}\right)^2 + c_2^2 - \frac{13}{12}c_2 + \left(\frac{13}{24}\right)^2 = c_1^2 + \frac{11}{12}c_1 + \left(\frac{11}{24}\right)^2 + c_2^2 + \frac{13}{12}c_2 + \left(\frac{13}{24}\right)^2 = 1.$$

Hence $\frac{11}{6}c_1 + \frac{13}{6}c_2 = 0 \implies c_2 = \frac{11}{13}c_1$. Solving equations yields: $c_1 \approx -1.2923, c_2 \approx 0.223908$. Thus, the circle has radius 1, lies in the plane $z = \frac{1}{4}$, and is centered at $C \approx (-1.2923, \frac{1}{4}, 0.223908)$. The curvature of this circle will match the curvature of the space curve near $t = 0$.

(6) We have $\overrightarrow{f'(t)} = v(t)\overrightarrow{T(t)}$, $\overrightarrow{f''(t)} = v'(t)\overrightarrow{T(t)} + v(t)\overrightarrow{T'(t)}$. Recall that the cross product of any vector with itself is zero. Further, since $\overrightarrow{T(t)}$ is perpendicular to $\overrightarrow{T'(t)}$, we see that $|\overrightarrow{T(t)} \times \overrightarrow{T'(t)}| = |\overrightarrow{T(t)}| \cdot |\overrightarrow{T'(t)}|$. Taking cross products yields:

$$|\overrightarrow{f'(t)} \times \overrightarrow{f''(t)}| = v(t)^2 |\overrightarrow{T(t)} \times \overrightarrow{T'(t)}| = v(t)^2 |\overrightarrow{T(t)}| \cdot |\overrightarrow{T'(t)}|.$$

Dividing both sides by $v(t)^3$ gives

$$\frac{|\overrightarrow{f'(t)} \times \overrightarrow{f''(t)}|}{v(t)^3} = \frac{|\overrightarrow{T'(t)}|}{v(t)} = \kappa(t).$$

(7) Apply the formula from problem **(6)** for the space curve $\overrightarrow{g(t)} = t\vec{i} + f(t)\vec{j}$, using the equalities $|\overrightarrow{g'(t)} \times \overrightarrow{g''(t)}| = |f''(t)|$ and $|\overrightarrow{g'(t)}| = 1 + (f'(t))^2$. Then substitute $x = t$.

(8) $\kappa(t) = \frac{e^x}{(1+e^{2x})^{\frac{3}{2}}}$. Maximum is at the point $x = -\frac{1}{2}\ln(2)$.

(9) Use the main property of the indefinite integral: $\frac{d}{dt}\int f(t)\,dt = f(t)$.

Solutions for §22.5

Recall that $\vec{a}(t) = a(t)\vec{T}(t) + k(t)v(t)^2\vec{N}(t)$. Then $a(t) = v'(t)$ is the tangential component of the acceleration vector and $k(t) \cdot v(t)^2$ is the normal component.

(1) We have

$$\begin{aligned}
\overrightarrow{f(t)} &= 8\vec{i} + (2t^2 - 3)\vec{j} + (t-2)\vec{k} \\
\overrightarrow{v(t)} &= \overrightarrow{f'(t)} = 4t\vec{j} + \vec{k} \implies v(t) = \sqrt{1 + 16t^2} \\
\overrightarrow{a(t)} &= \overrightarrow{v'(t)} = \overrightarrow{f''(t)} = 4\vec{j}.
\end{aligned}$$

Now, $a(t) = v'(t) = \frac{16t}{\sqrt{1+16t^2}}$ is the tangential component of acceleration. Further,

$$\overrightarrow{T(t)} = \frac{1}{\sqrt{1+16t^2}}(4t\vec{j} + \vec{k}) \implies \kappa(t) = \frac{|\overrightarrow{T'(t)}|}{v(t)} = \frac{4}{(1+16t^2)^{\frac{3}{2}}}.$$

Thus $\kappa(t)v(t)^2 = \frac{4}{\sqrt{1+16t^2}}$ is the normal component of acceleration.

(2) We have

$$\begin{aligned}
\overrightarrow{f(t)} &= 3\sin(t)\vec{i} - 5t\vec{j} + 3\cos(t)\vec{k} \\
\overrightarrow{v(t)} &= \overrightarrow{f'(t)} = 3\cos(t)\vec{i} - 5\vec{j} - 3\sin(t)\vec{k} \\
\overrightarrow{a(t)} &= \overrightarrow{v'(t)} = \overrightarrow{f''(t)} = -3\sin(t)\vec{i} - 3\cos(t)\vec{k}.
\end{aligned}$$

Now $v(t) = |\overrightarrow{v(t)}| = \sqrt{9\cos^2(t) + 25 + 9\sin^2(t)} = \sqrt{34}$. Thus, $a(t) = v'(t) = 0$ is the tangential component of acceleration. Further,

$$\overrightarrow{T(t)} = \frac{1}{\sqrt{34}}\left(3\cos(t)\vec{i} - 5\vec{j} + 3\sin(t)\vec{k}\right) \implies \kappa(t) = \frac{|\overrightarrow{T'(t)}|}{v(t)} = \frac{\frac{1}{\sqrt{34}} \cdot \sqrt{9\sin^2(t) + 9\cos^2(t)}}{\sqrt{34}} = \frac{3}{34}.$$

Hence, $\kappa(t) \cdot v(t)^2 = 3$ is the normal component of acceleration.

(3) We have

$$\begin{aligned}
\overrightarrow{f(t)} &= \sqrt{2} \cdot t\vec{i} + e^{-t}\vec{j} + e^t\vec{k} \\
\overrightarrow{v(t)} &= \overrightarrow{f'(t)} = \sqrt{2}\,\vec{i} - e^{-t}\vec{j} + e^t\vec{k} \\
\overrightarrow{a(t)} &= \overrightarrow{v'(t)} = \overrightarrow{f''(t)} = e^{-t}\vec{j} + e^t\vec{k}.
\end{aligned}$$

It follows that $v(t) = |\overrightarrow{v(t)}| = \sqrt{2 + e^{-2t} + e^{2t}} = \sqrt{(e^t + e^{-t})^2} = e^t + e^{-t}$. Hence, $a(t) = v'(t) = e^t + e^{-t}$ is the tangential component of acceleration. Further,

$$\overrightarrow{T(t)} = \frac{1}{e^t + e^{-t}}\left(\sqrt{2}\,\vec{i} - e^{-t}\vec{j} + e^t\vec{k}\right) \text{ and } \overrightarrow{T'(t)} = \frac{1}{(e^t + e^{-t})^2}\left(\sqrt{2}(e^{-t} - e^t)\,\vec{i} + 2\vec{j} + 2\vec{k}\right).$$

Consequently, $\kappa(t) = \frac{|\overrightarrow{T'(t)}|}{v(t)} = \frac{\sqrt{2}}{(e^t + e^{-t})^2}$ and, therefore, $\kappa(t) \cdot v(t)^2 = \sqrt{2}$ is the normal component of acceleration.

(4) Tangential component = 1, normal component = $\frac{1}{\sqrt{2t}}$.

(5) We have $\overrightarrow{f(t)} = 3\sqrt{2}\cos(\pi t)\vec{j} + 2\sqrt{2}\sin(\pi t)\vec{k}$ and $\overrightarrow{g(t)} = \left(t - \frac{5}{4}\right) + \left(t^2 - 3t - \frac{13}{16}\right)\vec{j} - \left(t + \frac{3}{4}\right)\vec{k}$.

The collision occurs for that value of t where $f(t) = g(t)$. Thus $t = \frac{5}{4}$ is the time of collision. As can be seen from the graphical display, the satellite has an elliptical orbit and the meteor collides with it at $t = \frac{5}{4}$.

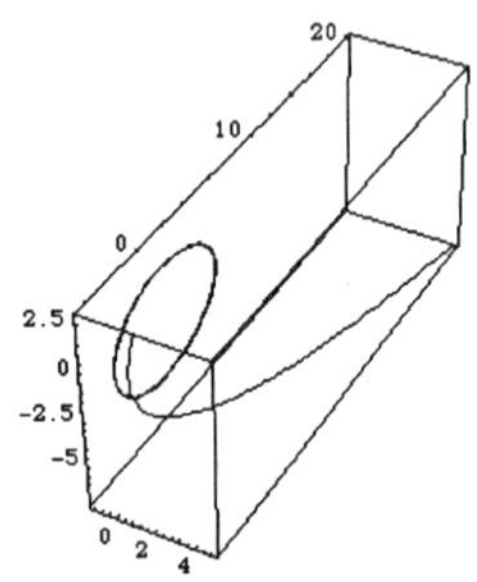

(a) The angle, θ, between the trajectories is the angle between the tangent vectors $\overrightarrow{f'\left(\frac{5}{4}\right)}$ and $\overrightarrow{g'\left(\frac{5}{4}\right)}$. Here $\overrightarrow{f'\left(\frac{5}{4}\right)} = 3\pi\vec{j} - 2\pi\vec{k}$ and $\overrightarrow{g'\left(\frac{5}{4}\right)} = \vec{i} - \frac{1}{2}\vec{j} - \vec{k}$. Since $\overrightarrow{f'\left(\frac{5}{4}\right)} \cdot \overrightarrow{g'\left(\frac{5}{4}\right)} = |\overrightarrow{f'\left(\frac{5}{4}\right)}| \cdot |\overrightarrow{g'\left(\frac{5}{4}\right)}|\cos(\theta)$, we see that $\theta = \cos^{-1}\left(\frac{1}{3\sqrt{13}}\right)$.

(b) The curvature of the trajectory of the meteor at time t is $\frac{4}{(11 - 12t + 4t^2)^3}$. Thus, the curvature at $t = \frac{5}{4}$ is $\frac{256}{729}$.

(c) The velocity vectors are: $\vec{\mathbf{v}}_{\text{satellite}} = \overrightarrow{f'\left(\frac{5}{4}\right)} = 3\pi\vec{j} - 2\pi\vec{k}$ and $\vec{\mathbf{v}}_{\text{meteor}} = \overrightarrow{g'\left(\frac{5}{4}\right)} = \vec{i} - \frac{1}{2}\vec{j} - \vec{k}$.

(d) The acceleration vectors are: $\vec{\mathbf{a}}_{\text{satellite}} = \overrightarrow{f''\left(\frac{5}{4}\right)} = 3\pi^2\vec{j} + 2\pi^2\vec{k}$ and $\vec{\mathbf{a}}_{\text{meteor}} = \overrightarrow{g''\left(\frac{5}{4}\right)} = 2\vec{j}$.

(6) The satellite makes a full rotation in $t = 24$ hours: $\frac{t\pi}{12} = 2\pi$.

We assume that the missile starts at $t = 0$ (and moves with a constant velocity). This means that the collision occurs at $t = 3$ at the point $(\frac{5\sqrt{2}}{2}, \frac{3\sqrt{2}}{2})$. The space curve for the missile is given by $\overrightarrow{g(t)} = \frac{5\sqrt{2}}{6}t\vec{\imath} + \frac{\sqrt{2}}{2}t\vec{k}$.

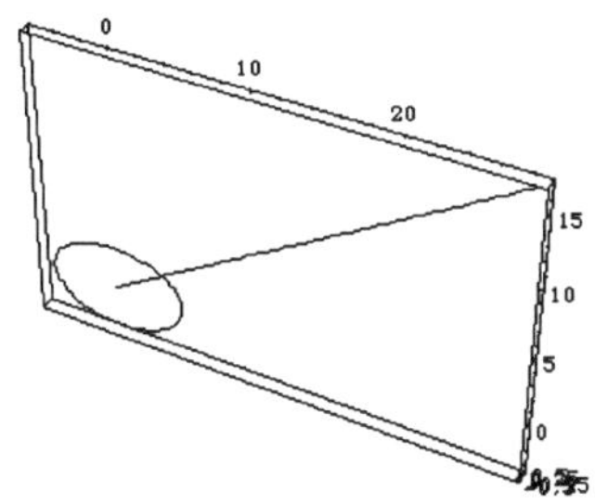

(a) The angle, θ, between the trajectories is the angle between the tangent vectors $\overrightarrow{f'(3)}$ and $\overrightarrow{g'(3)}$. Here $\overrightarrow{f'(3)} = \frac{\sqrt{2}\pi}{8}\left(-\frac{5}{3}\vec{\imath} + \vec{k}\right)$ and $\overrightarrow{g'(3)} = \frac{5\sqrt{2}}{6}\vec{\imath} + \frac{\sqrt{2}}{2}\vec{k}$. Since $\overrightarrow{f'(3)} \cdot \overrightarrow{g'(3)} = |\overrightarrow{f'(3)}| \cdot |\overrightarrow{g'(3)}| \cos(\theta)$, we see that $\theta = \cos^{-1}\left(-\frac{8}{17}\right)$.

(b) $\kappa_{sat}(t = 3) = \frac{1}{\sqrt{17}}$.

(c) The velocity vectors are: $\overrightarrow{\mathbf{v}}_{\text{satellite}} = \overrightarrow{f'(3)} = \frac{5\sqrt{2}}{6}\vec{\imath} + \frac{\sqrt{2}}{2}\vec{k}$ and $\overrightarrow{\mathbf{v}}_{\text{missile}} = \overrightarrow{g'(3)} = \frac{5\sqrt{2}}{6}\vec{\imath} + \frac{\sqrt{2}}{2}\vec{k}$.

(d) The acceleration vectors are: $\overrightarrow{\mathbf{a}}_{\text{satellite}} = \overrightarrow{f''(3)} = \frac{\pi^2\sqrt{2}}{288}(-5\vec{\imath} - 3\vec{\jmath})$ and $\overrightarrow{\mathbf{a}}_{\text{missile}} = \overrightarrow{g''(3)} = 0$.

(7) If the velocity of a particle is constant, then $\frac{d}{dt}|\overrightarrow{\mathbf{v}(t)}|^2 = 2\overrightarrow{\mathbf{v}(t)} \cdot |\overrightarrow{\mathbf{v}'(t)}| = \overrightarrow{\mathbf{v}(t)} \cdot \overrightarrow{\mathbf{a}(t)} = 0$, and one concludes that the acceleration only has a normal component, and is, therefore, in the direction of $\overrightarrow{N(t)}$.

(8) $|\mathbf{a}\vec{(t)}| = |a(t)T\vec{(t)} + \kappa(t)v(t)^2 N\vec{(t)}| = \sqrt{a(t)^2 + \kappa(t)^2 v(t)^4}$. (Note that the vectors $T\vec{(t)}$ and $N\vec{(t)}$ are orthogonal.)

Solutions for Additional Exercises for Chapter 22

(1) $|\overrightarrow{f'(t)}| = \sqrt{1 + 4t^2 + 9t^4}$. Since $\overrightarrow{(1, 2t, 3t^2)} \cdot \overrightarrow{(2, 2, 1)} = 0$ has no solution for $t \in \mathbb{R}$, there are no points on the curve at which the tangent vector is perpendicular to $\overrightarrow{(2, 2, 1)}$.

(2) We assume $\overrightarrow{f(t)} = \overrightarrow{(x(t), y(t), z(t))}$ is a vector valued differentiable function that is never zero for $u \leqslant t \leqslant v$.

(2a) We have $\vec{f} \cdot \overrightarrow{f'(t)} = \overrightarrow{(x, y, z)} \cdot \overrightarrow{(x', y', z')} = xx' + yy' + zz'$. Also, $|\vec{f}| = \sqrt{x^2 + y^2 + z^2}$, so $|\vec{f}| \cdot \frac{d|\vec{f}|}{dt} = \sqrt{x^2 + y^2 + z^2} \cdot \left(\frac{1}{2}\frac{2xx' + 2yy' + 2zz'}{\sqrt{x^2+y^2+z^2}}\right) = xx' + yy' + zz'$. Therefore $\vec{f} \cdot \frac{d\vec{f}}{dt} = |\vec{f}| \cdot \frac{d|\vec{f}|}{dt}$.

(2b) We have $|\vec{f}| = \sqrt{x^2 + y^2 + z^2}$ and $\frac{d|\vec{f}|}{dt} = \frac{1}{2}\frac{2xx' + 2yy' + 2zz'}{\sqrt{x^2+y^2+z^2}}$. So $|\vec{f}|$ is constant if and only if $\frac{d|\vec{f}|}{dt} = 0$ if and only if $xx' + yy' + zz' = 0$ if and only if $\vec{f} \cdot \vec{f'} = 0$.

(2c) We have $\vec{f} \times \vec{f'} = 0$ if and only if $xy' = yx'$, $xz' = zx'$, and $yz' = zy'$. This is true if and only if there is a function $w(t)$ and constants a, b, c so that $x = aw$, $y = bw$, and $z = cw$. Letting $\vec{u}$ denote the constant vector $\overrightarrow{(a, b, c)}$, we see that $\vec{f} \times \vec{f'} = 0$ if and only if $\overrightarrow{f(t)} = w(t)\vec{u}$ if and only if $\frac{\overrightarrow{f(t)}}{|\overrightarrow{f(t)}|} = \vec{u}$ if and only if $\vec{f}$ has constant direction.

(3) If the curve $x^5 = y^3$ is parameterized by $(x(t), y(t)) = \left(t^3, t^5\right)$, then $(x'(t), y'(t)) = \left(3t^2, 5t^4\right)$, so that $(x'(0), y'(0)) = (0,0)$, and there is no tangent vector at the origin. But, if the curve $x^5 = y^3$ is parameterized by $(x(t), y(t)) = \left(t, t^{\frac{5}{3}}\right)$, then $(x'(t), y'(t)) = \left(1, \frac{5}{3}t^{\frac{2}{3}}\right)$, so that $(x'(0), y'(0)) = (1, 0)$, and there is a tangent vector at the origin.

(4) If we write $\overrightarrow{g(t)} = \overrightarrow{(x(t), y(t), z(t))}$, then $\overrightarrow{g'(t)} = 0$ if and only if $x' = 0$, $y' = 0$, and $z' = 0$. This is true if and only if each coordinate function $x(t)$, $y(t)$, and $z(t)$ is constant on the interval $\{x \mid a \leqslant x \leqslant b\}$, which is then true if and only if $g(t)$ is constant on that interval as well.

(5) If $\overrightarrow{f(t)} = \overrightarrow{(\cos(t), \sin(t), 1)}$, then $\overrightarrow{f'(t)} = \overrightarrow{(-\sin(t), \cos(t), 0)}$. So the tangent vector $\overrightarrow{f'(t)}$ and the vector $\overrightarrow{(\cos(t), \sin(t), 0)}$ are perpendicular vectors lying in the xy-plane.

(6) Length of curve $= \int_0^3 \sqrt{4t^4 - 2t^2 + \frac{9}{4}}\, dt \approx 18.231$.

(7) Length of curve $= \int_{-1}^{2} \sqrt{16t^6 + 18t^4 + 144t^2}\, dt \approx 38.339$.

(8) Length of curve $= \int_0^{\pi} \sqrt{2}\, dt = \pi\sqrt{2}$.

(9) Let $x = t, y = t^{\frac{2}{3}}$. Then $\overrightarrow{f(t)} = t\vec{i} + t^{\frac{2}{3}}\vec{j}$ is a parameterization of our curve.
Length of curve $= \int_0^5 \sqrt{1 + \frac{4}{9}t^{\frac{2}{3}}}\, dt \approx 5.886$.

(10) Length of curve $= \int_0^{\log 2} 3e^{2t}\, dt = \frac{9}{2}$.

(11) The arclength $s(t)$ is given by $s(t) = \int_0^t \sqrt{a^2\cos^2(t) + a^2\sin^2(t)}\, dt = \int_0^t a\, dt = at$. Then $t = \frac{s}{a}$. A parameterization using arclength is given by $x = a\cos\left(\frac{s}{a}\right)$, $y = a\sin\left(\frac{s}{a}\right)$.

(12) Here $s(t) = \int_0^t \sqrt{a^2\cos^2(t) + a^2\sin^2(t) + c^2}\, dt = \sqrt{a^2 + c^2}\, t$. Thus $t = \frac{s}{\sqrt{a^2+c^2}}$. Parameterization using arclength is given by $x = a\cos\left(\frac{s}{\sqrt{a^2+c^2}}\right)$, $y = a\sin\left(\frac{s}{\sqrt{a^2+c^2}}\right)$, $z = \frac{cs}{\sqrt{a^2+c^2}}$.

(13) Here $s(t) = \int_0^t \sqrt{4e^{4t} \cdot 4}\, dt = \int_0^t 4e^{2t}\, dt = 2e^{2t}$. Parameterization using arclength is, therefore, given by $x = \sin\left(\frac{s}{2}\right)$, $y = \cos\left(\frac{s}{2}\right)$, $z = \frac{\sqrt{3}}{2}s$.

(14) Enclosed area $= \frac{1}{2}\int_0^{2\pi} (2 + 2\cos(\theta))^2\, d\theta - \frac{1}{2}\int_0^{2\pi} (2\cos(\theta))^2\, d\theta = 4\pi$.

(15) Enclosed area $= \frac{9}{4}\left(\frac{\pi}{2} - 1\right)$.

(16) See problem (6) in section 22.4.

(17) Curvature is $\kappa(t) = \frac{a}{a^2+c^2}$. (constant curvature)

Solutions for §23.1

(1) Domain$(f) = \mathbb{R} \times \mathbb{R}$, Range$(f) = \mathbb{R}$.

(2) Domain$(f) = \mathbb{R} \times \mathbb{R}$, Range$(f) = \{-3 \leq f < \infty\}$.

(3) Domain$(f) = \{(x, y) \mid y \neq 0\}$, Range$(f) = \mathbb{R}$.

(4) Domain$(f) = \{(x, y) \mid (x, y) \neq (0, 0)\}$, Range$(f) = \mathbb{R}$.

(5) Domain$(f) = \{(x, y, z) \mid x^2 + y^2 + z^2 \geq 11\}$, Range$(f) = \{0 \leq f < \infty\}$.

(6) Domain$(f) = \{(x \geq 0, y \neq 0, z \neq 0)\}$, Range$(f) = \mathbb{R}$.

(7) Domain$(f) = \{(x, y, z) \mid 9 > 3x^2 + y^2 + 3z^2\}$, Range$(f) = \mathbb{R}$.

(8) Domain$(f) = \{(x, y, z, w) \mid xy \geq 0, \quad z^2 + w^2 > 0\}$, Range$(f) = \{0 \leq f < \infty\}$.

Solutions for §23.2

(1) No singular points.

(2) No singular points.

(3) Singularity at $x = y = 0$.

(4) No singular points.

(5) Singularity at $x = 5$.

(6) Singularity is located on a line $x = y$.

(7)

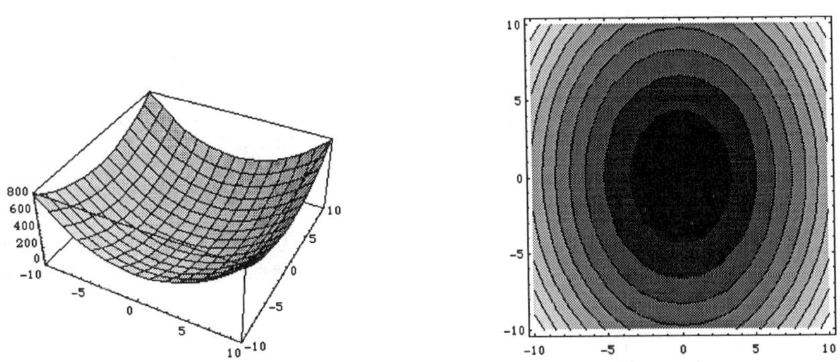

$$z = 5x^2 + 3y^2$$

(8)

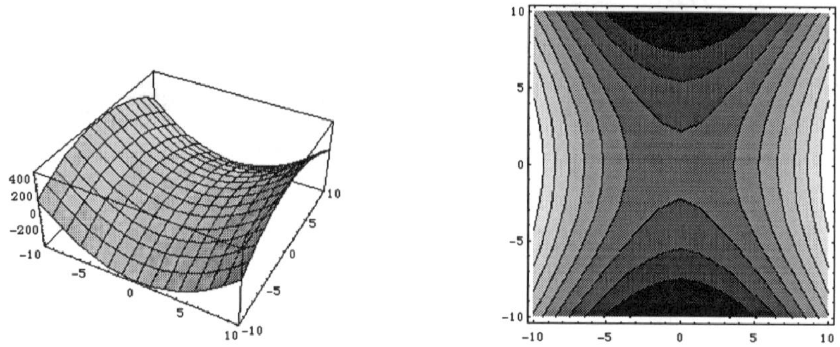

$$z = 5x^2 - 3y^2$$

(9)

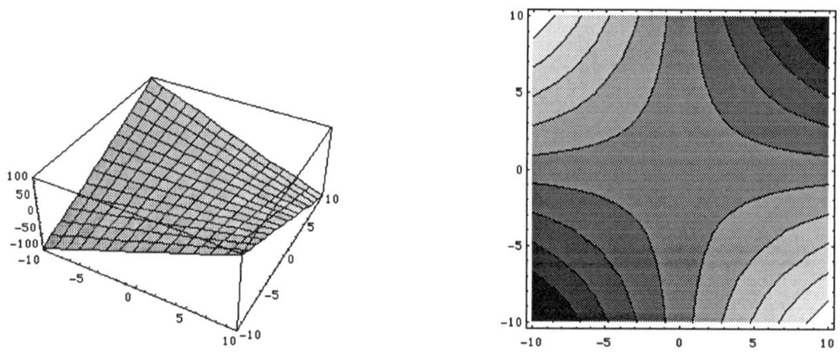

$$z = 1 - xy$$

(10)

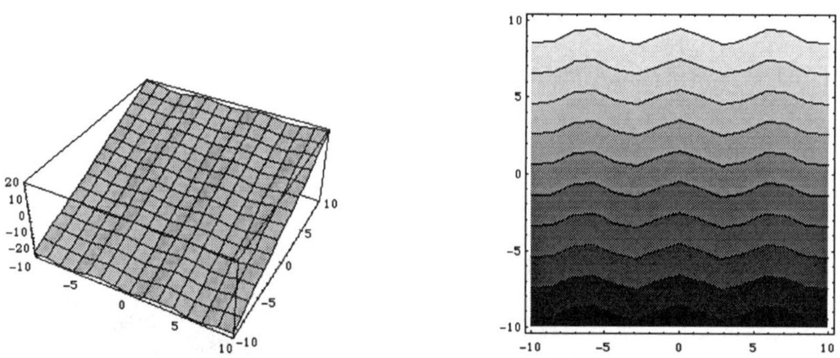

$$z = 2y - \cos(x)$$

(11)

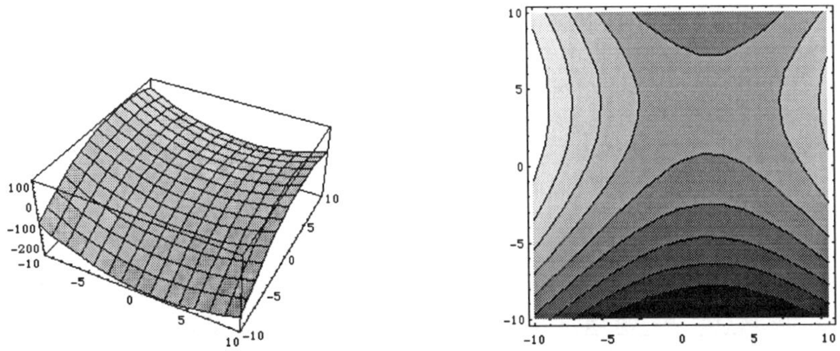

$$z = (x-2)^2 - (y-4)^2$$

(12)

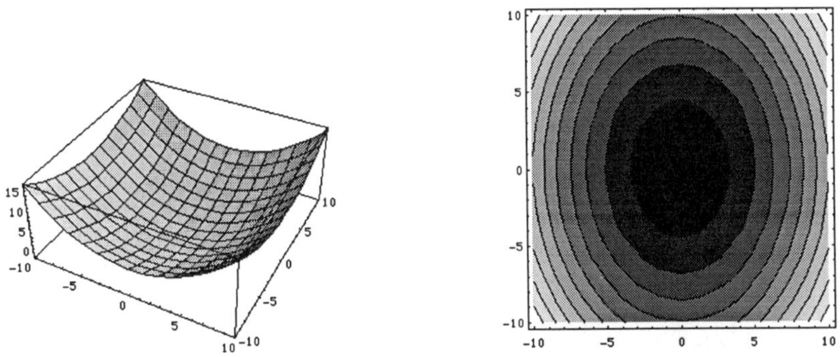

$$z = \frac{1}{9}x^2 + \frac{1}{16}y^2$$

Solutions for §23.3

(1) $\frac{\partial f}{\partial x} = 4x^3y^2$, $\frac{\partial f}{\partial y} = 2x^4y$, $\frac{\partial f}{\partial z} = 0$.

(2) $\frac{\partial f}{\partial x} = 4x^3y^2 + 15x^4$, $\frac{\partial f}{\partial y} = 2x^4y$, $\frac{\partial f}{\partial z} = 0$.

(3) $\frac{\partial f}{\partial x} = \frac{1}{x}$, $\frac{\partial f}{\partial y} = \frac{2}{y}$, $\frac{\partial f}{\partial z} = \frac{1}{z}$.

(4) $\frac{\partial f}{\partial x} = -2x$, $\frac{\partial f}{\partial y} = -6y^2z\sin(2y^3z)$, $\frac{\partial f}{\partial z} = -2y^3\sin(2y^3z)$.

(5) $\frac{\partial f}{\partial x} = 2xze^{x^2z}$, $\frac{\partial f}{\partial y} = 6yz^4$, $\frac{\partial f}{\partial z} = x^2e^{x^2z} + 12y^2z^3$.

(6) $\frac{\partial f}{\partial x} = \frac{yz}{\cos^2(xyz)}$, $\frac{\partial f}{\partial y} = \frac{xz}{\cos^2(xyz)}$, $\frac{\partial f}{\partial z} = \frac{xy}{\cos^2(xyz)}$.

(7) $\frac{\partial f}{\partial x} = yzx^{yz-1}$, $\frac{\partial f}{\partial y} = x^{yz}z\ln x$, $\frac{\partial f}{\partial z} = x^{yz}y\ln x$.

(8) $\frac{\partial f}{\partial x} = y + z$, $\frac{\partial f}{\partial y} = x + 3y^2$, $\frac{\partial f}{\partial z} = x$.

(9) $\frac{\partial f}{\partial x} = x\sinh(x-5y+z)$, $\frac{\partial f}{\partial y} = -5\sinh(x-5y+z)$, $\frac{\partial f}{\partial z} = \sinh(x-5y+z)$.

(10) $\overrightarrow{\nabla f} = y^2\vec{i} + 2xy\vec{j}$, $\overrightarrow{\nabla f(1,2,3)} = 4\vec{i} + 4\vec{j}$.

(11) $\overrightarrow{\nabla f} = 3x^2y^2z^4\vec{i} + 2x^3yz^4\vec{j} + 4x^3y^2z^3\vec{k}$, $\overrightarrow{\nabla f(1,2,3)} = 972\vec{i} + 324\vec{j} + 432\vec{k}$.

(12) $\overrightarrow{\nabla f} = -z^3\sin(xz^3)\vec{i} + 3y^2\vec{j} - 3xz^2\sin(xz^3)\vec{k}$, $\overrightarrow{\nabla f(1,2,3)} = -27\sin(27)\vec{i} + 12\vec{j} - 27\sin(27)\vec{k}$.

(13) $\overrightarrow{\nabla f} = 2xze^{x^2z}\vec{i} + x^2e^{x^2z}\vec{k}$, $\overrightarrow{\nabla f(1,2,3)} = 6e^3\vec{i} + e^3\vec{k}$.

(14) $\overrightarrow{\nabla f} = 2x\vec{i} + \frac{3y^2z^3}{\cos^2(y^3z^3)}\vec{j} + \frac{3y^3z^2}{\cos^2(y^3z^3)}\vec{k}$, $\overrightarrow{\nabla f(1,2,3)} = 2\vec{i} + \frac{324}{\cos^2(216)}\vec{j} + \frac{216}{\cos^2(216)}\vec{k}$.

(15) $\overrightarrow{\nabla f} = \frac{2x}{\ln(x^2+y^2)}\vec{i} + \frac{2y}{\ln(x^2+y^2)}\vec{j}$, $\overrightarrow{\nabla f(1,2,3)} = \frac{2}{\ln 5}\vec{i} + \frac{4}{\ln 5}\vec{j}$.

(16) $\frac{\partial^2 f}{\partial x^2} = 0$, $\frac{\partial^2 f}{\partial y^2} = 0$, $\frac{\partial^2 f}{\partial x \partial y} = 1$.

(17) $\frac{\partial^2 f}{\partial x^2} = 6xy^2$, $\frac{\partial^2 f}{\partial y^2} = 2x^3$, $\frac{\partial^2 f}{\partial x \partial y} = 6x^2y$.

(18) $\frac{\partial^2 f}{\partial x^2} = \frac{y^2}{(x^2+y^2)^{3/2}}$, $\frac{\partial^2 f}{\partial y^2} = \frac{x^2}{(x^2+y^2)^{3/2}}$, $\frac{\partial^2 f}{\partial x \partial y} = -\frac{xy}{(x^2+y^2)^{3/2}}$.

(19) $\frac{\partial^2 f}{\partial x^2} = -\frac{1}{(x+y)^2}$, $\frac{\partial^2 f}{\partial y^2} = -\frac{1}{(x+y)^2}$, $\frac{\partial^2 f}{\partial x \partial y} = -\frac{1}{(x+y)^2}$.

(20) $\frac{\partial^2 f}{\partial x^2} = y(y-1)x^{y-2}$, $\frac{\partial^2 f}{\partial y^2} = x^y\ln^2(x)$, $\frac{\partial^2 f}{\partial x \partial y} = (y\ln(x)+1)x^{y-1}$.

(21) $\frac{\partial^2 f}{\partial x^2} = -2\sin(x^2+y^2) - 4x^2\cos(x^2+y^2)$, $\frac{\partial^2 f}{\partial y^2} = -2\sin(x^2+y^2) - 4y^2\cos(x^2+y^2)$, $\frac{\partial^2 f}{\partial x \partial y} = -4xy\cos(x^2+y^2)$.

Solutions for §23.4

(1) $\frac{\partial f}{\partial \vec{u}} = \overrightarrow{\nabla f} \cdot \vec{u} = (2x\vec{i} + 6y\vec{j}) \cdot (\frac{3}{5}\vec{i} + \frac{4}{5}\vec{j}) = \frac{6}{5}x + \frac{24}{5}y = \frac{6}{5}(x+4y)$.

(2) $\frac{\partial f}{\partial \vec{u}} = \overrightarrow{\nabla f} \cdot \vec{u} = (2xy^4\vec{i} + (4x^2y^3+3)\vec{j}) \cdot (\frac{3}{5}\vec{i} - \frac{4}{5}\vec{j}) = \frac{6}{5}xy^2 - \frac{16}{5}x^2y^2 - \frac{12}{5} = \frac{2}{5}(3xy^4 - 8x^2y^3 - 6)$.

(3) $\frac{\partial f}{\partial \vec{u}} = \frac{2}{\sqrt{14}}(3xyz^4 + 3y\cos(xy) - x^2z^4 - 2x\cos(xy) + 2x^2yz^3)$.

(4) $\frac{\partial f}{\partial \vec{u}} = \frac{1}{3}(3x^2 - z + 4y - 2x)$.

(5) $\frac{\partial f}{\partial \vec{u}} = \frac{1}{3}(12xyz^4e^{3x^2y} + 3x^2z^4e^{3x^2y} + 8z^3e^{3x^2y})$.

(6) $\frac{\partial f}{\partial \vec{u}} = \frac{1}{\sqrt{14}}(yzx^{yz-1} - 2x^{yz}z\ln(x) + 3x^{yz}y\,ln(x))$.

(7) $\frac{\partial f}{\partial \vec{u}} = \frac{1}{\sqrt{2}}(2x\vec{i} - 4y\vec{j} + 3z^2\vec{k}) \cdot (\vec{i} - \vec{j}) = \frac{1}{\sqrt{2}}(2x+4y)$, and $\frac{\partial f}{\partial \vec{u}}(1,-1,2) = \frac{1}{\sqrt{2}}(2-4) = -\sqrt{2}$.

(8) $\frac{\partial f}{\partial \vec{u}} = x^2y^3$, $\frac{\partial f}{\partial \vec{u}}(-1,2,1) = 8$.

Solutions for §23.5

(1) $f(x, y) = 2x^2 - 3y^2$ and $x = t^4$, $y = t^2$. Thus

$$\begin{aligned}\frac{\partial f}{\partial t} &= \frac{\partial f}{\partial x}\frac{\partial x}{\partial t} + \frac{\partial f}{\partial y}\frac{\partial y}{\partial t}\\ &= 4x \cdot 4t^3 + (-6y) \cdot (2t)\\ &= 4t^4 \cdot 4t^3 - 6t^2 \cdot 2t = 16t^7 - 12t^3.\end{aligned}$$

(2) Here $f(x, y) = (x - y^2)^{10}$ and $x = \cos(t), y = t^3$, so

$$\begin{aligned}\frac{\partial f}{\partial t} &= \frac{\partial f}{\partial x}\frac{\partial x}{\partial t} + \frac{\partial f}{\partial y}\frac{\partial y}{\partial t}\\ &= 10(x - y^2)^9 \cdot (-\sin(t)) + 10(x - y^2)^9(-2y) \cdot (3t^2)\\ &= -10(\cos(t) - t^6)^9(\sin(t) + 6t^5).\end{aligned}$$

(3) $\frac{\partial f}{\partial t} = \frac{36t^3+10t^9+2t}{9t^4+t^{10}+t^2}$.

(4) $\frac{\partial f}{\partial t} = t^{t^5+4}(1 + 5\ln(t))$.

(5) $\frac{\partial f}{\partial t} = -(2t + 1 + \frac{1}{t} + 6t^5)\sin(t^2 + t + \ln(t) + t^6)$.

(6) $\frac{\partial g}{\partial s} = 12s^3 + 4st^2$, $\frac{\partial g}{\partial t} = 12s^2t + 4t^3$.

(7) $\frac{\partial g}{\partial s} = 5(st + \frac{s^3}{t^3})^4(t + 3\frac{s^2}{t^3})$, $\frac{\partial g}{\partial t} = 5(st + \frac{s^3}{t^3})^4(s - 3\frac{s^3}{t^4})$.

(8) $\frac{\partial g}{\partial s} = 16s(2s^2 + 2t^2 + t^6)^3$, $\frac{\partial g}{\partial t} = 8(2s^2 + 2t^2 + t^6)^3(2t + 3t^5)$.

(9) Given that $g(x, y, z) = \cos(x\,y\,z)$ and $x = s + t + u$, $y = s^2u^3$, and $z = u^4$ observe that both y and z are independent of the variable t. Thus $\frac{\partial g}{\partial t} = \frac{\partial g}{\partial x} \cdot \frac{\partial x}{\partial t} = -yz\sin(x\,y\,z) = -s^2u^7\sin(s^3u^7 + s^2tu^7 + s^2u^8)$. The other two partial derivatives are $\frac{\partial g}{\partial s} = -(3s^2u^7 + 2stu^7 + 2su^8)\sin(s^3u^7 + s^2tu^7 + s^2u^8)$, and $\frac{\partial g}{\partial u} = -(7s^3u^6 + 7s^2tu^6 + 8s^2u^7)\sin(s^3u^7 + s^2tu^7 + s^2u^8)$.

(10) Here we encounter a four variable function, $g(x, y, z, w) = \ln(x+y+z+w)$ where $x = stu$, $y = \frac{s-t}{u}$, and $z = s^3$. Notice that w is independent of t, and hence $\frac{\partial w}{\partial t} = 0$. The partial derivative with respect to s is given by

$$\begin{aligned}\frac{\partial g}{\partial s} &= \frac{\partial g}{\partial x}\frac{\partial x}{\partial t} + \frac{\partial g}{\partial y}\frac{\partial y}{\partial t} + \frac{\partial g}{\partial z}\frac{\partial z}{\partial t}\\ &= \frac{1}{x+y+z+w}\left(tu + \frac{1}{u} + 3s^2\right)\\ &= \frac{tu + \frac{1}{u} + 3s^2}{stu + \frac{s-t}{u} + s^3 + w}.\end{aligned}$$

The remaining derivatives are computed similarly: $\frac{\partial g}{\partial t} = \frac{su - \frac{1}{u}}{stu + \frac{s-t}{u} + s^3 + w}$, $\frac{\partial g}{\partial u} = \frac{st + \frac{t-s}{u^2}}{stu + \frac{s-t}{u} + s^3 + w}$.

Solutions for §23.6

(1) Given $z = 3x^2 + 2y^2$ and $P = (1, 2, 11)$ set $F(x, y, z) = 3x^2 + 2y^2 - z$. Then

$$\overrightarrow{\nabla F}(x, y, z) = 6x\vec{\imath} + 4y\vec{\jmath} - \vec{k},$$

$\overrightarrow{\nabla F}(1, 2, 11) = 6\vec{\imath} + 8\vec{\jmath} - \vec{k}$. Thus the equation of the tangent plane is given by

$$\overrightarrow{(x-1, y-2, z-11)} \cdot \overrightarrow{(6, 8, -1)} = 0$$

i.e., the equation of the tangent plane to $z = 3x^2 + 2y^2$ at $P = (1, 2, 11)$ is given by

$$6x + 8y - z = 11.$$

(2) $6x - 8y - z = -5$.

(3) Here $z = \cos(xy)$ and $P = (\pi, \frac{1}{2}, 0)$. Setting $F(x, y, z) = \cos(xy) - z$ we have $\overrightarrow{\nabla F}(x, y, z) = -y\sin(xy)\vec{\imath} - x\sin(xy)\vec{\jmath} - \vec{k}$ and $\overrightarrow{\nabla F}(\pi, \frac{1}{2}, 0) = -\frac{1}{2}\vec{\imath} - \pi\vec{\jmath} - \vec{k}$. Thus the equation of the tangent line is given by

$$\overrightarrow{(x-\pi, y-\frac{1}{2}, z)} \cdot \overrightarrow{(-\frac{1}{2}, -\pi, -1)} = 0,$$

i.e., the tangent plane is given by

$$2z + x + 2y\pi = 2\pi.$$

$x + 2\pi y + 2z = 2\pi$.

(4) $12x - 4y + z = -24$.

(5) $12x + 8y\ln 2 - z = 24(1 + \ln 2) - 8$.

(6) $(20x + 20y + ze^{200}) = 401$.

(7) $4x + 6y + 49z = 21$.

(8) $x - 2y - 4z = -4$.

(9) Consider the curve $y = e^{-x^2}$ as the intersection of the graph of $F(x, y, z) = e^{-x^2} - y$ and the plane $z = 0$. We have

$$\overrightarrow{\nabla F}(x, y, z) = -2xe^{-x^2}\vec{\imath} - \vec{\jmath},$$

and $\overrightarrow{\nabla F}(1, e^{-1}.0) = -2e^{-1}\vec{\imath} - \vec{\jmath}$. The equation of the tangent plane is given by

$$\overrightarrow{(x-1, y-e^{-1}, z-0)} \cdot \overrightarrow{(-2e^{-1}, -1, 0)} = 0 \implies 2xe^{-1} + y = 3e^{-1}.$$

Intersecting this plane with the plane $z = 0$ yields $2xe^{-1} + y = 3e^{-1}$ which is now viewed as the equation of the tangent line to the curve $y = e^{-x^2}$.

(10) Given that $y = x\ln(x)$ set $F(x, y, z) = x\ln(x) - y$. Then

$$\overrightarrow{\nabla F}(x, y, z) = (\ln(x) + 1)\vec{i} - \vec{j} + 0\vec{k},$$

and $\overrightarrow{\nabla F}(x, y, z) = 2\vec{i} - \vec{j}$. Hence the tangent plane is given by

$$\overrightarrow{(2, -1, 0)} \cdot \overrightarrow{(x - e, y - e, z)} = 0 \implies 2x - 2e - y + e = 0.$$

Intersecting this plane with the plane $z = 0$ we obtain the equation of the tangent line: $2x - y = e$.

Solutions for Additional Problems for Chapter 23

(1)

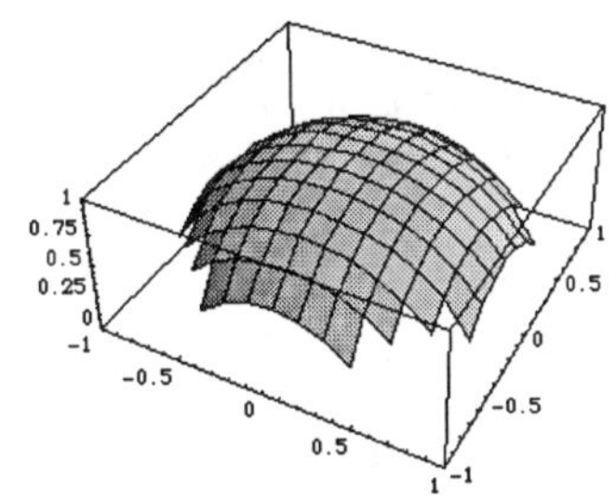

$$f(x, y) = \sqrt{1 - x^2 - y^2}$$

(2)

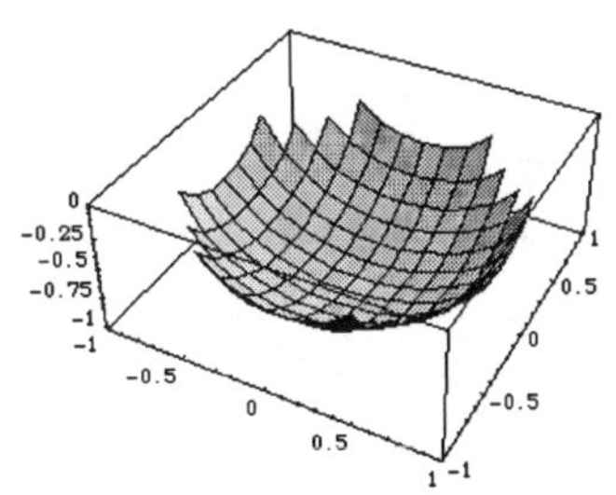

$$f(x, y) = -\sqrt{1 - x^2 - y^2}$$

(3)

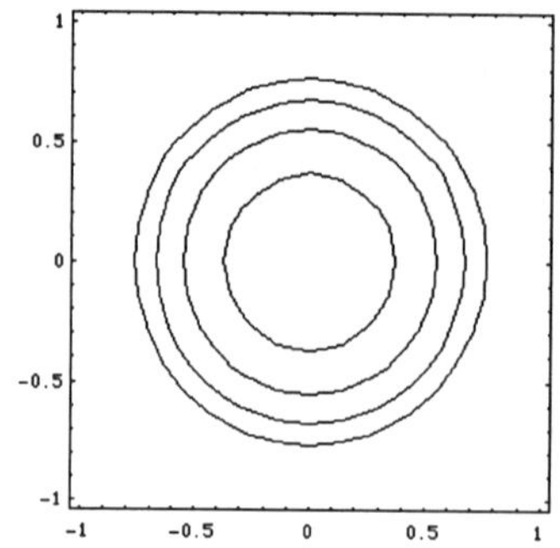

(4a) $x + y = 2\ln(3)$.

(4b) $x + y = 0$.

(4c) $x + y = \ln(2)$.

(5) The domain of f is $\{(x, y) \mid xy \neq 1\}$, and f is continuous and differentiable everywhere on its domain.

(6) The function

$$f(x,y) = \begin{cases} \frac{x^2y^2}{x^4+y^4} & \text{if } (x,y) \neq (0,0) \\ 0 & \text{if } (x,y) = (0,0) \end{cases}$$

is not continuous at $(0, 0)$, since:

(6a) $\lim_{h\to 0} f(h, 0) = \lim_{h\to 0} \frac{0}{h^4+0} = 0$,

(6b) $\lim_{h\to 0} f(0, h) = \lim_{h\to 0} \frac{0}{0+h^4} = 0$,

(6c) $\lim_{h\to 0} f(h, h^2) = \lim_{h\to 0} \frac{h^2h^4}{h^4+h^8} = \lim_{h\to 0} \frac{h^2}{1+h^4} = 0$, and

(6d) $\lim_{h\to 0} f(h, h) = \lim_{h\to 0} \frac{h^2h^2}{h^4+h^4} = \lim_{h\to 0} \frac{h^4}{2h^4} = \frac{1}{2}$.

(7a) Tangent plane to $z = x^2 + 3y^2 - x + 3y - 2$ at $(1, 3, 34)$ is
$z = 34 + (x - 1) + 21(y - 3)$, or $x + 21y - z = 30$.

(7b) Here $z = e^{-(x^2+y^2-2x)}$ and $(1, 0, e)$. Setting $F(x, y, z) = e^{-(x^2+y^2-2x)} - z$ we have

$$\overrightarrow{\nabla F}(x, y, z) = e^{-(x^2+y^2-2x)}(-2x + 2)\vec{\imath} + e^{-(x^2+y^2-2x)}(-2y)\vec{\jmath} - \vec{k},$$

and $\overrightarrow{\nabla F}(1, 0, e) = e \cdot 0\vec{\imath} + e \cdot 0\vec{\jmath} - \vec{k}$. The equation of the tangent plane is $\overrightarrow{(x - 1, y - 0, z - e)} \cdot \overrightarrow{(0, 0, -1)} = 0$
$\implies -z + e = 0 \implies z = e, \implies z = e + 0(x - 1) + 0(y - 0) \implies z = e$.

(7c) Tangent plane to $z = xy^2 - yx^2 - 2xy$ at $(1, -1, 4)$ is $z = 4 + 5(x - 1) - 5(y + 1)$, or $5x - 5y - z = 6$.

Solutions for §24.1

(1) Here $z = f(x,y) = 10000 - 3x^2 - y^4$. The highest point occurs at $z = 10000$. We now compute $\overrightarrow{\nabla f}(x,y) = -6x\vec{i} - 4y^3\vec{j}$ and $\overrightarrow{\nabla f}(10,5) = -60\vec{i} - 500\vec{j}$. This is the direction of maximal increase of f. The steepest descent occurs in the opposite direction, i.e. $60\vec{i} + 500\vec{j}$.

(2)

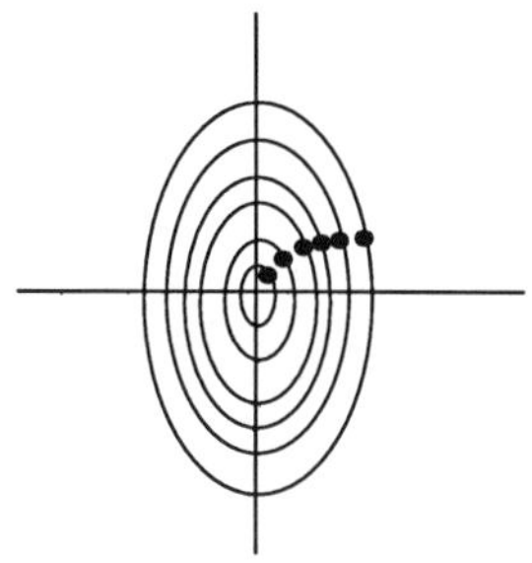

(3) $z = f(x,y) = 5x^2 + xy + y^3$. Then $\overrightarrow{\nabla f}(x,y) = \overrightarrow{(10x + y, x + 3y^2)}$ and $\overrightarrow{\nabla f}(1,-2) = \overrightarrow{(8,13)}$ is the direction of maximal increase.

(4) $a(x,y) = \tan\left(2\pi x + \frac{\pi}{3}y\right)$. Then

$$\overrightarrow{\nabla a}(x,y) = \overrightarrow{\left(\frac{2\pi}{\cos^2\left(2\pi x + \frac{\pi}{3}y\right)}, \frac{\pi}{3\cos^2\left(2\pi x + \frac{\pi}{3}y\right)}\right)} \implies \overrightarrow{\nabla a}(1.5,2) = \overrightarrow{\left(8\pi, \frac{4\pi}{3}\right)}$$

is the direction of maximal increase of elevation. A vector perpendicular to $\overrightarrow{(8\pi, \frac{4\pi}{3})}$ is $\overrightarrow{(1,-6)}$. So, if we travel in the direction $\pm\overrightarrow{(1,-6)}$, the elevation will not change.

(5) Depth is given by $d(x,y) = 150 - 0.01x^2 - 0.006y^2$.

(a) $d(10,20) = 146.6$ feet.

(b) $\overrightarrow{\nabla d}(x,y) = \overrightarrow{(-0.02x, -0.012y)} \implies \overrightarrow{\nabla d}(10,20) = \overrightarrow{(-0.2,-0.24)}$. Consequently, $\overrightarrow{(-0.2,-0.24)}$ is the direction of maximal increase of depth.

(c)

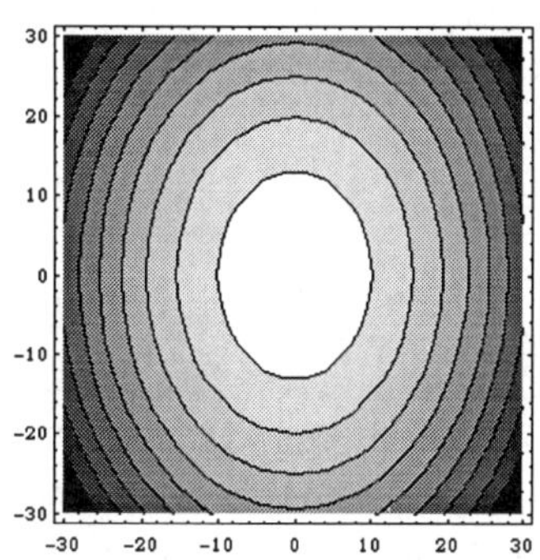

(6) $\overrightarrow{\nabla F}(x,y,z) = \left(\frac{\partial f(x,y)}{\partial x}, \frac{\partial f(x,y)}{\partial y}, -1\right)$ is orthogonal to the surface $z = f(x,y)$. The vector $\overrightarrow{\nabla f}(x,y)$ coincides with the projection of $\overrightarrow{\nabla F}(x,y,z)$ on the xy–plane and gives the direction (in the xy–plane) of the steepest increase of the z–coordinate of the surface.

Solutions for §24.2

(1) $f(x, y) = 20 - 2x^2 + 4x - y^2 + 4y$. Critical points occur for

$$\frac{\partial f}{\partial x} = -4x + 4 = 0, \qquad \frac{\partial f}{\partial y} = -2y + 4 = 0$$

$\implies$ $(1, 2)$ is the only critical point and $f(1, 2) = 26$. We now examine the behavior of f on the boundary.

For $y = 0$ and $0 \leq x \leq 3 \implies f(x, 0) = 20 - 2x^2 + 4x$ is minimized at $x = 3$ and maximized at $x = 1$. We have $f(3, 0) = 14$, $f(1, 0) = 22$.

For $y = 4$ and $0 \leq x \leq 3 \implies f(x, 4) = 20 - 2x^2 + 4x$ is minimized at $x = 3$ and maximized at $x = 1$. We have $f(3, 4) = 14$, $f(1, 4) = 22$.

For $x = 0$ and $0 \leq y \leq 4 \implies f(0, y) = 20 - y^2 + 4y$ is minimized at $y = 0, 4$ and maximized at $x = 1$. We have $f(0, 0) = f(0, 4) = 20$, $f(0, 2) = 24$.

For $x = 3$ and $0 \leq y \leq 4 \implies f(3, y) = 14 - y^2 + 4y$ is minimized at $y = 0, 4$ and maximized at $y = 2$. We have $f(3, 0) = f(3, 4) = 14$, $f(3, 2) = 18$.

We now see that f is maximized at the critical point $(1, 2)$ with $f(1, 2) = 26$ and is minimized at the boundary points $(3, 0), (3, 4)$ where it has the value 14.

(2) The minimum is at $(\sqrt[3]{\frac{3}{4}}, -1)$, where the value is $-\frac{9}{4}\sqrt[3]{\frac{3}{4}}$. The maximum value 4 is at the point $(1, 1)$.

(3) $f(x, y) = x^2 - xy - y^2 + 5y - 3$. Critical points occur for

$$\frac{\partial f}{\partial x} = 2x - y = 0, \qquad \frac{\partial f}{\partial y} = -x - 2y + 5 = 0$$

$\implies$ $(1, 2)$ is the only critical point and $f(1, 2) = -18$. We now examine f on the boundary of the rectangle.

For $-1 \leq x \leq 2, y = 0$ we have $f(x, 0) = x^2 \implies f(-1, 0) = 1, f(2, 0) = 4, f(0, 0) = 0$.
For $-1 \leq x \leq 2, y = 3$ we have $f(x, 3) = x^2 - 3x + 3 \implies f(-1, 3) = 1, f(2, 3) = 1, f\left(\frac{3}{2}, 3\right) = \frac{3}{4}$.
For $x = -1, 0 \leq y \leq 3$ we have $f(-1, y) = -y^2 + 6y - 2 \implies f(-1, 0) = -2, f(-1, 3) = 7$.
For $x = 2, 0 \leq y \leq 3$ we have $f(2, y) = -y^2 + 3y + 1 \implies f(2, 0) = 1, f(2, 3) = 1, f\left(2, \frac{3}{2}\right) = \frac{13}{4}$.

The maximum value is at $(-1, 3)$ and $f(-1, 3) = 7$. The minimum value is at the critical point $(1, 2)$ and $f(1, 2) = -18$.

(4) $f(x, y) = 3x^3y^2 - x^4y^2 - x^3y^3$. Critical points occur for

$$\frac{\partial f}{\partial x} = 9x^2y^2 - 4x^3y^2 - 3x^2y^3 = x^2y^2(9 - 4x - 3y) = 0$$
$$\frac{\partial f}{\partial y} = 6x^3y - 2x^4y - 3x^3y^2 = x^3y(6 - 2x - 3y) = 0.$$

So we have critical points at $(0, 0)$ and $\left(\frac{3}{2}, 1\right)$. Further, $f(0, y) = f(x, 0) = 0$ and $f\left(\frac{3}{2}, 1\right) = \frac{27}{16}$. When we examine the boundaries of the rectangle at $x = 10$ or $y = 10$, we see that f is negative. Thus, the minimum value of f is -17×10^5 at the point $(10, 10)$ and the maximum value is $\frac{27}{16}$ at the critical point $\left(\frac{3}{2}, 1\right)$.

(5) $f(x, y) = y^2 + \cos(\pi x) - y$. Critical points occur for

$$\frac{\partial f}{\partial x} = -\pi \sin(\pi x) = 0, \qquad \frac{\partial f}{\partial y} = 2y - 1 = 0.$$

The only critical point in the rectangle $-\frac{1}{2} \le x \le \frac{1}{2}, 0 \le y \le 1$, is at $\left(0, \frac{1}{2}\right)$. We have $f\left(0, \frac{1}{2}\right) = \frac{3}{4}$. When we examine the boundary, we find that f is maximized at $(0,0), (1,0)$ and $f(0,0) = f(1,0) = 0$. Further, f is minimized at $\left(\pm\frac{1}{2}, \frac{1}{2}\right)$ and $f\left(\pm\frac{1}{2}, \frac{1}{2}\right) = -\frac{1}{4}$.

(6) Let x, y, and z be the dimensions of the box. If the volume $V = xyz = 4$ cubic feet is fixed, then the area of the box $xy + 2xz + 2yz = xy + \frac{8}{x} + \frac{8}{y} = A(x,y)$ is minimal when $x = y = 2$; $z = \frac{V}{xy} = 1$.

(7) If the area $A = 2xy + 2xz + 2yz = 24$ square feet is fixed, then the largest volume $xyz = xy\frac{12-xy}{x+y} = V(x,y)$ is reached when $x = y = 2$; $z = \frac{12-xy}{x+y} = 2$.

(8) Minimizing the square of the distance $d^2 = (t-1)^2 + (3t-1)^2 + (2t+4)^2$, one gets that the closest point on the line corresponds to $t = -\frac{2}{7}$: $\left(-\frac{2}{7}, -\frac{13}{7}, \frac{17}{7}\right)$.

(9) Let (x,y,z) be a point on the plane $2x - 5y + z = 2$. We must minimize the square of the distance $(x-1)^2 + y^2 + (z+1)^2$. Substituting $z = 2 - 2x + 5y$, we see that we must minimize the function $f(x,y) = (x-1)^2 + y^2 + (3 - 2x + 5y)^2$. Critical points occur when

$$\frac{\partial f}{\partial x} = 2(x-1) - 4(3 - 2x + 5y) = 10x - 20y - 14 = 0$$
$$\frac{\partial f}{\partial y} = 2y + 10(3 - 2x + 5y) = -20x + 52y + 30 = 0.$$

Thus, $\left(\frac{16}{15}, -\frac{1}{6}\right)$ is the only critical point and the closest point is $\left(\frac{16}{15}, -\frac{1}{6}, \frac{23}{3}\right)$

(10) One should cut a piece of wire into three pieces of equal lengths.

(11) The critical points are on the lines $x = 0$ and $y = 0$. In the indicated region the function is not bounded either from above or from below.

(12) The square of a distance to the origin is given by $d^2(x,y) = x^2 + y^2 + \frac{4}{x^2y^2}$. Its minimum is at the points $(\pm\sqrt[3]{2}, \pm\sqrt[3]{2})$ (all choices of signs take place).

Solutions for §24.3

(1) $x^2 - 2xy + y^2$.

(2) 0.

(3) $\left(x + y - \frac{x^2}{4} - \frac{xy}{2} - \frac{y^2}{4}\right)\frac{\ln 2}{2}$.

(4) $1 + x - 3y + \frac{x^2}{2} - \frac{6xy}{2} + \frac{9y^2}{2}$.

(5) $1 - \frac{x^2}{2}$.

(6) $y - 2xy$.

(7) $f(x,y) = 1 - x^2 - y^2 + 2x + 6y$. Critical points occur for

$$\frac{\partial f}{\partial x} = -2x + 2 = 0, \qquad \frac{\partial f}{\partial y} = -2y + 6 = 0.$$

So $(1,3)$ is the only critical point. Now,

$$\frac{\partial^2 f}{\partial x^2} = -2, \quad \frac{\partial^2 f}{\partial y^2} = -2, \quad \frac{\partial^2 f}{\partial x \partial y} = 0, \quad D = \frac{\partial^2 f}{\partial x^2} \cdot \frac{\partial^2 f}{\partial y^2} - \left(\frac{\partial^2 f}{\partial x \partial y}\right)^2 = 4.$$

Since $\frac{\partial^2 f}{\partial x^2} < 0$ and $D > 0$ we see that $(1,3)$ is a local maximum.

(8) $(-4, 2)$ is a local minimum.

(9) $f(x,y) = xy - 3x - 3y + 9$. Critical points occur for $\frac{\partial f}{\partial x} = y - 3 = 0, \quad \frac{\partial f}{\partial y} = x - 3 = 0$. So $(3,3)$ is the only critical point. Now, $\frac{\partial^2 f}{\partial x^2} = 0, \; \frac{\partial^2 f}{\partial y^2} = 0, \; \frac{\partial^2 f}{\partial x \partial y} = 1, \; D = -1 \implies (3,3)$ is a saddle point.

(10) $(\pm\sqrt{\frac{2}{3}}, 0)$ are saddle points.

(11) $(0,0)$ is a saddle point, $(\frac{5}{3}, \frac{5}{3})$ is a local maximum.

(12) $(\sqrt{2}, 2)$ is a local maximum, $(\sqrt{2}, -2)$ is a saddle point, $(-\sqrt{2}, 2)$ is a saddle point, $(-\sqrt{2}, -2)$ is a local minimum.

(13) $f(x,y) = 2x^4 + y^4 - 2x^2 - 2y^2 + 8$. Critical points occur for

$$\frac{\partial f}{\partial x} = 8x^3 - 4x = 0 \implies x = 0, \pm 2^{-\frac{1}{2}}$$
$$\frac{\partial f}{\partial y} = 4y^3 - 4y = 0 \implies y = 0, \pm 1.$$

Further,

$$\frac{\partial^2 f}{\partial x^2} = 24x^2 - 4, \quad \frac{\partial^2 f}{\partial y^2} = 12y^2 - 4, \quad \frac{\partial^2 f}{\partial x \partial y} = 0, \quad D = (24x^2 - 4) \cdot (12y^2 - 4).$$

Thus, $\left(\pm 2^{-\frac{1}{2}}, \pm 1\right)$ are all local minima (for all choice of signs). Further, $(0, \pm 1)$, $\left(\pm 2^{-\frac{1}{2}}, 0\right)$ are saddle points while $(0,0)$ is a local maximum.

(14)

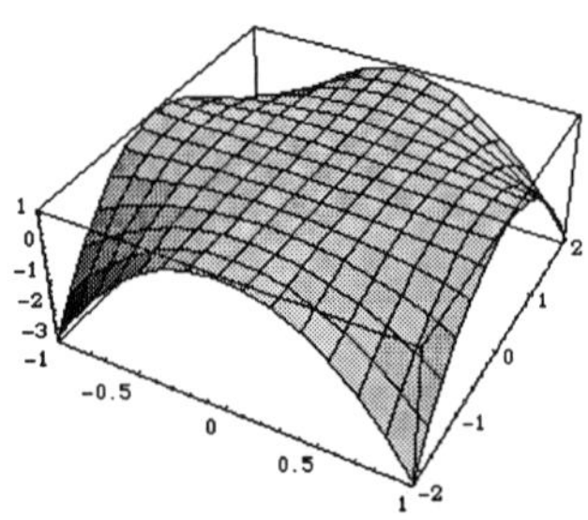

$f(x,y) = 1 - x^2 y^2$

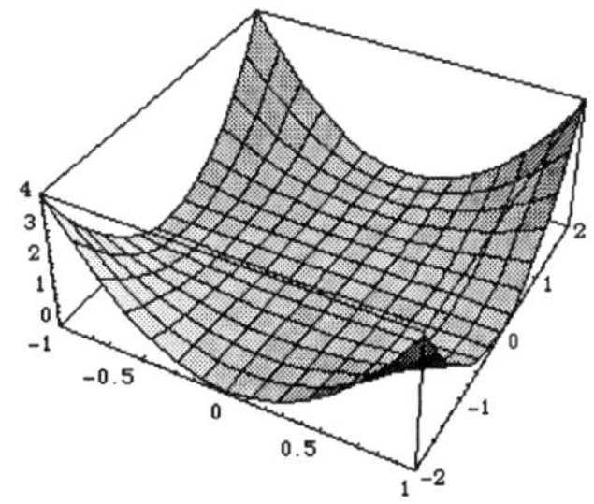

$f(x, y) = x^2y^2$

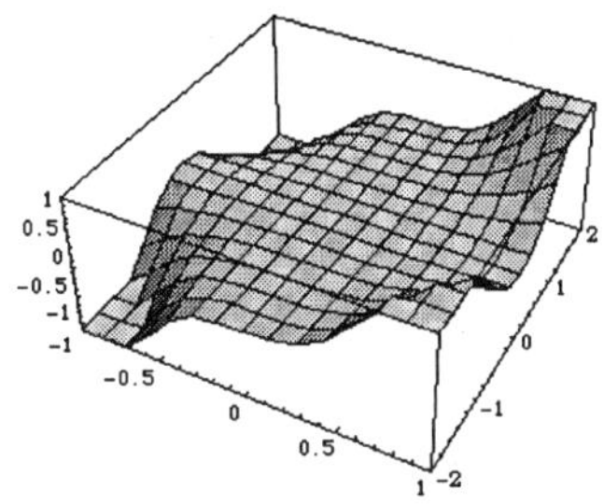

$f(x, y) = x^3y^2$

It is not possible to determine the nature of the critical points from the critical point test because the partial derivatives of the second order are zero at the origin. To conclude about the behavior at the origin, one needs to consider the derivatives of higher order.

Solutions for §24.4

(1) Let $x,\ y$ be the sides of a rectangle with area $A(x, y) = xy$. If the rectangle is inscribed in a circle of radius R

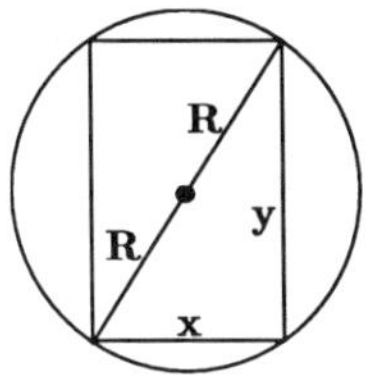

then $x^2 + y^2 = 4R^2$. Set $g(x, y) = x^2 + y^2 - 4R^2$. We must maximize $A(x, y)$ subject to the constraint $g(x, y) = 0$. By the method of Lagrange multipliers,

$$\overrightarrow{\nabla A}(x, y) = \overrightarrow{(y, x)} = \lambda \overrightarrow{\nabla g}(x, y) = \lambda \overrightarrow{(2x, 2y)}$$

$\Longrightarrow \quad y = 2\lambda x,\ x = 2\lambda y \quad \Longrightarrow \quad \lambda = \frac{y}{2x} = \frac{x}{2y} \quad \Longrightarrow \quad y^2 = x^2$. Thus, $x = y$, and the largest possible inscribed rectangle is a square.

(2) The square of the distance from (x, y, z) to $(1, 1, 2)$ is given by $f(x, y, z) = (x-1)^2 + (y-1)^2 + (z-2)^2$. We must minimize f subject to the constraint $g(x, y, z) = 2x - y + 3z - 2 = 0$. The method of Lagrange multipliers yields 4 equations in 4 unknowns:

$$\overrightarrow{\nabla f} = \overrightarrow{(2(x-1), 2(y-1), 2(z-2))} = \lambda \overrightarrow{\nabla g} = \overrightarrow{(2, -1, 3)}$$

$$g(x, y, z) = 0.$$

Thus $2(x-1) = 2\lambda,\ 2(y-1) = -\lambda,\ 2(z-2) = 3\lambda$ and this implies that $\lambda = x - 1,\ y = \frac{3-x}{2},\ z = \frac{3x+1}{2}$. Plugging this into the constraint equation $g(x, y, z) = 0$, yields $2x - \left(\frac{3-x}{2}\right) + 3\left(\frac{3x+1}{2}\right) - 2 = 0$. Hence, $x = \frac{2}{7},\ y = \frac{19}{14},\ z = \frac{13}{14}$.

(3) We want to minimize $f(x, y) = 2x^2 + 3y^2$ subject to the constraint $g(x, y) = 3x + 4y - 1 = 0$. Thus $\overrightarrow{\nabla f} = \overrightarrow{(4x, 6y)} = \lambda \overrightarrow{\nabla g} = \lambda \overrightarrow{(3, 4)}$. This implies that $4x = 3\lambda,\ 6y = 4\lambda \implies \lambda = \frac{4}{3}x$ and $y = \frac{8}{9}x$. Plugging this into the constraint equation $g(x, y) = 0$ yields $3x + \frac{32}{9}x = 1 \implies x = \frac{9}{59},\ y = \frac{8}{59}$.

(4) We want to optimize $f(x, y, z) = 3x - y - 4z$ subject to the constraint $g(x, y, z) = x^2 + y^2 + z^2 - 4 = 0$. By the method of Lagrange multipliers $\overrightarrow{\nabla f} = \overrightarrow{(3, -1, 4)} = \lambda \overrightarrow{\nabla g} = \lambda \overrightarrow{(2x, 2y, 2z)}$. This immediately implies that $3 = 2\lambda x,\ -1 = 2\lambda y,\ 4 = 2\lambda z$. Thus, $\lambda = \frac{3}{2\lambda}$ and $y = -\frac{x}{3},\ z = \frac{4x}{3}$. Plugging this into the constraint equation yields $x^2 + \frac{1}{9}x^2 + \frac{16}{9}x^2 = 4$. We obtain $x = \frac{6}{\sqrt{26}},\ y = -\frac{2}{\sqrt{26}},\ z = \frac{8}{\sqrt{26}}$.

(5) We must minimize $f(x, y) = 3x + 2y$ subject to the constraint $g(x, y) = xy - 5$. By the method of Lagrange multipliers, $\overrightarrow{\nabla f} = \lambda \overrightarrow{\nabla g} \implies \overrightarrow{(3, 2)} = \lambda \overrightarrow{(y, x)} \implies 3 = \lambda y,\ 2 = \lambda x$. It follows that $\lambda = \frac{2}{x}$ and $y = \frac{3}{2}x$. Plugging this into the constraint equation yields $\frac{3}{2}x^2 = 5 \implies x = \pm\sqrt{\frac{10}{3}},\ y = \pm\frac{3}{2}\sqrt{\frac{10}{3}}$. The function f is minimized at $\left(-\sqrt{\frac{10}{3}}, -\frac{3}{2}\sqrt{\frac{10}{3}}\right)$.

(6) We must maximize the function $(x-1)^2 + (y-2)^2 + (z-3)^2 = f(x, y, z)$ subject to the constraint $g(x, y, z) = (x-1)^2 + y^2 + (z+1)^2 - 1 = 0$. Then

$$\overrightarrow{\nabla f} = \lambda \overrightarrow{\nabla g} \implies 2(x-1) = 2\lambda(x-1),\ 2(y-2) = 2\lambda y,\ 2(z-3) = 2\lambda(z+1).$$

The assumption $\lambda = 1$ leads to a contradiction. So we must have $x = 1 \implies \lambda = \frac{y-2}{y},\ z = \frac{3+\lambda}{1-\lambda} = 2y - 1$. Plugging these values into the constraint equation yields $y^2 + 4y^2 = 1 \implies y = \pm\frac{1}{\sqrt{5}}$. Thus, f is maximized at $\left(1, -\frac{1}{\sqrt{5}}, \frac{-2\sqrt{5}+5}{5}\right)$.

(7) We must minimize the square of the distance from (x, y, z) to the origin, which is $f(x, y, z) = x^2 + y^2 + z^2$, subject to the constraint $g(x, y, z) = \frac{1}{4}x^2 + \frac{1}{9}y^2 + z^2 - 1 = 0$. Setting $\overrightarrow{\nabla f} = \lambda \overrightarrow{\nabla g}$ yields the three equations:

$$2x = \frac{1}{4}2x \cdot \lambda, \quad 2y = \frac{1}{9}2y \cdot \lambda, \quad 2z = 2z \cdot \lambda.$$

The possible solutions are $x = 0,\ y = 0,\ \lambda = 1$ or $x = 0,\ \lambda = 9,\ z = 0$ or $\lambda = 4,\ y = 0,\ z = 0$. When we plug these three possibilities into $g(x, y, z) = 0$, we obtain three possible sets of solutions:

$$x = 0,\ y = 0,\ z = 1, \quad x = 0,\ y = 3,\ z = 0, \quad x = 2,\ y = 0,\ z = 0.$$

Thus, f is minimized for $x = 0,\ y = 0,\ z = 1$.

(8) We must minimize $f(x,y,z) = x^2 - y + 2z^2 - 1$ subject to the constraint $g(x,y,z) = x^4 + y^4 - z^2 = 0$. Then $\overrightarrow{\nabla f} = \lambda \overrightarrow{\nabla g} \implies 2x = 4\lambda x^3,\ -1 = 4\lambda y^3,\ 4z = -2\lambda z$. If $\lambda = -2$, then $x = 0,\ y = \frac{1}{2},\ z = \pm\frac{1}{4}$. If $\lambda \neq -2 \implies z = 0 \implies x = 0,\ y = 0,\ z = 0$ since $x^4 + y^4 = 0$ only has the solution $x = 0,\ y = 0$. Thus, f is minimized for $x = 0,\ y = \frac{1}{2},\ z = \pm\frac{1}{4}$.

(9) Let $x =$ length, $y =$ width, $z =$ depth. We must maximize the volume $V(x,y,z) = xyz$ subject to the constraint $x + y + z \leq 100$. Set $S_u(x,y,z) = x + y + z - u$ with $0 \leq u \leq 100$. We may rewrite the constraint condition as $S_u(x,y,z) = 0$ where u is an undetermined constant. Then

$$\overrightarrow{\nabla V} = \lambda \overrightarrow{\nabla S_u} \implies yz = \lambda,\ xz = \lambda,\ xy = \lambda \implies yz = xz = xy \implies x = y = z.$$

Thus, the volume is maximized when $x = y = z = \frac{u}{3}$ and $u = 100$. The corresponding volume is $\frac{10^6}{27}$.

(10) The function $f(x,y,z) = 2x^2 + y^2 + 3z^2$ is minimized at $(x,y,z) = \left(\pm\frac{\sqrt[6]{6}}{\sqrt{2}}, \pm\sqrt[6]{6}, \frac{\pm\sqrt[6]{6}}{\sqrt{3}}\right)$. The constraint condition is obeyed if either none or two of the coordinates are negative.

(11) We minimize the square of the distance to the origin $f(x,y,z) = x^2 + y^2 + z^2$ subject to the two constraints: $g(x,y,z) = x + y - 2 = 0$ and $h(x,y,z) = x^2 + (y-3)^2 + z^2 - 1 = 0$. The method of Lagrange multipliers yields: $\overrightarrow{\nabla f} = \lambda_1 \overrightarrow{\nabla g} + \lambda_2 \overrightarrow{\nabla h}$. Consequently, $\overrightarrow{(2x, 2y, 2z)} = \lambda_1 \overrightarrow{(1,1,0)} + \lambda_2 \overrightarrow{(2x, 2(y-3), 2z)}$

$$\implies \quad \begin{aligned} 2x &= \lambda_1 + 2\lambda_2 x \\ 2y &= \lambda_1 + 2\lambda_2 (y-3) \\ 2z &= 2\lambda_2 z. \end{aligned}$$

Thus $\lambda_2 \neq 1$ and z must be zero. When we plug into the constraints, we find two possible solutions: $x = 0,\ y = 2,\ z = 0$ or $x = -1,\ y = 3,\ z = 0$. These points are nearest to and furthest from the origin, respectively.

(12) We must maximize $f(x_1, x_2, \ldots, x_K) = \frac{1}{K}\sum_{i=1}^{K} x_i$ subject to $g(x_1, x_2, \ldots, x_K) = \sum_{i=1}^{K} x_i^2 - 1 = 0$. By the method of Lagrange multipliers, we have

$$\overrightarrow{\nabla f} = \lambda \overrightarrow{\nabla g} \implies \frac{1}{K}\overrightarrow{(1, 1, \ldots, 1)} = 2\lambda \overrightarrow{(x_1, x_2, \ldots, x_K)} \implies x_1 = x_2 = \cdots = x_K = \frac{1}{2\lambda K}.$$

Plugging into the constraint equation $g(x_1, \ldots, x_K) = 0$ implies $x_1 = x_2 = \cdots = x_K = \frac{1}{\sqrt{K}}$.

(13) Minimizing the function $(x - x_0)^2 + (y - y_0)^2 + (z - z_0)^2$ subject to the constraint $ax + by + cz - d = 0$, we obtain the minimum value $\frac{1}{\sqrt{a^2+b^2+c^2}} |ax_0 + by_0 + cz_0 - d|$.

One can as well obtain this formula projecting the vector $\overrightarrow{(x - x_0, y - y_0, z - z_0)}$ onto the normal direction to the plane, which is given by the unit vector parallel to the gradient $\overrightarrow{(a, b, c)}$ of the constraint function $ax + by + cz - d$.

(14) We must minimize $f(x,y,z) = x^2 + y^2 + z^2$ subject to $g(x,y,z) = 2 - \sin(x) - \cos(y) - z = 0$. Then

$$\overrightarrow{\nabla f} = \lambda \overrightarrow{\nabla g} \implies 2x = -\lambda \cos(x),\ 2y = \lambda \sin(y),\ 2z = -\lambda.$$

If $y = 0 \implies \frac{x}{\cos(x)} = z = 1 - \sin(x) \implies x = \cos(x)(1 - \sin(x))$. This has the solution $x \approx 0.479$. Thus, $x \approx 0.479,\ y = 0,\ z \approx 0.539$ and the minimum of f is approximately 0.52.

Solutions for Additional Exercises for Chapter 24

(1) Direction of fastest increase is $\overrightarrow{\nabla f}(2,3) = 84\overrightarrow{(9,4)}$. Directional derivative in direction $\overrightarrow{(4,-3)}$ is $\overrightarrow{\nabla} f(2,3)\cdot \overrightarrow{\left(\frac{4}{5},-\frac{3}{5}\right)}$ (normalize!) $= \frac{2016}{5} = 403.2$.

(2) $\overrightarrow{\nabla f}(x,y) = \overrightarrow{(e^y - ye^x,\, xe^x - e^x)}$. Direction of fastest increase is $\overrightarrow{\nabla f}(2,3) = \overrightarrow{(e^3 - 3e^2,\, 2e^3 - e^2)}$. The directional derivative in the direction $2\vec{i} + \vec{j}$ is $\overrightarrow{\nabla f}(2,3)\cdot \overrightarrow{\left(\frac{2}{\sqrt{5}}, \frac{1}{\sqrt{5}}\right)} = \frac{1}{\sqrt{5}}\left(4e^3 - 7e^2\right)$.

(3) $\overrightarrow{\nabla f}(x,y) = \overrightarrow{\left(\frac{y}{(x+y)^2}, \frac{-x}{(x+y)^2}\right)}$. Direction of fastest increase is $\overrightarrow{\nabla f}(1,0) = \overrightarrow{(0,-1)}$. The directional derivative in the direction $\overrightarrow{(1,0)}$ is $\overrightarrow{\nabla f}(1,0)\cdot\overrightarrow{(1,0)} = 0$.

(4) $\overrightarrow{\nabla f}(x,y) = \overrightarrow{\left(\frac{2x}{1+x^2+y}, \frac{1}{1+x^2+y}\right)}$. Direction of fastest increase is $\overrightarrow{\nabla f}(1,2) = \overrightarrow{\left(\frac{2}{4},\frac{1}{4}\right)}$. The directional derivative in the direction $\overrightarrow{(1,1)}$ is $\overrightarrow{\nabla f}(1,2)\cdot\overrightarrow{\left(\frac{1}{\sqrt{2}},\frac{1}{\sqrt{2}}\right)} = \frac{3\sqrt{2}}{2}$.

(5) Direction of fastest increase is $\overrightarrow{\nabla f}(-2,3) = \overrightarrow{\left(-\frac{1}{2},\frac{1}{2}\right)}$. Directional derivative in direction $\overrightarrow{(1,0)}$ is $\overrightarrow{\nabla f}(-2,3)\cdot\overrightarrow{(1,0)} = -\frac{1}{2}$.

(6) $|\overrightarrow{\nabla f}(x,y)| = 10$ if and only if $100 = 9x^4y^2 + \left(x^3 + 8y\right)^2$.

(7) First, we calculate $\overrightarrow{\nabla f}(2,10) = \overrightarrow{(-4,-60)}$. If the climber walks due west (in direction $-\vec{i}$), then the directional derivative is $\overrightarrow{\nabla f}(2,10)\cdot\overrightarrow{(-1,0)} = 4$, so he will be ascending. If, on the other hand, the climber walks northeast (in direction $\frac{1}{\sqrt{2}}\vec{i} + \frac{1}{\sqrt{2}}\vec{j}$), then the directional derivative is $\overrightarrow{\nabla f}(2,10)\cdot\overrightarrow{\left(\frac{1}{\sqrt{2}},\frac{1}{\sqrt{2}}\right)} = -32\sqrt{2}$, so he will be descending. If the climber walks in direction $15\vec{i} - \vec{j}$ he will remain (for the moment) on a level path.

(8) Critical point is $(-1,-1)$ (minimum).

(9) Critical points are (letting $\alpha = \arctan\left(\frac{1}{2}\right) \approx 0.4636$): $\left(\frac{\pi}{2},\frac{\pi}{2}\right)$ (saddle point), $\left(\frac{\pi}{2},-\frac{\pi}{2}\right)$ (minimum), $\left(-\frac{\pi}{2},\frac{\pi}{2}\right)$ (maximum), $\left(-\frac{\pi}{2},-\frac{\pi}{2}\right)$ (saddle point), $\left(\frac{\pi}{2}+\alpha,\alpha\right)$ (maximum), $\left(-\frac{\pi}{2}+\alpha,-\pi+\alpha\right)$ (minimum), $\left(-\frac{\pi}{2}-\alpha,-\alpha\right)$ (minimum), $\left(\frac{\pi}{2}-\alpha,\pi-\alpha\right)$ (maximum).

(10) Critical points are: $(0,0)$ (minimum), $(2,1)$ (saddle point), $(-2,1)$ (saddle point).

(11) Critical point at $(0,0)$. Saddle point if $\frac{4}{a^5b^5} < 1$. Maximum if $\frac{4}{a^5b^5} > 1$ and $a > 0$. Minimum if $\frac{4}{a^5b^5} > 1$ and $a < o$.

(12) Critical point is $\left(1,-\frac{3}{2}\right)$ (maximum).

(13) Critical points are $(x_0, 0)$, where x_0 is an arbitrary number between 0 and 1. All critical points are minima.

(14) $x = y = z = \frac{100}{3}$.

(15) $x = y = z = \frac{100}{3}$.

(16) The maximum range value of the function is $\frac{8\sqrt{3}}{9}$. It is attained at the points $\left(\frac{2\sqrt{3}}{3}, \frac{2\sqrt{3}}{3}\right)$ and $\left(\frac{2\sqrt{3}}{3}, -\frac{2\sqrt{3}}{3}\right)$.

(17) The point on the paraboloid closest to the plane is $\left(-\frac{1}{2}, -\frac{1}{2}, \frac{5}{2}\right)$. The distance to the plane at this point is $\frac{\sqrt{3}}{2}$.

Solutions for §25.2

(1)

$$\begin{aligned}\int_1^3\int_0^1 (x^2y - y^2 + y)\,dx\,dy &= \int_1^3 \left(\frac{x^3}{3}y - y^2x + yx\right)\Big]_{x=0}^{1} dy\\ &= \int_1^3 \left(\frac{y}{3} - y^2 + y\right) dy\\ &= \frac{y^2}{6} - \frac{y^3}{3} + \frac{y^2}{2}\Big]_{y=1}^{3}\\ &= \left(\frac{9}{6} - \frac{27}{3} + \frac{9}{2}\right) - \left(\frac{1}{6} - \frac{1}{3} + \frac{1}{2}\right) = -\frac{10}{3}\end{aligned}$$

(2) $\int_{\frac{1}{2}}^{1}\int_1^3 \left(1 + \frac{y}{x}\right)\,dx\,dy = 1 + \frac{3}{8}\ln 3$.

(3)

$$\begin{aligned}\int_0^\pi\int_0^1 y\sin(xy)\,dx\,dy &= \int_0^\pi -\cos(xy)\Big]_0^1 dy\\ &= \int_0^\pi (-\cos(y) + 1)\,dy\\ &= \sin(y) + y\Big]_{y=0}^{\pi} = \pi.\end{aligned}$$

(4) $\int_0^\pi\int_0^1 (e^x\cos(y))\,dx\,dy = 0$.

(5)

$$\int_0^1\int_0^2 (16 - x^2 - y^2)\,dx\,dy = \int_0^1 16x - \frac{x^3}{3} - y^2x\Big]_0^2 dy$$
$$= \int_0^1 \left(32 - \frac{8}{3} - 2y^2\right) dy$$
$$= \frac{88}{3}y - \frac{2}{3}y^3\Big]_0^1 = \frac{88}{3} - \frac{2}{3} = \frac{86}{3}.$$

(9) We must compute the following sum of integrals:

$$\int_{-\frac{1}{2}}^{\frac{1}{2}}\int_{-1}^{0} dx\,dy \;+\; \int_{-\frac{1}{2}}^{\frac{1}{2}}\int_{1}^{2} dx\,dy \;=\; 2.$$

(11)

$$\int_0^{\frac{\pi}{4}}\int_0^{\frac{\pi}{4}} (\cos(x))\cdot(\sin(x))\,dx\,dy = \int_0^{\frac{\pi}{4}} \cos(x)\,dx \cdot \int_0^{\frac{\pi}{4}} \sin(y)\,dy$$
$$= \left(\sin(x)\Big]_{x=0}^{\frac{\pi}{4}}\right)\cdot\left(-\cos(y)\Big]_{y=0}^{\frac{\pi}{4}}\right)$$
$$= \frac{\sqrt{2}}{2}\cdot\left(-\frac{\sqrt{2}}{2}+1\right) = \frac{\sqrt{2}-1}{2}.$$

(12) $\int_0^1\int_1^e (y\cdot\ln(x))\,dx\,dy = \frac{1}{2}$.

(13) We may express

$$\int_0^2\int_0^1 (x-2y)\,dx \;\approx\; \sum_{m=1}^{N}\sum_{n=1}^{2N}\left(\frac{m}{N} - 2\frac{n}{N}\right)\cdot\frac{1}{N^2}.$$

For $N = 3$, the double integral is approximated by $-\frac{10}{3}$, while the actual value is -3.

(14) We may express

$$\int_0^2\int_0^1 (x^2-y^2)\,dx \;\approx\; \sum_{m=1}^{N}\sum_{n=1}^{2N}\left[\left(\frac{m}{N}\right)^2 - \left(\frac{n}{N}\right)^2\right]\cdot\frac{1}{N^2}.$$

For $N = 3$, the double integral is approximated by $-\frac{7}{3}$, while the actual value is -2.

Solutions for §25.3

(1) Here $f(x,y) = xy$ and the region is given by $-1 \leq x \leq 2$, $3x \leq y \leq 6$. Thus

$$\int_{-1}^{2}\int_{3x}^{6} (xy)\,dy\,dx = \int_{-1}^{2} x \cdot \frac{y^2}{2}\Big]_{3x}^{6}\,dx$$
$$= \int_{-1}^{2} \left(18x - \frac{9x^3}{2}\right)\,dx$$
$$= 9x^2 - \frac{9}{8}x^4\Big]_{x=-1}^{2} = \frac{81}{8}.$$

(2) Here $f(x,y) = x^2 + y^2$ and $0 \leq x \leq 2$, $0 \leq y \leq x+1$. Thus

$$\int_{0}^{2}\int_{0}^{x+1} (x^2+y^2)\,dy\,dx = \int_{0}^{2} x^2y + \frac{y^3}{3}\Big]_{y=0}^{x+1}\,dx$$
$$= \int_{0}^{2} \left(x^2(x+1) + \frac{(x+1)^3}{3}\right)\,dx$$
$$= \int_{0}^{2} \left(x^3 + x^2 + \frac{1}{3}(x+1)^3\right)\,dx$$
$$= \frac{x^4}{4} + \frac{x^3}{3} + \frac{(x+1)^4}{12}\Big]_{x=0}^{2} = \frac{40}{3}.$$

(3) On the interval $0 \leq x \leq 1$ we have that $x^3 \leq y \leq x^2 + 2$ and our integral is computed as follows:

$$\int_{0}^{1}\int_{x^3}^{x^2+2} (x+y)\,dy\,dx = \int_{0}^{1} xy + \frac{y^2}{2}\Big]_{x^3}^{x^2+2}\,dx$$
$$= \int_{0}^{1} \left(x(x^2+2) + \frac{(x^2+2)^2}{2}\right) - \left(x^4 + \frac{x^6}{2}\right)\,dx$$
$$= \frac{x^4}{4} + \frac{2x^3}{3} + x^2 + 2x - \frac{x^5}{10} - \frac{x^7}{14}\Big]_{0}^{1} = \frac{1573}{420}.$$

(4) On the interval $0 \leq x \leq 1$ we have $x^4 \leq y \leq x^3 + 3$ and

$$\int_{0}^{1}\int_{x^4}^{x^3+3} (x^2y - x)\,dy\,dx = \frac{236}{495}.$$

(5) On the interval $0 \le y \le 1$ we see that $y^2 \le x \le y^2+2$ and the computation proceeds as follows:

$$\int_0^1 \int_{y^2}^{y^2+2} (2x^3y + xy)\,dx\,dy = \int_0^1 \frac{x^4}{2}y + \frac{x^2}{2}y\Big]_{x=y^2}^{y^2+2} dy$$

$$= \int_0^1 \left(\frac{(y^2+2)^4}{2} \cdot y + \frac{(y^2+2)^4}{2} \cdot y - \frac{y^9}{2} - \frac{y^5}{2} \right) dy$$

$$= \frac{(y^2+2)^5}{20} + \frac{(y^2+2)^3}{12} - \frac{y^{10}}{20} - \frac{y^6}{12}\Big]_0^1 = 12.$$

(6) Here we consider the interval $1 \le y \le 2$ where $y^3 \le x \le y^4+3$. The integral is computed as follows:

$$\int_1^2 \int_{y^3}^{y^4+3} (2x - y)\,dx\,dy = \int_1^2 x^2 - xy\Big]_{x=y^3}^{y^4+3} dy$$

$$= \int_1^2 \left((y^4+3)^2 - (y^4+3)y - y^6 + y^4 \right) dy = \frac{23951}{315}.$$

(7) The region in this example is specified by $0 \le x \le \pi$ and $0 \le y \le \sin(x)$. Thus

$$\int_0^\pi \int_0^{\sin(x)} (4y)\,dy\,dx = \int_0^\pi 2y^2\Big]_0^{\sin(x)} dx$$

$$= \int_0^\pi 2\sin^2(x)\,dx$$

$$= \int_0^\pi (1 - \cos(2x))\,dx$$

$$= x - \frac{1}{2}\sin(2x)\Big]_0^\pi = \pi.$$

(8) On the interval $0 \le x \le 1$ we have $x^3 \le y \le x^2$ and thus

$$\int_0^1 \int_{x^3}^{x^2} (x+y)\,dy\,dx = \frac{11}{140}.$$

(9) This is entirely analogous to problem (8) above: $\int_0^1 \int_{y^3}^{y^2} (x+y)\,dx\,dy = \frac{11}{140}$.

(10) The region in this case is a circle which can be looked at as the union of two semi–circles which are specified by $-3 \le x \le 3$, $0 \le y \le \sqrt{9-x^2}$ and $-3 \le x \le 3$, $-\sqrt{9-x^2} \le y \le 0$. Thus the integral breaks up into two integrals as follows:

$$\iint_{x^2+y^2\le 9} (3x)\,dy\,dx = \int_{-3}^{3}\int_{0}^{\sqrt{9-x^2}} (3x)\,dy\,dx + \int_{-3}^{3}\int_{-\sqrt{9-x^2}}^{0} (3x)\,dy\,dx.$$

Now

$$\int_{-3}^{3}\int_{0}^{\sqrt{9-x^2}} (3x)\,dy\,dx = 3\int_{-3}^{3} x\sqrt{9-x^2}\,dx = -\frac{3}{2}\cdot\frac{2}{3}(9-x^2)^{\frac{3}{2}}\Big]_{x=-3}^{3} = 0,$$

and similarly

$$\int_{-3}^{3}\int_{-\sqrt{9-x^2}}^{0} (3x)\,dy\,dx = 0.$$

Thus $\iint_{x^2+y^2\le 9} (3x)\,dy\,dx = 0.$

(11) This is entirely analogous to problem (10) above: $\iint_{x^2+y^2\le 9} (3y)\,dy\,dx = 0.$

(12) That $\iint_{x^2+y^2\le 9} (x+y)\,dy\,dx = 0$ follows from (10) and (11) above.

(13) $\int_{0}^{1}\left(\int_{\sqrt[4]{x}}^{1} x\,dy\right)dx = \frac{1}{18}.$

(14) The region $1 \le y \le 3,\ y \le x \le y^2$ can alternatively be expressed as the union of the two regions $1 \le x \le 3,\ \sqrt{x} \le y \le x$ and $3 \le x \le 9,\ \sqrt{x} \le y \le 3$, as can be observed from the figure below.

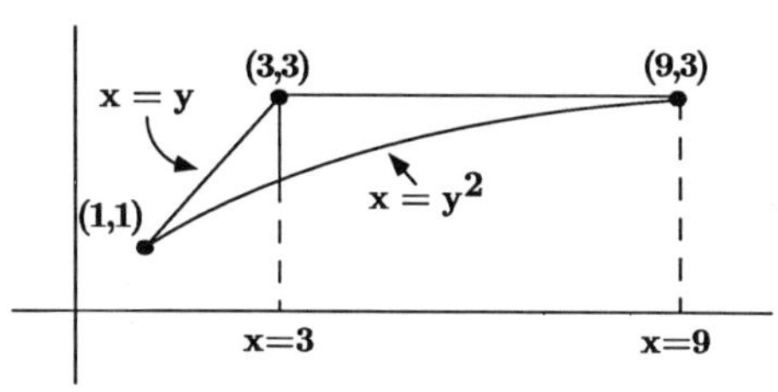

Consequently, $\int_{1}^{3}\left(\int_{\sqrt{x}}^{x}(x^2-y)\,dy\right)dx + \int_{3}^{9}\left(\int_{\sqrt{x}}^{3}(x^2-y)\,dy\right)dx = \frac{1808}{21}.$

(15) The region $0 \le x \le 2,\ x \le y \le 2x^3$ can alternatively be expressed as the union of the two regions $0 \le y \le 2,\ \sqrt[3]{\frac{y}{2}} \le x \le y$ and $2 \le y \le 16,\ \sqrt[3]{\frac{y}{2}} \le x \le 2$, as can be observed from the figure below.

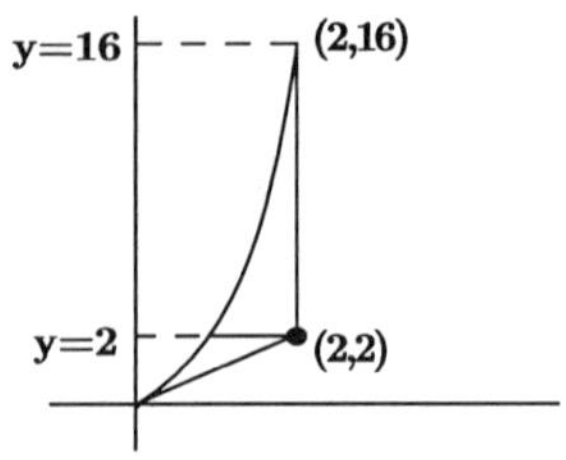

Consequently, $\int_0^2 \left(\int_{\sqrt[3]{\frac{y}{2}}}^{y} y\,dx \right) dy + \int_2^{16} \left(\int_{\sqrt[3]{\frac{y}{2}}}^{2} y\,dx \right) dy = \frac{740}{21}$.

(16)
$\int_1^0 \left(\int_{x^2/2}^{x} dx \right) dy + \int_2^1 \left(\int_{x^2/2}^{1} dx \right) dy = -\frac{5}{6}$.

Solutions for §25.5

(1) Here the region $\mathcal{R}$ is described in polar form by $0 \le r \le 2$ and $\frac{\pi}{2} \le \theta \le \pi$. Thus

$$\iint_{\mathcal{R}} (x^2 + y^2)\,dy\,dx = \int_{\frac{\pi}{2}}^{\pi} \int_0^2 r^2 \cdot r\,dr\,d\theta$$
$$= \int_{\frac{\pi}{2}}^{\pi} \frac{r^4}{4}\Big]_{r=0}^{2} d\theta = 2\pi.$$

(2) The region $\mathcal{R}$ in this example is bounded by the circle $(x-2)^2 + y^2 = 4$. When we switch to polar coordinated this circle becomes

$$(t\cos(\theta) - 2)^2 - r\sin(\theta) = 2^2 \implies r = 4\cos(\theta).$$

Thus the region is specified by $-\frac{\pi}{2} \le \theta \le \frac{\pi}{2}$ and $0 \le r \le 4\cos(\theta)$ and the integral is computed as follows:

$$\int_{-\frac{\pi}{2}}^{\frac{\pi}{2}} \int_0^{4\cos(\theta)} (r\sin(\theta))r\,dr\,d\theta = \int_{-\frac{\pi}{2}}^{\frac{\pi}{2}} \sin(\theta)\,\frac{r^3}{3}\Big]_0^{4\cos(\theta)} d\theta$$
$$= \int_{-\frac{\pi}{2}}^{\frac{\pi}{2}} \frac{4^3}{3} \sin(\theta)\cos^3(\theta)\,d\theta$$
$$= -\frac{4^3}{3} \cdot \frac{\cos^4(\theta)}{4}\Big]_{-\frac{\pi}{2}}^{\frac{\pi}{2}} = 0.$$

(3) Here $f(x,y) = xy = r^2 \sin(\theta)\cos(\theta)$ and the region is a disk of radius 3 centered at the origin.

$$\int_0^{2\pi}\int_0^{3} r^2 \sin(\theta)\cos(\theta)\cdot r\,dr\,d\theta = \int_0^{2\pi} \sin(\theta)\cos(\theta)\left.\frac{r^4}{4}\right]_0^3 d\theta$$

$$= \frac{81}{4}\left.\frac{\sin^2(\theta)}{2}\right]_0^{2\pi} = 0.$$

(4) The region in this example is bounded by the unit circle centered at the origin,line $y = \sqrt{3}x$, and the x–axis. In order to describe the region in polar form we must first compute where the curves intersect:

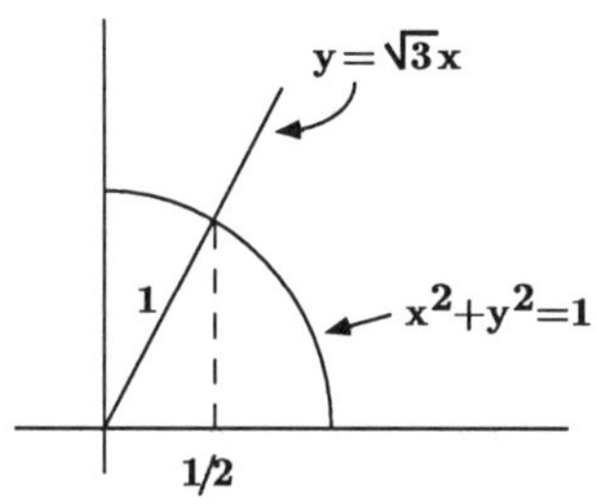

$$y = \sqrt{1-x^2} = \sqrt{3}x \implies 1 = 4x^2 \implies x = \frac{1}{2}.$$

Now this intersection point lies on the circle, hence $r = 1$, and $\cos(\theta) = \frac{1}{2} \implies \theta = \frac{\pi}{3}$. Thus the region is specified by $0 \le r \le 1$ and $0 \le \theta \le \frac{\pi}{3}$ and the integral is computed as follows:

$$\int_0^{\frac{\pi}{3}}\int_0^{1} r\cos(\theta)\cdot r\,dr\,d\theta = \int_0^{\frac{\pi}{3}} \frac{1}{3}\cos(\theta)\,d\theta = \frac{\sqrt{3}}{6}.$$

(5) Here $f(x,y) = \sqrt{x^2+y^2} = \sqrt{r^2} = r$ and

$$\int_0^{2\pi}\int_0^{1+\sin(\theta)} r\cdot r\,dr\,d\theta = \int_0^{2\pi}\left.\frac{r^3}{3}\right]_0^{1+\sin(\theta)} = \frac{5\pi}{3}.$$

(6) The integral in this case is given by

$$\int_0^{2\pi}\int_0^{3+2\sin(\theta)} r\sin(\theta)\cdot r\,dr\,d\theta = \int_0^{2\pi} \sin(\theta)\cdot\frac{(3+2\cos(\theta))^3}{3}\,d\theta$$

$$= \frac{1}{6}\left.\frac{(3+2\cos(\theta))^4}{4}\right]_0^{2\pi} = 0.$$

Solutions for §25.6

(1) The curves $y = x^2$ and $y = x^4$ intersect when $x^2 = x^4 \implies x^2(x^2-1) = 0 \implies x = 0$ or $x = 1$. Since $x^4 \le x^2$ on the interval $0 \le x \le 1$ the area of the region is given by

$$\int_0^1 \int_{x^4}^{x^2} dy\,dx = \int_0^1 (x^2 - x^4)\,dx$$

$$= \frac{x^3}{3} - \frac{x^5}{5}\Big]_0^1 = \frac{1}{3} - \frac{1}{5} = \frac{2}{15}.$$

(2) This problem is entirely analogous to (1) above: Area $= \frac{2}{15}$.

(3) The figure below demonstrates that on the interval $-2 \le x \le -1$ we have $x^2 \le -3x - 2$.

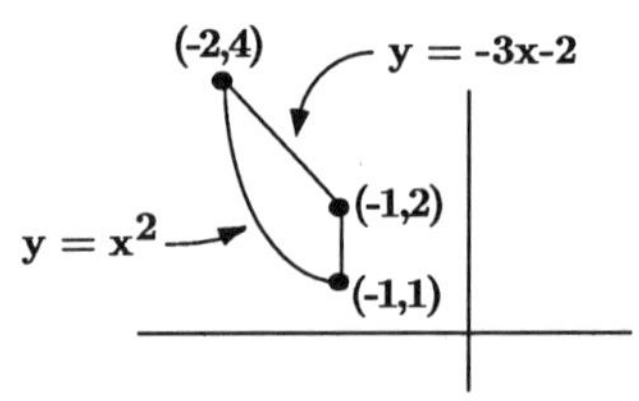

Hence the area of the region is given by

$$\int_{-2}^{-1} \int_{x^2}^{-3x-2} dy\,dx = \int_{-2}^{-1} (-3x - 2 - x^2)\,dx = -\frac{3}{2}x^2 - 2x - \frac{x^3}{3}\Big]_{-2}^{-1} = \frac{1}{6}.$$

(4) In this example we express the curves as functions of y: $x = \sqrt{4 - y^2}$ and $x = \sqrt{4 - (y-2)^2}$. The curves intersect at the points $(\pm\sqrt{3}, 1)$. From the graph below we see that the intersection of the two disks consists of four equivalent regions. We have

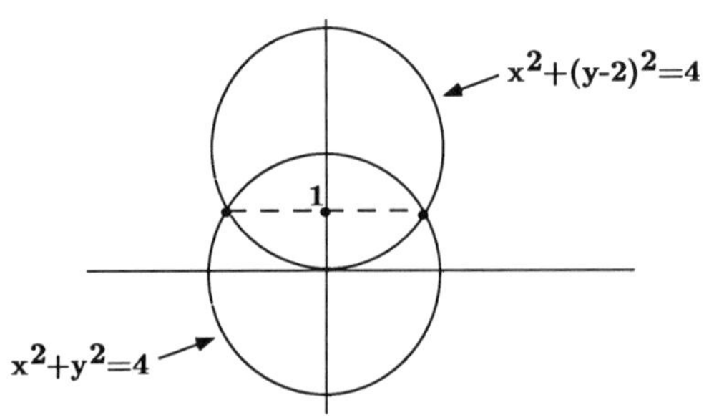

$$\text{Area} = 4 \cdot \int_0^1 \int_0^{\sqrt{4-(y-2)^2}} dx\,dy = 4 \cdot \int_0^1 \sqrt{4 - (y-2)^2}\,dy = \frac{8\pi}{3} - 2\sqrt{3}.$$

(5) Here again we express the curves as functions of y: $x = y$ and $x = 2 - y^2$.

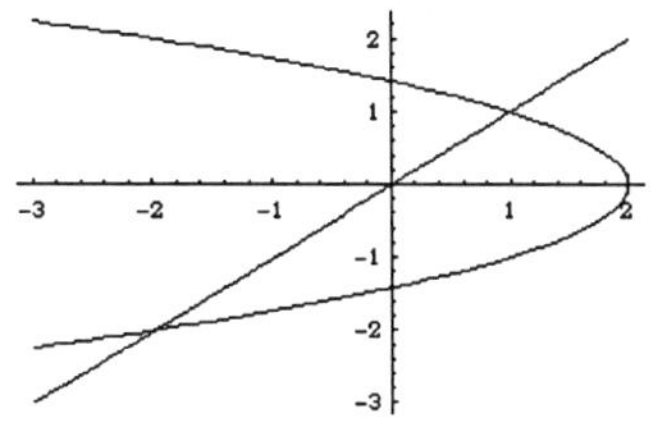

The curves intersect when $2 - y^2 = y \implies y^2 + y - 2 = (y+2)(y-1) = 0 \implies y = -2, y = 1$. Thus the region is specified by $-2 \le y \le 1$ and $y \le 2 - y^2$, and the area is given by

$$\int_{-2}^{1}\int_{y}^{2-y^2} dx\,dy = \int_{-2}^{1} (2 - y^2 - y)\,dy$$

$$= 2y - \frac{y^3}{3} - \frac{y^2}{2}\Big]_{-2}^{1} = \frac{9}{2}.$$

(6) From the figure below we see that the area is infinite.

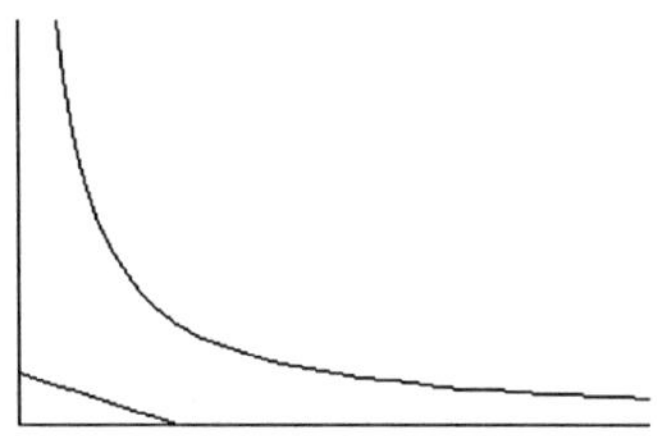

(7) Here we describe the region in polar form: $0 \le \theta \le 2\pi$ and $0 \le r \le r$, and thus the area is given by

$$\int_{0}^{2\pi}\int_{0}^{r} r\,dr\,d\theta = \pi r^2.$$

(8) Since $z = 2x + y - 2$ the volume is given by

$$\int_{1}^{2}\int_{0}^{1} (2x + y - 2)\,dy\,dx = \int_{1}^{2} 2xy + \frac{y^2}{2} - 2y\Big]_{0}^{1} dx$$

$$= \int_{1}^{2} \left(2x - \frac{3}{2}\right) dx$$

$$= x^2 - \frac{3}{2}x\Big]_{1}^{2} = \frac{3}{2}.$$

(9) Here $z = \frac{4-x+2y}{3}$ and the region is given by $-1 \leq x \leq 1$ and $0 \leq y \leq 2$. Thus the volume is

$$\int_0^2 \int_{-1}^1 \left(\frac{4-x+2y}{3}\right) dx\, dy = 8.$$

(10) Here the volume is given by

$$\begin{aligned}\int_0^1 \int_0^1 (x^2 y + e^x)\, dy\, dx &= \int_0^1 \left. x^2 \frac{y^2}{2} + e^x y \right]_0^1 dx \\ &= \int_0^1 \left(\frac{x^2}{2} + e^x\right) dx \\ &= \frac{1}{6} + e - 1 = e - \frac{5}{6}.\end{aligned}$$

(11) $\int_0^{\frac{1}{2}} \int_0^{\frac{1}{2}} (\sin(\pi(x+y)))\, dy\, dx = \frac{2}{\pi^2}$.

(12) The triangular region described here can be looked at as the region bounded by the x–axis, the y–axis, and the line $y = 1 - x$, i.e., $0 \leq x \leq 1$ and $0 \leq y \leq 1 - x$. Hence the volume in question is given by

$$\begin{aligned}\int_0^1 \int_0^{1-x} (x^2 + 2y^2)\, dy\, dx &= \int_0^1 \left. x^2 y + \frac{2}{3} y^3 \right]_0^{1-x} dx \\ &= \int_0^1 \left(x^2(1-x) + \frac{2}{3}(1-x)^3\right) dx = \frac{1}{4}.\end{aligned}$$

(13) The region in the $x\,y$–plane is specified by $0 \leq x \leq 1$ and $x^2 \leq y \leq x$ and

$$\int_0^1 \int_{x^2}^x (x^2 y)\, dy\, dx = \frac{1}{35}.$$

Solutions for §25.7

(1) $x = r\sin(\theta)$, $y = r\cos(\theta)$ and thus

$$\begin{aligned}\frac{\partial(x,y)}{\partial(r,\theta)} &= \frac{\partial x}{\partial r}\frac{\partial y}{\partial \theta} - \frac{\partial x}{\partial \theta}\frac{\partial y}{\partial r} \\ &= -\sin(\theta)\cdot r\sin(\theta) - r\cos(\theta)\cdot\cos(\theta) \\ &= -r(\sin^2(\theta)+\cos^2(\theta)) = -r.\end{aligned}$$

(2) $v = 3u + 2v$, $y = u - v$, thus

$$\begin{aligned}\frac{\partial(x,y)}{\partial(u,v)} &= \frac{\partial x}{\partial u}\frac{\partial y}{\partial v} - \frac{\partial x}{\partial v}\frac{\partial y}{\partial u} \\ &= 3\cdot(-1) - 2\cdot 1 = -5.\end{aligned}$$

(3) $x = ve^u$, $y = ue^v$, and

$$\begin{aligned}\frac{\partial(x,y)}{\partial(u,v)} &= \frac{\partial x}{\partial u}\frac{\partial y}{\partial v} - \frac{\partial x}{\partial v}\frac{\partial y}{\partial u} \\ &= ve^u\cdot ue^v - e^u\cdot e^v \\ &= e^{(u+v)}(u\cdot v - 1).\end{aligned}$$

(4) $x = \cos(u = v)$, $y = \sin(u - v)$, and

$$\begin{aligned}\frac{\partial(x,y)}{\partial(u,v)} &= \frac{\partial x}{\partial u}\frac{\partial y}{\partial v} - \frac{\partial x}{\partial v}\frac{\partial y}{\partial u} \\ &= (-\sin(u+v))\cdot(-\cos(u-v)) - (-\sin(u+v))\cdot(\cos(u-v)) \\ &= 2\sin(u+v)\cdot\cos(u-v).\end{aligned}$$

(5) The Jacobian of the transformation $x = \frac{1}{3}(u+v)$, $y = \frac{1}{3}(v - 2u)$ is

$$\frac{\partial(x,y)}{\partial(u,v)} = \frac{1}{3}\cdot\frac{1}{3} - \frac{1}{3}\left(-\frac{2}{3}\right) = \frac{1}{3}.$$

In the new variables the region is specified by $0 \le v \le 2$ and $0 \le u \le 1$, and we have

$$\begin{aligned}\iint\limits_{\substack{0\le x-y\le 1\\ 0\le 2x+y\le 2}} (2x)\,dx\,dy &= \int_0^1\int_0^2 \frac{2}{3}(u+v)\cdot\frac{1}{3}\,du\,dv \\ &= \frac{2}{9}\int_0^1 \frac{u^2}{2} + vu\Big]_0^2\,dv = \frac{2}{9}\int_0^1 (2+2v)\,dv \\ &= \frac{2}{9}\left(2v + v^2\Big]_0^2\right) = \frac{2}{3}.\end{aligned}$$

(6) In this case the Jacobian is $-\frac{1}{5}$ and we have

$$\iint\limits_{\substack{0\le x+2y\le 2\\ -1\le x-3y\le 1}} (x+y)\,dx\,dy = \int_0^2\int_{-1}^1 \frac{3u+2v}{5}+\frac{u-v}{5}\cdot\left(-\frac{1}{5}\right)du\,dv$$

$$= \frac{1}{25}\int_0^2\int_{-1}^1 (4u-v)\,dv\,du = \frac{16}{25}.$$

(7) Here we set $u=3x-2y$ and $v=x+3y$. Solve for x,y in terms of u,v we see that $y=\frac{3v-u}{11}$ and $x=\frac{2v+3u}{11}$. The Jacobian of this transformation turns out to be $\frac{1}{11}$ and the integral becomes

$$\iint\limits_{\substack{0\le 3x-2y\le 2\\ -1\le x+3y\le 1}} (xy)\,dx\,dy = \int_1^2\int_2^3 \left(\frac{-3u^2+7uv+6v^2}{121}\right)\cdot\left(\frac{1}{11}\right)du\,dv = \frac{1}{11^3}\cdot\frac{229}{4} = \frac{229}{5324}.$$

(8) Here we use the transformation $u=x-2y$, $y=2x+y$ which yields $x=\frac{2v+u}{5}$, $y=\frac{v-2u}{5}$. The Jacobian is $\frac{1}{5}$ and the integral becomes

$$\iint\limits_{\substack{0\le x-2y\le 2\\ 0\le 2x+y\le 1}} e^{2x+y}\,dx\,dy = \int_0^1\int_0^2 \left(e^v\cdot\frac{1}{5}\right)du\,dv = \frac{1}{5}\int_0^1 (2e^v)\,dv = \frac{2}{5}(e-1).$$

(9) The Jacobian of the transformation $x=u+v$, $y=\frac{v}{u+v}$ is $\frac{1}{u+v}$, and the integral becomes

$$\iint\limits_{\substack{1\le x(1-y)\le 3\\ 1\le xy\le 2}} 3x\,dx\,dy = \int_1^2\int_1^3 3(u+v)\cdot\frac{1}{u+v}\,du\,dv = 6.$$

(10) The Jacobian of the transformation $x=2r\cos(\theta)$, $y=3r\sin(\theta)$ is

$$(2\cos(\theta))\cdot(3r\cos(\theta)) - (-2r\sin(\theta))\cdot(3\sin(\theta)) = 6r.$$

Thus the integral is computed as follows:

$$\iint\limits_{\frac{x^2}{4}+\frac{y^2}{9}\le 1} (y^2)\,dx\,dy = \int_0^{2\pi}\int_0^1 9r^2\sin^2(\theta)\cdot 6r\,dr\,d\theta = 54\int_0^{2\pi} \sin^2(\theta)\frac{r^4}{4}\Big]_0^1\,d\theta = \frac{27}{2}\pi.$$

Solutions for Additional Exercises for Chapter 25

(1) $\int_{-1}^{1}\int_{-2}^{2}(3x^2y-4yx^2)\,dy\,dx = 0 = \int_{-1}^{1} 3x^2\frac{y^2}{2}-x^2 2y^2\Big]_{-2}^{2}\,dx = \int_{-1}^{1}\left((6x^2-8x^2)-(6x^2-8x^2)\right)\,dx = 0.$

(2) $\int_{0}^{2}\int_{0}^{3}\frac{xy}{\sqrt{x^2+y^2+2}}\,dy\,dx = \frac{1}{3}\left(2\sqrt{2}-6\sqrt{6}-11\sqrt{11}+15\sqrt{15}\right).$

(3) $\int_{0}^{1}\int_{0}^{x^2}(\ln(xy)+x^2)\,dy\,dx = \int_{0}^{1}\int_{0}^{x^2}(\ln(x)+\ln(y)+x^2)\,dy\,dx = -\frac{7}{15}.$

(4) The curves in question intersect when $x = 3x - x^2 \implies x^2 - 2x = 0 \implies x = 0, 2$. Thus the region in question is given by $0 \le x \le 2$, $x \le y \le 3x - x^2$ and

$$\int_{0}^{2}\int_{x}^{3x-x^2}(x^2 - xy^3)\,dy\,dx = -\frac{72}{35}.$$

(5) The curves $y = \frac{1}{x}$ and $y = 2x$ intersect when $\frac{1}{x} = 2x \implies x = \frac{\sqrt{2}}{2}$. Furthermore, $y = \frac{1}{x}$ intersect $y = x$ when $x = 1$ since $x \ge 0$.

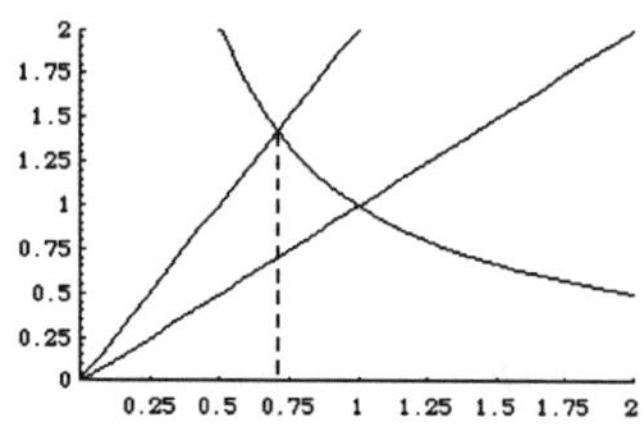

Thus the integral splits into the sum of two integrals as follows:

$$\int_{0}^{\frac{\sqrt{2}}{2}}\int_{x}^{2x}(x^2y^3)\,dy\,dx + \int_{\frac{\sqrt{2}}{2}}^{1}\int_{x}^{\frac{1}{x}}(x^2y^3)\,dy\,dx = \frac{2(\sqrt{2}-1)}{7}.$$

(6) $\int_{-2}^{2}\int_{-\sqrt{4-x^2}}^{\sqrt{4-x^2}}(4-y)\,dy\,dx = \int_{-2}^{2}\int_{-\sqrt{4-y^2}}^{\sqrt{4-y^2}}(4-y)\,dx\,dy = 16\pi.$

(7) $\int_{\frac{1}{4}}^{1}\int_{x^2}^{x}\sqrt{\frac{x}{y}}\,dy\,dx = \int_{\frac{1}{4}}^{1}\int_{y}^{\sqrt{y}}\sqrt{\frac{x}{y}}\,dx\,dy + \int_{\frac{1}{16}}^{\frac{1}{4}}\int_{y}^{\frac{1}{4}}\sqrt{\frac{x}{y}}\,dx\,dy = \frac{13}{80}.$

(8) $\int_0^{\frac{\pi}{2}} \int_0^{y^2} e^{\frac{x}{y^2}} \, dx \, dy = \int_0^{\frac{\pi^2}{4}} \int_{\sqrt{x}}^{\frac{\pi}{2}} e^{\frac{x}{y^2}} \, dy \, dx = \frac{(e-1)\pi^3}{24}.$

(9) $\int_0^{\pi} \int_0^{6\sin(3\theta)} r \, dr \, d\theta = 9\pi.$

(10) $\int_0^{\pi} \int_0^{6\sin(3\theta)} \frac{r}{1+r^2} \, dr \, d\theta = \frac{1}{2} \int_0^{\pi} \ln\left(1 + 36\sin^2(3\theta)\right) \, d\theta.$

(11) $\int_0^{\pi} \int_0^{7\sin(5\theta)} r^4 \cos^2(\theta)\sin(\theta) \, dr \, d\theta = \frac{16807}{5} \int_0^{\pi} \sin^5(5\theta)\cos^2(\theta)\sin(\theta) \, d\theta.$

(12) $\int_0^{2\pi} \int_0^{\frac{2}{3}\cos(4\theta)} \frac{r^3\cos^2(\theta)}{1+r^2\sin^2(\theta)} \, dr \, d\theta.$

(13) $\int_0^{2\pi} \int_0^{4} r \, dr \, d\theta = 16\pi. \; \frac{255\pi}{4}.$

(14) $\int_0^{1} \int_0^{1} (x^4 + y^4) \, dy \, dx = \frac{2}{5}.$

(15) $\frac{\partial(x,y)}{\partial(u,v)} = -2e^v\left(u^2 + v\right).$

(16) $\frac{\partial(x,y)}{\partial(u,v)} = \frac{1}{2}.$

(17) To compute this integral we set $u = 2x - xy$ and $v = 2xy$. Thus

$$u = 2x - \frac{v}{2}, y = \frac{2v}{2u+v} \implies \frac{\partial(x,y)}{\partial(u,v)} = \frac{1}{2u+v} \implies 4y\frac{\partial(x,y)}{\partial(u,v)} = \frac{8v}{(2u+v)^2},$$

and

$$\iint_{\substack{1\le x(2-y)\le 4\\ 1\le 2xy\le 3}} 4y \, dx \, dy = \int_1^3 \int_1^4 \left(\frac{8v}{(2u+v)^2}\right) du \, dv = -8(-\ln(3) + \ln(5)) + 32(-\ln(9) + \ln(11)).$$

(18) Here we set $u = x - 3y$ and $v = 2x + 1$ which yields

$$x = \frac{v-1}{2}, y = \frac{v-2u-1}{6}, \implies \frac{\partial(x,y)}{\partial(u,v)} = \frac{1}{6}.$$

$$\iint_{\substack{0\le x-3y\le 4\\ 0\le 2x+1\le 4}} \exp(3x+2y) \, dx \, dy = \int_0^4 \int_0^4 e^{\frac{11v-4u-11}{6}} \, du \, dv = \frac{3}{22}\left(e^{-\frac{9}{2}} - e^{-\frac{-11}{6}} - e^{\frac{17}{6}} + e^{\frac{11}{2}}\right).$$

Solutions for §26.1

(1)

$$\int_0^1\int_0^1\int_0^1 (xy^2z)\,dx\,dy\,dz = \int_0^1\int_0^1 \left(y^2z\frac{x^2}{2}\Big]_0^1\right)\,dy\,dz = \frac{1}{2}\int_0^1\int_0^1 (y^2z)\,dy\,dz$$

$$= \frac{1}{2}\int_0^1 \left(z\frac{y^3}{3}\Big]_0^1\right)\,dz = \frac{1}{6}\int_0^1 z\,dz = \frac{1}{12}.$$

(2) Here we have

$$\int_0^1\int_{-1}^0\int_{-1}^2 (x^2+y^2+z^2)\,dx\,dy\,dz = \int_0^1\int_{-1}^0 \left(y^2x + z^2x + \frac{x^3}{3}\Big]_{-1}^2\right)\,dy\,dz$$

$$= \int_0^1\int_{-1}^0 \left(2y^2+2z^2+\frac{8}{3}\right) - \left(-y^2-z^2-\frac{1}{3}\right)\,dy\,dz$$

$$= \int_0^1\int_{-1}^0 (3y^2+3z^2+3)\,dy\,dz$$

$$= \int_0^1 \left(y^3+3z^2y+3y\Big]_{-1}^0\right) = \int_0^1 -(-1-3z^2-3)\,dz$$

$$= \int_0^1 (4+3z^2)\,dz = 4z+z^3\Big]_0^1 = 5.$$

(3)

$$\int_{-1}^1\int_2^3\int_0^1 (x^2y+xz^3)\,dx\,dy\,dz = \int_{-1}^1\int_2^3 \left(\frac{x^3}{3}y + \frac{x^2}{2}z^3\Big]_0^1\right)\,dy\,dz$$

$$= \int_{-1}^1\int_2^3 \left(\frac{y}{3}+\frac{z^3}{2}\right)\,dy\,dz$$

$$= \int_{-1}^1 \left(\frac{y^2}{6}+\frac{z^3}{2}y\Big]_2^3\right)\,dz$$

$$= \int_{-1}^1 \left(\frac{5}{5}-\frac{1}{2}z^3\right)\,dz = \frac{5}{3}.$$

(4)

$$\int_0^1\int_1^3\int_{-1}^1 (e^x(y+z))\,dx\,dy\,dz = \int_0^1\int_1^3 \left((y+z)e^x\Big]_{-1}^1\right)dy\,dz$$

$$= (e+e^{-1})\int_0^1 \left(\frac{y^2}{2}+zy\Big]_1^3\right)dz$$

$$= (e+e^{-1})\int_0^1 (4+2z)\,dz = 5(e+e^{-1}).$$

(5) $\int_3^4\int_2^3\int_0^1 \left((y^2+z)x^{-\frac{1}{2}}\right)dx\,dy\,dz = \frac{59}{3}$.

(6) $\int_0^2\int_0^1\int_{-2}^1 = \left(yz\cos\left(\frac{\pi xy}{2}\right)\right)dx\,dy\,dz = \frac{16}{\pi^2}$.

Solutions for §26.2

(1) $\int_{-\frac{1}{2}}^{\frac{1}{2}}\int_{-\frac{1}{2}}^{\frac{1}{2}}\int_{-\frac{1}{2}}^{\frac{1}{2}} (x^2+y)^2z^2\,dx\,dy\,dz = \frac{1}{144}$.

(2) $\int_0^\pi\int_0^1\int_1^3 (x-y)\sin(z)\,dx\,dy\,dz = 6$.

(3) The solid Ω may be expressed as the set of points (x,y,z) satisfying $0 \le x \le 1$, $0 \le y \le 3$ and $2 \le z \le s_2(x,y)$ where $z = s_2(x,y)$ is the plane going through the points $(0,0,2)$, $(1,0,2)$, $(1,3,4)$. The equation of this plane is given by $z = 2+\frac{2}{3}y$. Our integral is:

$$\int_0^3\int_0^1\left(\int_2^{2+\frac{2}{3}y} 2x\,dz\right)dx\,dy = \int_0^3\int_0^1 \frac{4}{3}xy\,dx\,dy = 3.$$

(4) $\int_0^1\int_{-1}^1\left(\int_0^{x^2+1} xyz\,dz\right)dx\,dy = \int_0^1\int_{-1}^1 \frac{1}{2}xy(x^2+1)^2\,dx\,dy = \frac{1}{4}\int_{-1}^1 x(x^2+1)^2\,dx = 0.$

(5) $\int_2^3\int_{-1}^2\left(\int_0^{x^2+y^2} (y+2z)\,dz\right)dx\,dy = \frac{4549}{20}$.

(6) The solid Ω may be expressed as the set of points (x, y, z) satisfying $0 \le x \le 2$, $0 \le y \le 3$ and $s_1(x, y) \le z \le s_2(x, y)$ where $z = s_1(x, y)$ is the plane going through $(2, 0, 4)$, $(0, 0, 4)$, $(0, 3, 3)$ given by $z = s_1(x, y) = -\frac{1}{3}y + 4$, and $z = s_2(x, y)$ is the plane going through $(2, 0, 4)$, $(0, 0, 4)$, $(0, 3, 7)$ given by $z = s_2(x, y) = y + 4$. Our integral is:

$$\int_0^3 \int_0^2 \left(\int_{-\frac{1}{3}y+4}^{y+4} yz\, dz \right) dx\, dy = \int_0^3 \int_0^2 y \cdot \frac{(y+4)^2 - \left(-\frac{1}{3}y + 4\right)^2}{2}\, dx\, dy = 114.$$

(7) The upper and lower surfaces intersect when $\sqrt{1 - x^2 - y^2} = \sqrt{2x^2 + 2y^2}$, i.e., $x^2 + y^2 = \frac{1}{3}$. Thus we may express the solid region as $x^2 + y^2 \le \frac{1}{3}$, $\sqrt{2x^2 + 2y^2} \le z \le \sqrt{1 - x^2 - y^2}$, and our integral is

$$\iint_{x^2+y^2 \le \frac{1}{3}} \left(\int_{\sqrt{2x^2+2y^2}}^{\sqrt{1-x^2-y^2}} 2\, dz \right) dx\, dy = \iint_{x^2+y^2 \le \frac{1}{3}} \left(2\sqrt{1 - x^2 - y^2} - 2\sqrt{2x^2 + 2y^2} \right) dx\, dy$$

$$= \int_0^{\frac{1}{3}} \int_0^{2\pi} \left(2\sqrt{1 - r^2} - 2\sqrt{2r^2} \right) r\, dr\, d\theta$$

$$= 4\pi \int_0^{\frac{1}{3}} \left(\sqrt{1 - r^2} - \sqrt{2}r \right) r\, dr = \frac{4\pi(27 - 17\sqrt{2})}{81}.$$

(8) The surfaces $z = x^2 + y^2$ and $z = 2 - x^2 - y^2$ intersect when $x^2 + y^2 = 2 - x^2 - y^2 \implies x^2 + y^2 = 1$. The solid region Ω may be expressed as $x^2 + y^2 \le 1$, $x^2 + y^2 \le z \le 2 - x^2 - y^2$and our integral is

$$\iint_{x^2+y^2 \le 1} \left(\int_{x^2+y^2}^{2-x^2-y^2} 3\, dz \right) dx\, dy = 3\pi.$$

(9) The solid tetrahedron Ω is the set of points (x, y, z) so that $x + y \le 1, x \ge 0, y \ge 0, 0 \le z \le 1 - x - y$.

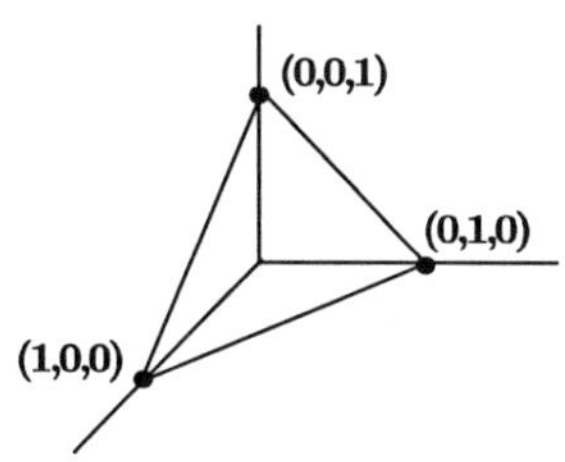

Our integral is

$$\iint_{\substack{x+y \le 1 \\ 0 \le x, y}} \left(\int_0^{1-x-y} 3\, dz \right) dx\, dy = \iint_{\substack{x+y \le 1 \\ 0 \le x, y}} 3(1 - x - y)\, dx\, dy = \int_0^1 \int_0^{1-y} 3(1 - x - y)\, dx\, dy = \frac{1}{2}.$$

(10) The volumes are **(4)** $\frac{8}{3}$, **(5)** 22, **(6)** 12, **(7)** $\frac{2\pi(27-17\sqrt{2})}{81}$, **(8)** π, **(9)** $\frac{1}{6}$.

Solutions for §26.3

We recall that the Jacobian of the transformation $x = g_1(u,v,w), y = g_2(u,v,w), z = g_3(u,v,w)$ *is given by*

$$\frac{\partial(x,y,z)}{\partial(u,v,w)} = \left|\frac{\partial x}{\partial u}\left(\frac{\partial y}{\partial v}\cdot\frac{\partial z}{\partial w} - \frac{\partial z}{\partial v}\cdot\frac{\partial y}{\partial w}\right) - \frac{\partial y}{\partial u}\left(\frac{\partial x}{\partial v}\cdot\frac{\partial z}{\partial w} - \frac{\partial z}{\partial v}\cdot\frac{\partial x}{\partial w}\right) + \frac{\partial z}{\partial u}\left(\frac{\partial x}{\partial v}\cdot\frac{\partial y}{\partial w} - \frac{\partial y}{\partial v}\cdot\frac{\partial x}{\partial w}\right)\right|$$

(1) $\frac{\partial(x,y,z)}{\partial(u,v,w)} = 11.$

(2) $\frac{\partial(x,y,z)}{\partial(u,v,w)} = 56.$

(3) $\frac{\partial(x,y,z)}{\partial(u,v,w)} = |a_1b_2c_3 + a_2b_3c_1 + a_3b_1c_2 - a_3b_2c_1 - a_2b_1c_3 - a_1b_3c_2|.$

(4) $\frac{\partial(x,y,z)}{\partial(u,v,w)} = |4(u^2v^2 + u^2w^2 - v^2w^2)|.$

(5) $\frac{\partial(x,y,z)}{\partial(u,v,w)} = |2uw - w^2|.$

(6) The Jacobian of the given change of variable is $\frac{\partial(x,y,z)}{\partial(u,v,w)} = 30$ and thus

$$\iiint\limits_{\frac{x^2}{4}+\frac{y^2}{25}+\frac{z^2}{9}\leq 1} dx\,dy\,dz = \iiint\limits_{u^2+v^2+w^2\leq 1} 30\,du\,dv\,dw.$$

The latter integral is visibly the volume of the sphere of radius 1 times 30, i.e., the integral is 40π.

(7) In this case $\frac{\partial(x,y,z)}{\partial(u,v,w)} = 1$ and the integral is

$$\int_0^2\int_0^2\int_0^1 (uvw - v^2w - vw^2)\,du\,dv\,dw = -\frac{26}{3}.$$

(8) Here $\frac{\partial(x,y,z)}{\partial(u,v,w)} = \frac{1}{3}$, and the integral is given by

$$\int_{-1}^0\int_0^1\int_1^2 \left(\frac{2u + 2w - v}{3}\cdot\frac{1}{3}\right) du\,dv\,dw = \frac{1}{6}.$$

Solutions for §26.4

(1) $\int_0^{2\pi}\int_0^2\int_2^3 r^3\cos^2(\theta)\,dz\,dr\,d\theta = \frac{75\pi}{8}.$

(2) $\int\limits_0^{2\pi}\int\limits_0^1\int\limits_{-1}^1 r^2\sin(\theta)\,dx\,dr\,d\theta = 0.$

(3) $\int\limits_0^{2\pi}\int\limits_0^{\pi}\int\limits_1^{\sqrt{2}} r^2\sin(\phi)\,dr\,d\phi\,d\theta = \frac{4\pi}{3}(2\sqrt{2}-1).$

(4) $\int\limits_0^{2\pi}\int\limits_0^{\pi}\int\limits_1^{\sqrt{2}} r^3\sin^2(\phi)\cos(\theta)\sin(\theta)\cos(\phi)\cdot r^2\sin(\phi)\,dr\,d\phi\,d\theta = 0.$

(5) A pipe of radius 2 and of height 2.

(6) A segment of a pipe of height 1.

(7) The plane given by the equation $x+y+z=1$.

(8) The bottom part of the solid obtained by revolving the parabola around the vertical axis, of height 4.

(9) The bottom part of the solid obtained by revolving the parabola around the vertical axis, of height 1.

(10) The plane defined by the equation $x=1$.

(11) The sphere of radius 1, centered at the origin.

(12) The segment of the sphere of radius 3, corresponding to positive values of x and y.

(13) The intersection of the cone $z=\sqrt{x^2+y^2}$ with a unit ball centered at the origin.

(14) The unit ball centered at the origin, except for the points inside the cone $z=-\sqrt{x^2+y^2}$.

(15) The plane defined by the condition $z=1$.

(16) The unit sphere centered at the point $(0,0,1)$.

Solutions for Additional Exercises for Chapter 26

(1) $\int\limits_{-1}^1\int\limits_0^1\left(\int\limits_{-1}^1 x\frac{\cos^2(\pi zy)}{x^2+y^2}\,dx\right)dy\,dz = \int\limits_{-1}^1\int\limits_0^1\left(\cos^2(\pi zy)\cdot\frac{\ln(x^2+y^2)}{2}\Big]_{-1}^1\right)dy\,dz = 0.$

(2) $\int\limits_1^3\int\limits_x^{x^2}\left(\int\limits_0^{\ln x} xe^y\,dy\right)dz\,dx = \int\limits_1^3\left(\int\limits_x^{x^2}(x^2-x)\,dz\right)dx = \int\limits_1^3(x^2-x)^2\,dx = \frac{256}{15}.$

(3) $\int\limits_0^1\int\limits_0^1\left(\int\limits_{-1}^{e^y}(z\ln(x))\,dx\right)dy\,dz = \int\limits_0^1\int\limits_0^1\left(z(x\ln(x)-x)\Big]_1^{e^y}\right)dy\,dz = \int\limits_0^1\int\limits_0^1 z(ye^y-e^y+1)\,dy\,dz = 1+\frac{1-e}{2}.$

(4) We set $y = r\sin(\theta)$, $z = r\cos(\theta)$, $x = x$. The solid wedge ω depicted below can be be expressed as the set of all points (x, r, θ) where $0 \le \theta \le \frac{\pi}{2}$, $0 \le r \le \frac{\sqrt{2}}{2}$, $0 \le x \le r\sin(\theta)$.

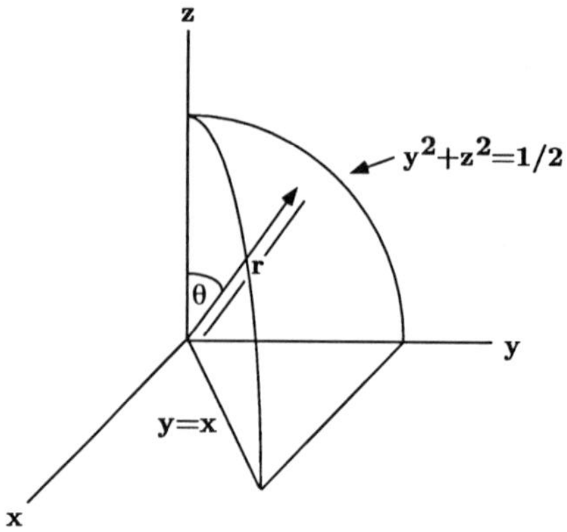

$$\text{Volume} = \int_0^{\frac{\sqrt{2}}{2}} \int_0^{\frac{\pi}{2}} \int_0^{r\sin(\theta)} r\,dx\,dr\,d\theta = \frac{\sqrt{2}}{12}.$$

(5) The parabaloids $z = 4x^2 + 4y^2$ and $z = 6 - 7x^2 - y^2$ intersect when $4x^2 + 4y^2 = 6 - 7x^2 - y^2 \implies 11x^2 + 5y^2 = 6$. Thus

$$\begin{aligned}\text{Volume} &= \iint_{11x^2+5y^2\le 6} \left(\int_{z=4x^2+4y^2}^{6-7x^2-y^2} dz \right) dx\,dy \\ &= \iint_{11x^2+5y^2\le 6} (6 - 11x^2 - 5y^2)\,dx\,dy \\ &= \frac{1}{\sqrt{55}} \iint_{x^2+y^2\le 6} (6 - x^2 - y^2)\,dx\,dy = \frac{18\pi}{\sqrt{55}}.\end{aligned}$$

(6) The parabaloids $z = 4x^2 + y^2$ and $z = 4 - 3y^2$ intersect when $x^2 + y^2 = 1$. Then

$$\begin{aligned}\text{Volume} &= \iint_{x^2+y^2\le 1} \left(\int_{z=4x^3+y^2}^{4-3y^2} dz \right) dx\,dy \\ &= 4 \iint_{x^2+y^2\le 1} (1 - x^2 - y^2)\,dx\,dy = 2\pi.\end{aligned}$$

(7)

$$\begin{aligned}\text{Volume} &= \iint_{x^2+y^2\le 9} \left(\int_{z=0}^{9-x^2-y^2} dz \right) dx\,dy \\ &= \int_0^{2\pi} \int_0^{3} \int_0^{9-r^2} r\,dz\,dr\,d\theta = \frac{81}{2}\pi.\end{aligned}$$

(**8**) Volume $= \int_0^{2\pi}\int_0^{1}\int_{r^2}^{10} r\,dz\,dr\,d\theta = \frac{19\pi}{2}$.

(**9**) Volume $= \int_0^{2\pi}\int_0^{\sqrt{2}}\int_{r^2}^{\sqrt{10-r^2}} r\,dz\,dr\,d\theta = 2\pi\left[\frac{\sqrt{2}}{6}\left(10^{\frac{3}{2}} - 8^{\frac{3}{2}}\right) - 1\right]$.

(Above the limit $\sqrt{2}$ the page also prints $\frac{1}{\sqrt{2}}$.)

(**10**) Volume $= \int_0^{2\pi}\int_0^{a}\int_{\frac{5r}{a}}^{5} r\,dz\,dr\,d\theta = \frac{5}{3}\pi a^2$.

(**11**) Volume $= \int_0^{\frac{\pi}{2}}\int_0^{4}\int_{\frac{\pi}{4}}^{\frac{\pi}{3}} \rho^2\sin(\phi)\,d\phi\,d\rho\,d\theta = \frac{16}{3}\left(\sqrt{2}-1\right)\pi$.

(**12**) Assuming $a < \frac{1}{2}$, volume $= \int_0^{2\pi}\int_{\frac{\pi}{3}}^{2\pi}\int_0^{2a} \rho^2\sin(\phi)\,d\rho\,d\phi\,d\theta + \int_0^{2\pi}\int_0^{\frac{\pi}{3}}\int_0^{a\sec(\phi)} \rho^2\sin(\phi)\,d\rho\,d\phi\,d\theta = 9\pi a^3$.

(**13**)

$$\overline{x} = \frac{1}{\text{Volume}(\Omega)}\iint_{11x^2+5y^2\le 6}\left(\int_{z=4x^2+4y^2}^{6-7x^2-y^2} dz\right) x\,dx\,dy$$

$$= \frac{1}{\left(\frac{18\pi}{\sqrt{55}}\right)}\iint_{11x^2+5y^2\le 6}(6-11x^2-5y^2)\cdot x\,dx\,dy$$

$$= \frac{\sqrt{55}}{18\pi}\cdot\frac{1}{\sqrt{55}}\iint_{x^2+y^2\le 6}(6-11x^2-5y^2)\cdot\frac{1}{\sqrt{11}}\,dx\,dy$$

$$= \frac{1}{18\sqrt{11}\cdot\pi}\int_0^{2\pi}\int_0^{\sqrt{6}}(6-r^2)r\cos(\theta)\,r\,dr\,d\theta = 0.$$

Similarly, $\overline{y} = 0$ while

$$\overline{z} = \sqrt{55}\cdot 18\pi\iint_{x^2+y^2\le 6}\left(\int_{z=4x^2+4y^2}^{6-7x^2-y^2} z\,dz\right)dx\,dy$$

$$= \sqrt{55}\cdot 18\pi\iint_{x^2+y^2\le 6}\left(\frac{(6-7x^2-y^2)^2}{2} - \frac{(4x^2+4y^2)^2}{2}\right)dx\,dy = \frac{174}{55}.$$

(**14**) $\overline{x} = \overline{y} = 0$, and $\overline{z} = \frac{2}{81\pi}\int_0^{2\pi}\int_0^{3}\int_0^{9-r^2} rz\,dz\,dr\,d\theta = 3$.

(15) Here, we have $\overline{x} = \overline{y} = \frac{3}{16(\sqrt{2}-1)\pi} \int\limits_0^{\frac{\pi}{2}} \int\limits_0^4 \int\limits_{\frac{\pi}{4}}^{\frac{\pi}{3}} \rho^3 \sin^2(\phi)\cos(\theta)\, d\phi\, d\rho\, d\theta = \frac{\sqrt{2}+1}{2} + \frac{3(\sqrt{2}+1)}{\pi}\left(1 - \frac{\sqrt{3}}{2}\right).$

$\overline{z} = \frac{3}{16(\sqrt{2}-1)\pi} \int\limits_0^{\frac{\pi}{2}} \int\limits_0^4 \int\limits_{\frac{\pi}{4}}^{\frac{\pi}{3}} \rho^3 \sin(\phi)\cos(\phi)\, d\phi\, d\rho\, d\theta = \frac{3}{4}(\sqrt{2}+1).$

(16) $\bar{x} = \bar{y} = 0.$

$\bar{z} = \frac{1}{2\pi\left[\frac{\sqrt{2}}{6}\left(10^{3/2}-8^{3/2}\right)-1\right]} \int\limits_0^{2\pi} \int\limits_0^{\sqrt{2}} \int\limits_0^{\sqrt{5-\frac{1}{2}r^2}} rz\, dz\, dr\, d\theta = \frac{9}{8\left[\frac{\sqrt{2}}{6}\left(10^{\frac{3}{2}}-8^{\frac{3}{2}}\right)-1\right]}.$

(17) Volume $= \frac{h}{3}\pi a^2$. $\overline{x} = \overline{y} = 0$. $\overline{z} = \frac{3}{h\pi a^2} \int\limits_0^{2\pi} \int\limits_0^a \int\limits_{\frac{hr}{a}}^h (rz)\, dz\, dr\, d\theta = \frac{3}{4}h.$

Solutions for §27.1

(1)

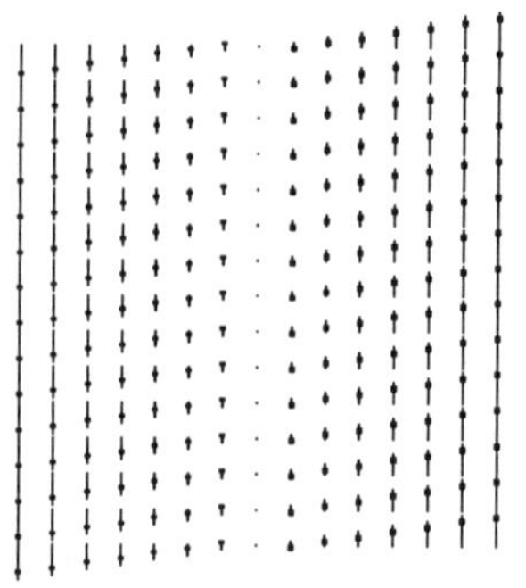

$\vec{v} = x\vec{j}$

(2)

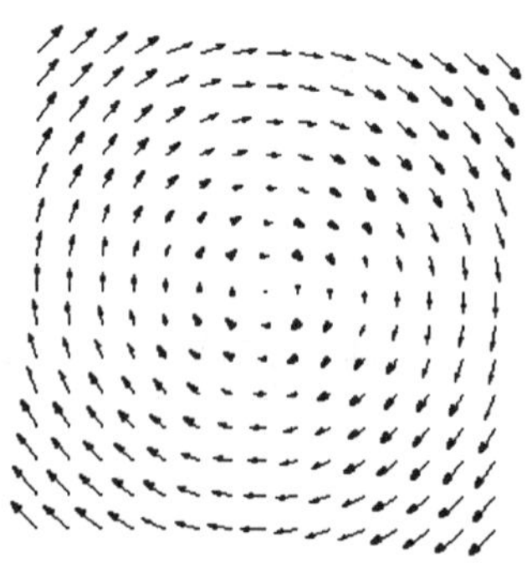

$\vec{v} = y\vec{i} - x\vec{j}$

(3)

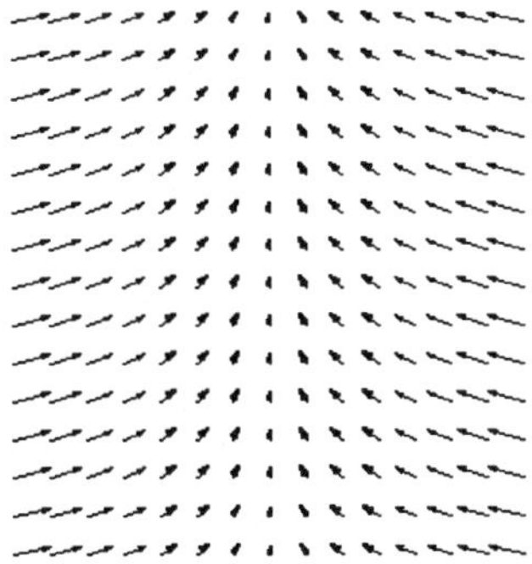

$$\vec{v} = -2x\vec{i} + \vec{j}$$

(4)

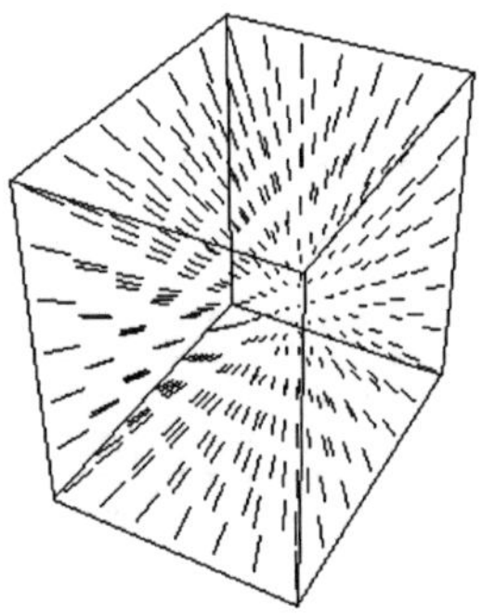

$$\vec{v} = x\vec{i} + y\vec{j} + z\vec{k}$$

(5)

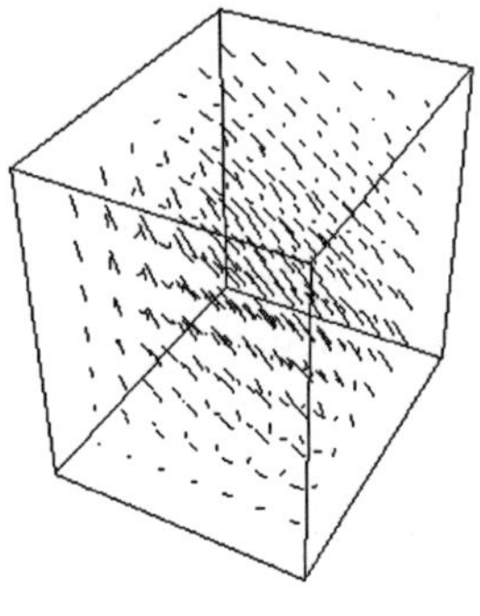

$$\vec{v} = (z-y)\vec{i} + (x-z)\vec{j} + (y-x)\vec{k}$$

(6 a)

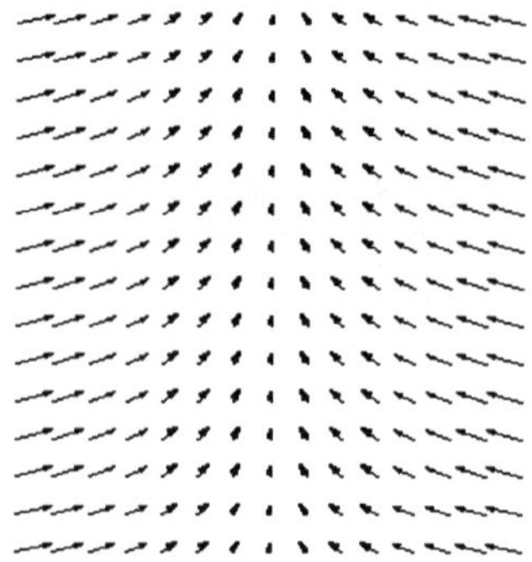

$$\vec{v} = \overrightarrow{\nabla}(y - x^2) = -2x\vec{i} + \vec{j}$$

(6 b)

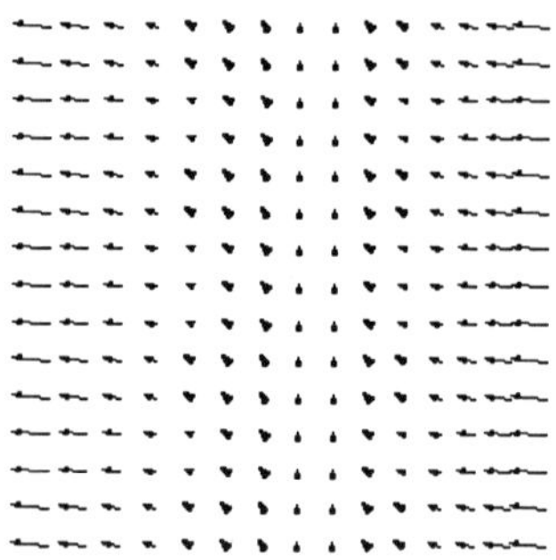

$$\vec{v} = \overrightarrow{\nabla}(y - x^3) = -3x^2\vec{i} + \vec{j}$$

(6 c)

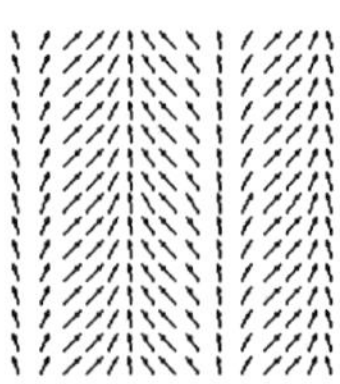

$$\vec{v} = \overrightarrow{\nabla}(y - \sin(x)) = -\cos(x)\vec{i} + \vec{j}$$

(7)

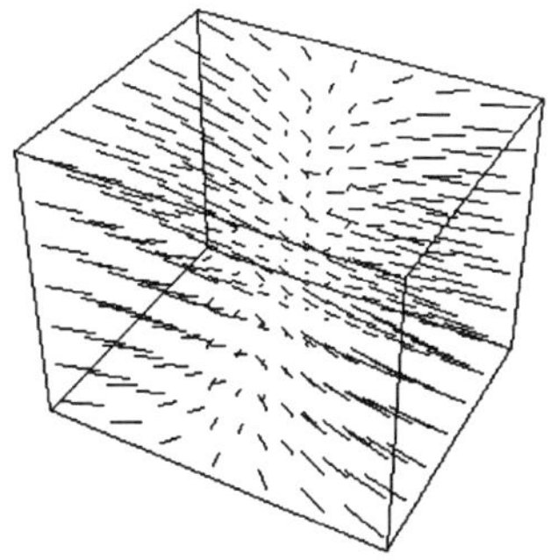

$$\vec{v} = \vec{\nabla}(x^2 + \frac{y^2}{4} + \frac{z^2}{9}) = 2x\vec{i} + \frac{1}{2}y\vec{j} + \frac{2}{9}z\vec{k}$$

(8)

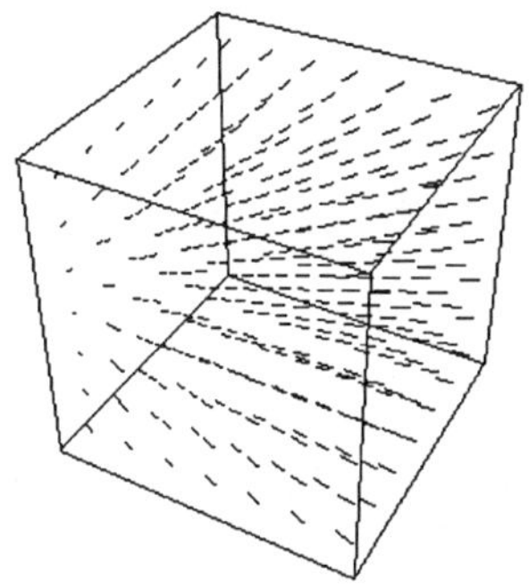

$$\vec{v} = \vec{\nabla}(x - y + z) = \vec{i} - \vec{j} + \vec{k}$$

(9) Since $\frac{\partial}{\partial y}y^2 = \frac{\partial}{\partial x}2xy \implies \vec{v} = y^2\vec{i} + 2xy\vec{j}$ is a gradient vector field. To compute a potential function, ϕ, we write

$$\phi(x, y) = \int \frac{\partial \phi}{\partial x}\, dx = \int y^2\, dx = y^2 x + c(y).$$

Then $\frac{\partial \phi}{\partial y} = 2xy + c'(y) \implies c'(y) = 0 \implies c(y) = C$ a constant. Thus, $\phi(x, y) = y^2 x + C$.

(10) Since $\frac{\partial}{\partial y}3x^2 y = 3x^2 \neq \frac{\partial}{\partial x}2x^3 \implies \vec{v}$ is not a gradient vector field.

(11) $\frac{\partial}{\partial y}(6x) = \frac{\partial}{\partial x}(-4y) = 0 \implies \vec{v} = 6x\vec{i} - 4y\vec{j}$ is a gradient vector field. Then $\phi(x, y) = \int \frac{\partial \phi}{\partial x}\, dx = \int 6x\, dx = 3x^2 + c(y)$. Further $\frac{\partial \phi}{\partial y} = c'(y) = -4y \implies c(y) = -2y^2 + C$. Thus $\phi(x, y) = 3x^2 - 2y^2 + C$.

(12) $\frac{\partial}{\partial y}(y\cos(xy)) = \frac{\partial}{\partial x}(x\cos(xy)) \implies \vec{v} = y\cos(xy)\vec{i} + x\cos(xy)\vec{j}$ is a gradient vector field. Then, $\phi(x, y) = \int \frac{\partial \phi}{\partial x}\, dx = \int y\cos(xy)\, dx = \sin(xy) + c(y)$. Further, $\frac{\partial}{\partial y} = x\cos(xy) + c'(y) = x\cos(xy) \implies c(y) = C$ and $\phi(x, y) = \sin(xy) + C$.

(13) Not a gradient vector field.

(14) $\phi(x, y) = y^2 \ln(x) + C$.

(15) Not a gradient vector field.

(16) Here $\vec{v} = yz\vec{i} + xz\vec{j} + xy\vec{k}$. Since $\text{curl}(\vec{v}) = 0 \implies \vec{v}$ is a gradient vector field. Let $\phi(x, y, z)$ be a potential function for $\vec{v}$. Since $\frac{\partial\phi}{\partial x} = yz \implies \phi(x, y, z) = xyz + c(y, z)$. But $\frac{\partial\phi}{\partial y} = xz \implies xz + \frac{\partial c(yz)}{\partial y} = xz \implies \frac{\partial c(y,z)}{\partial y} = 0 \implies c(y, z) = c(z)$. So $\phi(x, y, z) = xyz + c(z)$. Finally, $\frac{\partial\phi}{\partial z} = xy \implies c(z) = C$ and $\phi(x, y, z) = xyz + C$.

(17) $\phi(x, y, z) = x^2 + yz + C.$

(18) $\phi(x, y, z) = y^2 z \sin(x) + C.$

Solutions for §27.2

Let $\vec{\gamma} = \overrightarrow{(\gamma_1(t), \gamma_2(t), \gamma_3(t))}$ *be a space curve for* $a \leq t \leq b$. *Let* $\vec{F} = f_1\vec{i} + f_2\vec{j} + f_3\vec{k}$ *where* f_1, f_2, f_3 *are functions of* x, y, z. *Then the line integral of* $\vec{F}$ *over gamma is defined to be*

$$\int_{\gamma} \vec{F} \cdot \vec{dr} = \int_a^b \vec{F}(\gamma(t)) \cdot \vec{\gamma}'(t)\, dt.$$

(1) $\vec{F} = x^2y\vec{i} + 2xyz\vec{k}$, $\overrightarrow{\gamma(t)} = \overrightarrow{(-1, t, 0)}$ for $0 \leq t \leq 1$. We compute

$$\int_{\gamma} \vec{F} \cdot \vec{dr} = \int_0^1 \overrightarrow{(t, 0, 0)} \cdot \overrightarrow{(0, 1, 0)}\, dt = 0.$$

(2) $\vec{F} = x^2y\vec{i} + 2xyz\vec{k}$, $\overrightarrow{\gamma(t)} = \overrightarrow{(t, t^2, 1)}$ for $0 \leq t \leq 1$. We compute

$$\int_{\gamma} \vec{F} \cdot \vec{dr} = \int_0^1 \overrightarrow{(t^4, 0, 2t^3)} \cdot \overrightarrow{(1, 2t, 0)}\, dt = \int_0^1 t^4\, dt = \frac{1}{5}.$$

(3) $\vec{F} = 3xz\vec{i} + x^2\vec{j} + 2yz\vec{k}$, $\overrightarrow{\gamma(t)} = \overrightarrow{(t, t, t)}$ for $1 \leq t \leq 2$. We compute

$$\int_{\gamma} \vec{F} \cdot \vec{dr} = \int_0^1 \overrightarrow{(3t^2, t^2, 2t^2)} \cdot \overrightarrow{(1, 1, 1)}\, dt = \int_0^1 6t^2\, dt = 14.$$

(4) $\vec{F} = 3xzz\vec{i} + x^2\vec{j} + 2yz\vec{k}$, $\overrightarrow{\gamma(t)} = \overrightarrow{(3, t, 2t^2)}$ for $0 \leq t \leq 1$. We compute

$$\int_{\gamma} \vec{F} \cdot \vec{dr} = \int_0^1 \overrightarrow{(18t^2, 9, 4t^3)} \cdot \overrightarrow{(0, 1, 4t)}\, dt = \int_0^1 (9 + 16t^4)\, dt = \frac{61}{5}.$$

(5) $\vec{F} = \overrightarrow{(y\sqrt{x}, -z\sqrt{y}, x)}$, $\overrightarrow{\gamma(t)} = \overrightarrow{(1, t^3, 5)}$ for $0 \le t \le 4$. Thus

$$\int_\gamma \vec{F} \cdot \vec{dr} = \int_0^4 \overrightarrow{(t^3, -5t^{\frac{3}{2}}, 1)} \cdot \overrightarrow{(0, 3t^2, 0)}\, dt = \frac{5120}{3}.$$

(6) Here $\vec{F} = \overrightarrow{(3xz, x^2, 2yz)}$ and $\vec{\gamma}(t)$ consists of the four sides of the square: for $-\frac{1}{2} \le t \le \frac{1}{2}$,

$$\vec{\gamma_1}(t) = \overrightarrow{(\frac{1}{2}, t, 0)} \text{ for } -\frac{1}{2} \le t \le \frac{1}{2} \qquad \vec{\gamma_2}(t) = \overrightarrow{(t, \frac{1}{2}, 0)} \text{ for } \frac{1}{2} \ge t \ge -\frac{1}{2}$$

$$\vec{\gamma_3}(t) = \overrightarrow{(-\frac{1}{2}, t, 0)} \text{ for } \frac{1}{2} \ge t \ge -\frac{1}{2} \qquad \vec{\gamma_4}(t) = \overrightarrow{(t, -\frac{1}{2}, 0)} \text{ for } -\frac{1}{2} \le t \le \frac{1}{2}$$

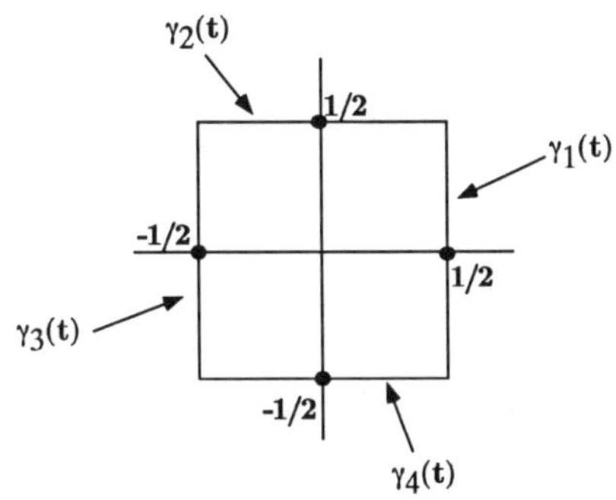

Thus

$$\int_\gamma \vec{F} \cdot \vec{dr} = \sum_{i=1}^{4} \int_{\gamma_i} \vec{F} \cdot \vec{dr}$$

$$= \int_{-\frac{1}{2}}^{\frac{1}{2}} \overrightarrow{(0, \frac{1}{4}, 0)} \cdot \overrightarrow{(0, 1, 0)}\, dt + \int_{\frac{1}{2}}^{-\frac{1}{2}} \overrightarrow{(0, t^2, 0)} \cdot \overrightarrow{(1, 0, 0)}\, dt$$

$$+ \int_{\frac{1}{2}}^{-\frac{1}{2}} \overrightarrow{(0, \frac{1}{4}, 0)} \cdot \overrightarrow{(0, 1, 0)}\, dt + \int_{-\frac{1}{2}}^{\frac{1}{2}} \overrightarrow{(0, t^2, 0)} \cdot \overrightarrow{(1, 0, 0)}\, dt = 0.$$

(7) Let γ denote the line segment joining $(1, 3)$ and $(-2, 4)$. We may write $\gamma(t) = (at + b, ct + d)$ with $\gamma(0) = (1, 3)$ and $\gamma(1) = (-2, 4)$. Thus, $b = 1, d = 3, a = -3, c = 1$, and $\gamma(t) = (-3t + 1, t + 3)$. Consequently,

$$\int_\gamma (3x^3 y\, dx + x^3\, dy) = \int_0^1 \overrightarrow{(3(-3t+1)^2(t+3), (-3t+1)^3)} \cdot \overrightarrow{(-3, 1)}\, dt = -35.$$

(8) Let $\gamma(t) = (at + b, ct + d, et + f)$ satisfy $\gamma(0) = (1, -1, 0)$ and $\gamma(1) = (0, 2, -1)$. It easily follows that $\gamma(t) = (1 - t, 3t - 1, -t)$ and $\gamma'(t) = (-1, 3, -1)$. Then

$$\int_\gamma (2xy\, dx + 3x^2y^3\, dy - xz\, dz) = \int_0^1 \overrightarrow{(2(1-t)(3t-1), 3(1-t)^2(3t-1)^3, (1-t)t)} \cdot \overrightarrow{(-1, 3, -1)}\, dt$$

$$= -\frac{7}{15}.$$

(9) We may express the parabola $y = x^2$ for $0 \leq x \leq 2$ as the parametric curve $\gamma(t) = (t, t^2)$ for $0 \leq t \leq 2$. Then

$$\text{Work} = \int_0^2 \overrightarrow{(-t^2, t)} \cdot \overrightarrow{(1, 2t)}\, dt = \int_0^2 t^2\, dt = \frac{8}{3}.$$

(10) The unit circle (counterclockwise) may be parameterized by $\gamma(t) = (\cos(t), \sin(t))$ for $0 \leq t \leq 2\pi$. Then

$$\text{Work} = \int_0^{2\pi} \overrightarrow{(-\sin(t), \cos(t))} \cdot \overrightarrow{(-\sin(t), \cos(t))}\, dt = 2\pi.$$

The work done in moving the particle in a clockwise motion is -2π.

(11) $\vec{F} = \overrightarrow{(2xy, x^2)}$, $\gamma(t) = \overrightarrow{(\cos(t), \sin(t))}$ for $0 \leq t \leq 2\pi$. We have

$$\text{Work} = \int_0^{2\pi} \overrightarrow{(2\cos(t)\sin(t), \cos^2(t))} \cdot \overrightarrow{(-\sin(t), \cos(t))}\, dt = \int_0^{2\pi} \left(-2\cos(t)\sin^2(t) + \cos^3(t)\right)\, dt = 0.$$

(12) Work $= 0$.

Solutions for §27.3

(1) Take $\phi(x, y) = xy + y^3$. Then $\int\limits_{(0,0)}^{(1,1)} \left(y\, dx + (x + 3y^2)\, dy\right) = \phi(1, 1) - \phi(0, 0) = 2.$

(2) Take $\phi(x, y) = x^2y^3$. Then $\int\limits_{(1,-1)}^{(0,3)} \left(2xy^3\, dx + 3x^2y^2)\, dy\right) = \phi(0, 3) - \phi(1, -1) = -1.$

(3) Take $\phi(x, y) = x\sin(y)$. Then $\int\limits_{(0,0)}^{(1,\frac{\pi}{3})} \left(\sin(y)\, dx + x\cos(y))\, dy\right) = \phi(1, \frac{\pi}{3}) - \phi(0, 0) = \frac{\sqrt{3}}{2}.$

(4) Take $\phi(x, y) = x\ln(y)$. Then $\int\limits_{(1,1)}^{(2,e)} \left(\ln(y)\, dx + \frac{x}{y}\, dy\right) = \phi(2, e) - \phi(1, 1) = 2.$

(5) Take $\phi(x, y) = 2y\sqrt{xy}$. Then $\int\limits_{(1,1)}^{(4,9)} \left(y \cdot \sqrt{\frac{y}{x}}\, dx + 3\sqrt{xy}\, dy\right) = \phi(4, 9) - \phi(1, 1) = 106.$

(6) Take $\phi(x, y, z) = x^2yz$. Then $\int\limits_{(0,0,0)}^{(1,1,1)} \left(2xyz\, dx + x^2z\, dy + x^2y\, dz\right) = \phi(1, 1, 1) - \phi(0, 0, 0) = 1.$

(7) Take $\phi(x, y, z) = \cos(x) + yz$. Then $\int\limits_{(0,1,1)}^{(\frac{\pi}{2},1,2)} \left(-\sin(x)\, dx + z\, dy + y\, dz\right) = \phi(\frac{\pi}{2}, 1, 2) - \phi(0, 1, 1) = 0.$

(8) Take $\phi(x, y, z) = e^x + y^2z$. Then $\int\limits_{(0,0,0)}^{(0,1,1)} \left(e^x\, dx + 2yz\, dy + y^2\, dz\right) = \phi(0, 1, 1) - \phi(0, 0, 0) = 1.$

(9) Take $\phi(x,y,z) = xz$. Then $\int_{(1,1,1)}^{(2,2,2)} (z\,dx + x\,dz) = \phi(2,2,2) - \phi(1,1,1) = 3.$

Solutions for §27.4

(1) By graphing for $0 \le t \le \frac{4\pi}{3}$ we observe a counterclockwise orientation.

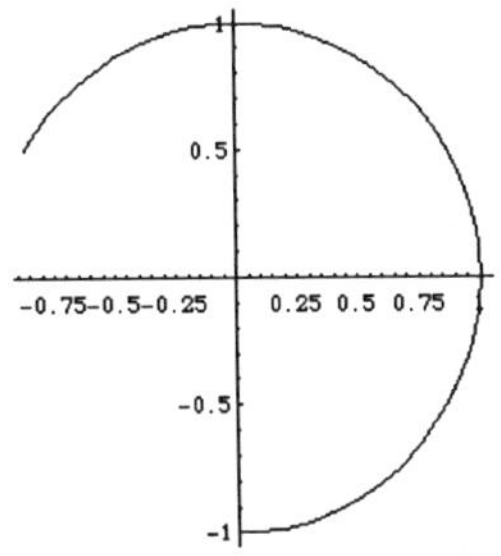

$$\gamma(t) = (\sin(t), -\cos(t)) \text{ for } 0 \le t \le \frac{4\pi}{3}$$

(2) By graphing for $0 \le t \le \frac{4\pi}{3}$ we observe a clockwise orientation.

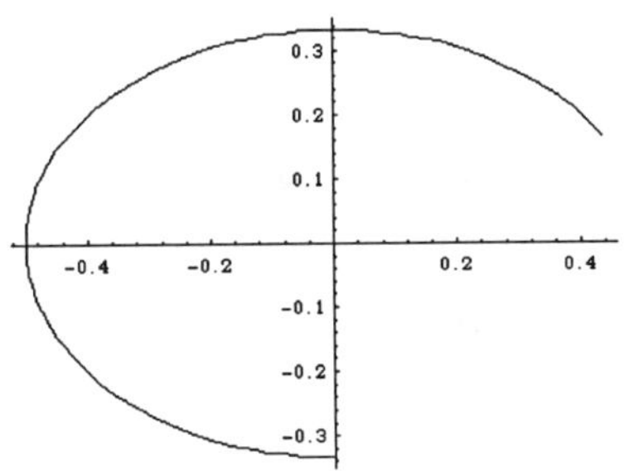

$$\gamma(t) = \left(-\frac{\sin(t)}{2}, -\frac{\cos(t)}{3}\right) \text{ for } 0 \le t \le \frac{4\pi}{3}$$

(3) By graphing for $0 \le t \le \frac{4\pi}{3}$ we observe a clockwise orientation.

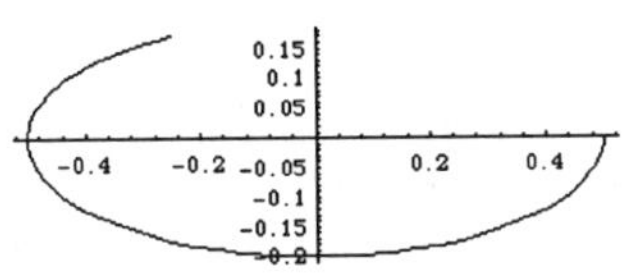

$$\gamma(t) = \left(\frac{\cos(t)}{2}, -\frac{\sin(t)}{5}\right) \text{ for } 0 \le t \le \frac{4\pi}{3}$$

(4) By graphing for $0 \le t \le \frac{4\pi}{3}$ we observe a clockwise orientation.

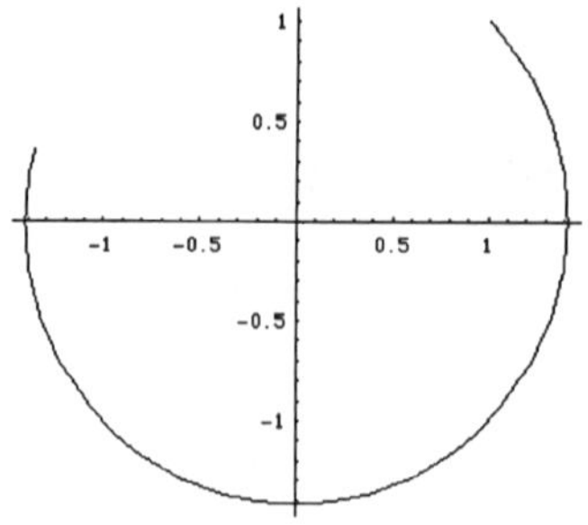

$$\gamma(t) = \Big(\cos(t) + \sin(t), \cos(t) - \sin(t)\Big) \text{ for } 0 \le t \le \frac{4\pi}{3}$$

Green's theorem asserts that $\oint\limits_{\partial\mathcal{R}} (P\,dx + Q\,dy) = \iint\limits_{\mathcal{R}} \left(\frac{\partial Q}{\partial x} - \frac{\partial P}{\partial y}\right) dx\,dy$ *where* $\partial\mathcal{R}$ *is a simple closed curve with counterclockwise orientation which forms the boundary of* $\mathcal{R}$.

(5) Let $\mathcal{R}$ be the rectangle $0 \le x \le 1,\ 0 \le y \le 2$. By Green's theorem,

$$\oint\limits_{\partial\mathcal{R}} \Big((x^2 + y^2)\,dx - xy\,dy\Big) = \int_0^2 \int_0^1 (-y - 2y)\,dx\,dy = -6.$$

(6) $\mathcal{D}$ = the disk $x^2 + y^2 \le 4$. By Green's theorem

$$\oint\limits_{\partial\mathcal{D}} \Big((3x^2y + y^3)\,dx - y^{10}\,dy\Big) = \iint\limits_{x^2+y^2\le 4} (-3x^2 - 3y^2)\,dx\,dy = -24\pi.$$

(7) $\mathcal{T}$ = the triangle with vertices at $(0,0)$, $(0,-1)$, $(1,0)$.

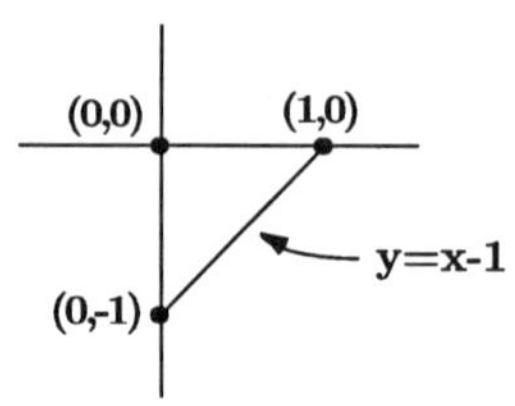

By Green's theorem,

$$\oint\limits_{\partial\mathcal{T}} \Big((e^x + y)\,dx - (x - e^y)\,dy\Big) = \iint\limits_{\mathcal{T}} -2\,dx\,dy = \int_0^1 \int_0^{x-1} -2\,dy\,dx = 1.$$

(8)The ellipse $\gamma(t) = \left(\frac{\cos(t)}{3}, -\frac{\sin(t)}{2}\right)$ for $0 \le t \le 2\pi$ has clockwise orientation as t goes from 0 to 2π. Since Green's theorem requires counterclockwise orientation, we must introduce an extra minus sign. The interior of the ellipse consists of all points (x, y) satisfying $9x^2 + 4y^2 \le 1$. By Green's theorem

$$\oint\limits_{\gamma} \Big((y^2 + x)\,dx + (xy - y)\,dy\Big) = -\iint\limits_{9x^2+4y^2\le 1} (y - 2y)\,dx\,dy = 0.$$

(9) The circle $\gamma(t) = 3\big(-\cos(t), \sin(t)\big)$ for $0 \le t \le 2\pi$ has clockwise orientation as t goes from 0 to 2π. Since Green's theorem requires counterclockwise orientation, we must introduce an extra minus sign. The interior of the circle consists of all points (x, y) satisfying $x^2 + y^2 \le 9$. By Green's theorem

$$\oint_\gamma \Big((y^2 + x^2)\,dx + (2x - y^2)\,dy\Big) = -\iint_{x^2+y^2\le 9} (2 - 2y)\,dx\,dy = -18\pi.$$

Let $\mathcal{R}$ be a region whose boundary $\partial\mathcal{R}$ is a simple closed curve with counterclockwise orientation. Then Area$(\mathcal{R}) = \int_{\partial\mathcal{R}} x\,dy$.

(10) $\gamma =$ the ellipse $\frac{x^2}{4} + \frac{y^2}{9} = 1$. We may parameterize the ellipse by $x = 2\cos(t)$, $y = 3\sin(t)$, and $dy = 3\cos(t)\,dt$. We have

$$\text{Area(ellipse)} = \oint_\gamma x\,dy = \int_0^{2\pi} 2\cos(t)\cdot 3\cos(t)\,dt = 6\pi.$$

(11) The ellipse $\gamma(t) = \Big(-\frac{\cos(t)}{4}, \frac{\sin(t)}{25}\Big)$ has clockwise orientation as t goes from 0 to 2π $\implies$ we need an extra minus sign in Green's theorem. Set $x = -\frac{\cos(t)}{4}$, $y = \frac{\sin(t)}{25}$, $dy = \frac{\cos(t)}{25}\,dt$. Then γ is the ellipse $(4x)^2 + (25y)^2 = 1$.

$$\text{Area(ellipse)} = \oint_\gamma x\,dy = -\int_0^{2\pi} -\frac{\cos(t)}{4}\cdot\frac{\cos(t)}{25}\,dt = \frac{\pi}{100}.$$

(12) The curve $\gamma(t) = \big(\cos^3(t), \sin^3(t)\big)$ has counterclockwise orientation for $0 \le t \le 2\pi$. Set $x = \cos^3(t)$, $y = \sin^3(t)$, $dy = 3\sin^2(t)\cos(t)\,dt$. We have

$$\text{Area} = \oint_\gamma x\,dy = \int_0^{2\pi} \cos^3(t)\cdot 3\sin^2(t)\cos(t)\,dt = \frac{3\pi}{8}.$$

(13) The circle $\gamma(t) = \big(\cos(t) + \sin(t), \cos(t) - \sin(t)\big)$ has clockwise orientation for $0 \le t \le 2\pi$, so we must insert an extra minus sign in Green's theorem. Set $x = \cos(t) + \sin(t)$, $y = \cos(t) - \sin(t)$. Then

$$\text{Area} = -\oint_\gamma x\,dy = -\int_0^{2\pi} \big(\cos(t) + \sin(t)\big)\cdot\big(-\sin(t) - \cos(t)\big)\,dt = 2\pi.$$

(14) Area = 3.

(15) Area = 6.

Solutions for Additional Exercises for Chapter 27

(1)

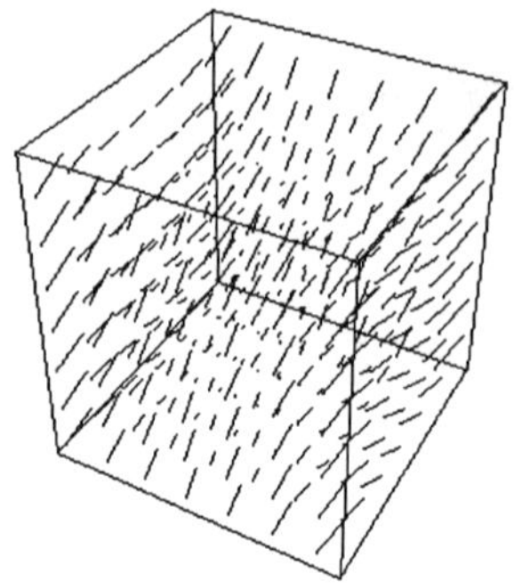

$$\vec{v} = x^2\vec{i} + y^2\vec{j} + (x^2 + y^2)\vec{k}$$

(2)

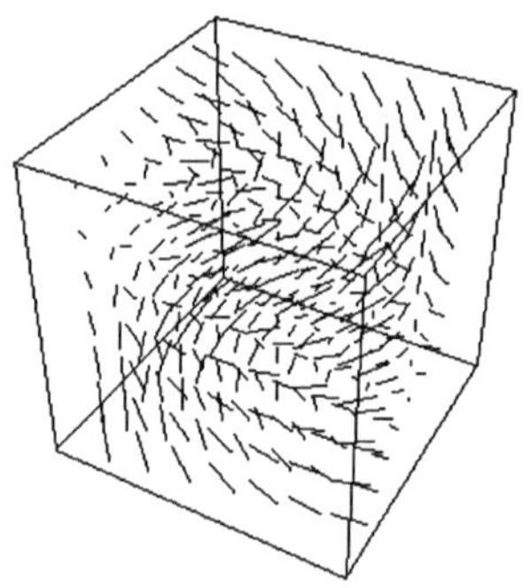

$$\vec{v} = -y\vec{i} + x\vec{j} + z\vec{k}$$

(3)

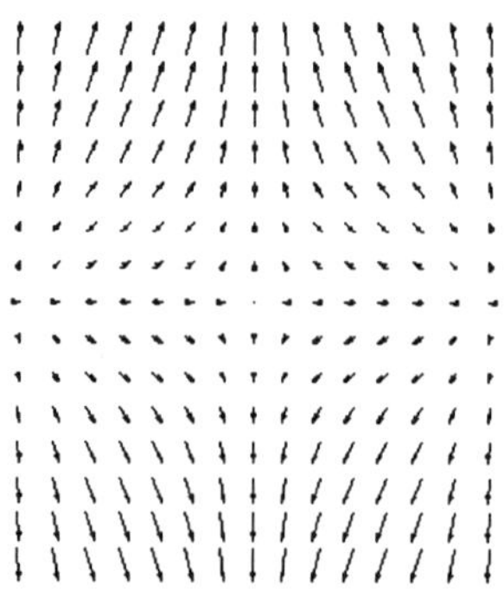

$$\vec{v} = \overrightarrow{\nabla}\phi \text{ with } \phi(x,y) = \cos(x) + \frac{1}{2}y^2$$

(4) $\vec{v}$ is NOT a gradient vector field.

(5) $\vec{v}$ is a gradient vector field; potential function is $\phi(x,y) = e^{xy}$.

(6) $\vec{v}$ is a gradient vector field; potential function is $\phi(x,y) = x^3y^2z$.

(7) $\vec{v}$ is a gradient vector field; potential function is $\phi(x,y) = \frac{1}{3}\left(x^3 + y^3 + z^3\right) + xyz$.

(8) Work $= \int\limits_0^{10} (t^2 + e^{3t})\ dt = \frac{1}{3}\left(999 + e^{30}\right)$.

(9) Work $= 24 \int\limits_0^{1} (t^2 - t)\ dt = -4$.

(10) Work $= 0$, since $\vec{F}$ is a gradient vector field and γ is a closed loop.

(11) Work $= 0$.

(12) Work $= \int\limits_0^{5} (t^2 + 2t^5)\ dt = 5250$.

(13) There is no potential function for this vector field.

(14) Potential function is $\phi(x,y) = e^x \ln y - e^y \ln x$. So $\int\limits_{(1,0)}^{(3,3)} \left(e^x \ln y - \frac{e^y}{x}\right)\ dx + \left(\frac{e^y}{x} - e^y \ln x\right)\ dy = \infty$.

(15) $g(x) = e^x$.

(16) $\oint \left(\left(x^2 - y^2\right)\ dx + x\ dy\right) = \iint\limits_{\text{small circle}} (1 + 2y)\ dx\ dy = -80\pi$.

(17) $\oint \left(\ln(1+y)\ dx - \frac{xy}{1+y}\ dy\right) = \iint\limits_{\text{small triangle}} -1\ dx\ dy = -5$.

(18) Area $= \int\limits_0^1 t\ dt + \int\limits_0^1 \frac{\sqrt{10}-1}{\sqrt{10}+(1-\sqrt{10})t}\ dt + \int\limits_0^1 (t-1)\ dt = \ln\sqrt{10}$.

(19) Area $= \int\limits_0^{2\pi} 15 \sin^2(t) \cos^4(t)\ dt = \frac{15\pi}{8}$.

(20) Area $= \int\limits_0^1 12\sqrt{3}(2t-1)^2\ dt = 4\sqrt{3}$.

Solutions for §28.1

The area of a surface $z = f(x,y)$ for $x, y \in \mathcal{R}$ is given by

$$\iint\limits_{\mathcal{R}} \sqrt{1 + \left(\frac{\partial f}{\partial x}\right)^2 + \left(\frac{\partial f}{\partial x}\right)^2}\, dx\, dy.$$

(1) $z = \sqrt{x^2+y^2}$ with $0 \le z \le 1 \implies x^2 + y^2 \le 1$. Thus,

$$\text{Surface Area} = \iint\limits_{x^2+y^2\le 1} \sqrt{1 + \left(\frac{x^2}{x^2+y^2}\right) + \left(\frac{y^2}{x^2+y^2}\right)}\, dx\, dy = \pi\sqrt{2}.$$

(2) $z = \frac{1}{3}(4 - 2x + y)$ for $x^2 + \frac{y^2}{4} \le 2$. Thus,

$$\text{Surface Area} = \iint\limits_{x^2+\frac{y^2}{4}\le 2} \sqrt{1 + \left(\frac{-2}{3}\right)^2 + \left(\frac{1}{3}\right)^2}\, dx\, dy = \frac{4\pi}{3}\sqrt{14}.$$

(3) $z = xy$ for $x^2 + y^2 \le 4$. Thus,

$$\text{Surface Area} = \iint\limits_{x^2+y^2\le 4} \sqrt{1 + y^2 + x^2}\, dx\, dy = \int_0^{2\pi}\int_0^2 \sqrt{1+r^2}\, r\, dr\, d\theta = \frac{2\pi}{3}(5\sqrt{5} - 1).$$

(4) The sphere $x^2 + y^2 + z^2 = 1$ intersects the plane $z = \frac{\sqrt{2}}{2}$ when $x^2 + y^2 = \frac{1}{2}$. So we must find the surface area of $z = \sqrt{1 - x^2 - y^2}$ for $x^2 + y^2 \le \frac{1}{2}$.

$$\begin{aligned}\text{Surface Area} &= \iint\limits_{x^2+y^2\le \frac{1}{2}} \sqrt{1 + \left(\frac{x^2}{1-x^2-y^2}\right) + \left(\frac{y^2}{1-x^2-y^2}\right)}\, dx\, dy \\ &= \iint\limits_{x^2+y^2\le \frac{1}{2}} \frac{1}{\sqrt{1-x^2-y^2}}\, dx\, dy = \pi(2 - \sqrt{2}).\end{aligned}$$

(5) Consider the upper half of the ellipsoid $\frac{x^2}{a^2} + \frac{y^2}{b^2} + \frac{z^2}{c^2} = 1$. Then $z = |c| \cdot \sqrt{1 - \frac{x^2}{a^2} - \frac{y^2}{b^2}}$ and $\frac{x^2}{a^2} + \frac{y^2}{b^2} \le 1$. It follows that

$$\begin{aligned}\text{Surface Area} &= \iint\limits_{\frac{x^2}{a^2}+\frac{y^2}{b^2}\le 1} \sqrt{1 + \frac{c^2x^2}{a^4\left(1 - \frac{x^2}{a^2} - \frac{y^2}{b^2}\right)} + \frac{c^2y^2}{b^4\left(1 - \frac{x^2}{a^2} - \frac{y^2}{b^2}\right)}}\, dx\, dy \\ &= |ab| \iint\limits_{x^2+y^2\le 1} \sqrt{1 + \frac{c^2x^2}{a^2(1-x^2-y^2)} + \frac{c^2y^2}{b^2(1-x^2-y^2)}}\, dx\, dy.\end{aligned}$$

(6) Here $z = \sqrt{4 - x^2 - y^2}$ and

$$\text{Surface Area} = \iint\limits_{x^2+\frac{y^2}{4}\leq 1} \sqrt{1 + \frac{x^2}{4-x^2-y^2} + \frac{y^2}{4-x^2-y^2}}\,dx\,dy = \iint\limits_{x^2+\frac{y^2}{4}\leq 1} \frac{2}{\sqrt{4-x^2-y^2}}\,dx\,dy.$$

Solutions for §28.2

Let $\vec{F} = P\vec{\imath} + Q\vec{\jmath} + R\vec{k}$ where P, Q, R are functions of x, y, z. Let S be a surface over a region $\mathcal{R}$ in the $x\,y$–plane. Assume $z = f(x, y)$. We use the formulae

$$\iint\limits_S \vec{F} \cdot \overrightarrow{d\sigma}^{\pm} = \pm \iint\limits_{\mathcal{R}} \left(-\frac{\partial f}{\partial x} P - \frac{\partial f}{\partial y} Q + R \right) dx\,dy.$$

(1) $\vec{F} = xz\vec{\imath} - \vec{\jmath} + y\vec{k}$ and S is the plane $z = f(x,y) = \frac{1}{3}(2 - 2x + y)$ lying over $0 \leq x \leq 1, 0 \leq y \leq 2$. We have

$$\begin{aligned}\iint\limits_S \vec{F} \cdot \overrightarrow{d\sigma}^{\pm} &= \pm \int_0^1 \int_0^2 \left(\frac{2}{3} \cdot x \cdot z - \frac{1}{3}(-1) + y \right) dx\,dy \\ &= \pm \int_0^1 \int_0^2 \left(\frac{2}{9} x(2 - 2x + y) + \frac{1}{3} + y \right) dx\,dy = \pm \frac{82}{27}.\end{aligned}$$

(2) $\vec{F} = 2y\vec{\imath} - 2x\vec{\jmath} + \vec{k}$ and S is the cone $z = f(x,y) = 4 - x^2 - y^2$ lying above the $x\,y$–plane. The cone S intersects the $x\,y$-plane when $z = 0 = 4 - x^2 - y^2 \implies x^2 + y^2 = 4$.

$$\begin{aligned}\iint\limits_S \vec{F} \cdot \overrightarrow{d\sigma}^{\pm} &= \pm \iint\limits_{x^2+y^2\leq 4} (-(-2x)\cdot(2y) - (-2y)\cdot(-2x) + 1)\,dx\,dy \\ &= \pm \iint\limits_{x^2+y^2\leq 4} (1 + 8xy)\,dx\,dy = \pm 4\pi.\end{aligned}$$

(3) $\vec{F} = -x\vec{\imath} + y\vec{\jmath} + x^2\vec{k}$ and S is the portion of the surface $z = f(x,y) = xy$ lying above $-1 \leq x \leq 1$, $0 \leq y \leq 2$. We have

$$\begin{aligned}\iint\limits_S \vec{F} \cdot \overrightarrow{d\sigma}^{\pm} &= \pm \int_{-1}^1 \int_0^2 (-y(-x) - xy + x^2)\,dx\,dy \\ &= \pm \int_{-1}^1 \int_0^2 (x^2 - 2xy)\,dx\,dy = \pm \frac{4}{3}.\end{aligned}$$

(4) $\vec{F} = x\vec{i} + y\vec{j} + z\vec{k}$ and S is the cone $z = f(x,y) = \sqrt{x^2+y^2}$ for $0 \le z \le 1 \implies x^2+y^2 \le 1$. We have

$$\iint_S \vec{F} \cdot \overrightarrow{d\sigma}^{\pm} = \pm \iint_{x^2+y^2\le 1} \left(\frac{-x}{\sqrt{x^2+y^2}} \cdot x - \frac{y}{\sqrt{x^2+y^2}} \cdot y + \sqrt{x^2+y^2} \right) dx\,dy$$
$$= \pm \iint_{x^2+y^2\le 1} \left(\sqrt{x^2+y^2} - \frac{x^2+y^2}{\sqrt{x^2+y^2}} \right) dx\,dy = 0.$$

(5) $\vec{F} = x\vec{i} + y\vec{j} + z\vec{k}$ and S is the portion of the surface $z = f(x,y) = x^2 - xy + y^2$ lying above $0 \le x \le 1$, $1 \le y \le 2$. We have

$$\iint_S \vec{F} \cdot \overrightarrow{d\sigma}^{\pm} = \pm \int_0^1 \int_1^2 \left(-2(2x-y) \cdot x - (2y-x) \cdot y + (x^2 - xy + y^2)\right) dx\,dy = \mp \frac{23}{12}.$$

(6) $\vec{F} = xz\vec{i} + yz\vec{j} + x^2\vec{k}$ and S is the surface $z = f(x,y) = -\sqrt{1-x^2-y^2}$ for $x^2+y^2 \le 1$. We have

$$\iint_S \vec{F} \cdot \overrightarrow{d\sigma}^{\pm} = \pm \iint_{x^2+y^2\le 1} \left(\frac{-x}{\sqrt{1-x^2-y^2}} \cdot xz - \frac{y}{\sqrt{1-x^2-y^2}} \cdot yz + x^2 \right) dx\,dy$$
$$= \pm \iint_{x^2+y^2\le 1} \left(x^2+y^2+x^2\right) dx\,dy = \pm \frac{3\pi}{4}.$$

(7) $\vec{F} = x^2\vec{i} + y^2\vec{j} + z\vec{k}$ and S is the triangle which passes through the points $(1,0,0)$, $(0,1,0)$, $(0,0,2)$.

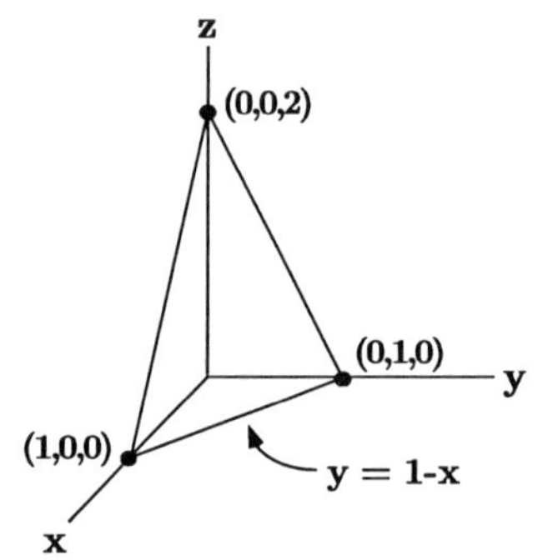

Thus S is the portion of the plane $z = f(x,y) = 2 - 2x - 2y$ lying above $0 \le x \le 1, 0 \le y \le 1-x$. Hence

$$\iint_S \vec{F} \cdot \overrightarrow{d\sigma}^{\pm} = \pm \int_0^1 \int_1^{1-x} \left(-(-2 \cdot x^2 - (-2) \cdot y^2 + (2 - 2x - 2y)\right) dx\,dy = \pm \frac{2}{3}.$$

(8) $\vec{F} = 3xz\vec{i} + \vec{j} - y^2\vec{k}$ and S is the portion of the plane $z = f(x,y) = 5 - 4x - 9y$ lying above the region $\{0 \le x \le 1, 0 \le y \le \frac{1}{2}(1-x)\} \cup \{-1 \le x \le 0, -x \le y \le \frac{1}{2}(1-x)\} = \mathcal{R}$.

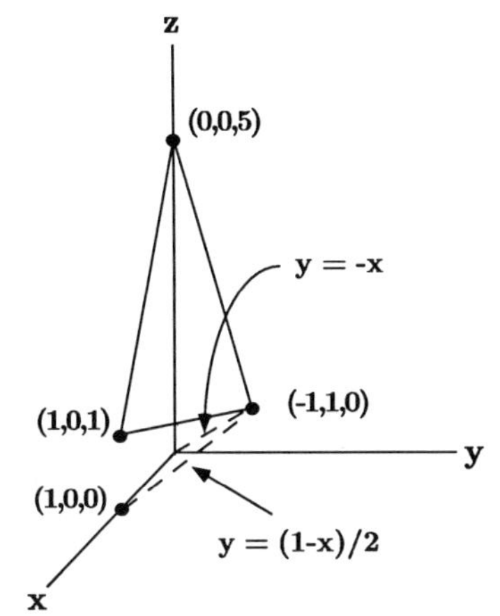

Hence

$$\iint_S \vec{F} \cdot \overrightarrow{d\sigma}^{\pm} = \pm \int_{\mathcal{R}} \left(4 \cdot 3x(5 - 4x - 9y) + 9 - y^2\right)\, dx\, dy = \pm\frac{59}{12}.$$

Solutions for §28.3

(1) The six faces of the a cube are given by:

$$S_1 : z = 0, \quad 0 \le x \le 1,\ 0 \le y \le 1 \implies \overrightarrow{d\sigma_1} = -\vec{k}\, dx\, dy.$$
$$S_2 : z = 1, \quad 0 \le x \le 1,\ 0 \le y \le 1 \implies \overrightarrow{d\sigma_2} = \vec{k}\, dx\, dy.$$
$$S_3 : y = 0, \quad 0 \le x \le 1,\ 0 \le z \le 1 \implies \overrightarrow{d\sigma_3} = -\vec{j}\, dx\, dz.$$
$$S_4 : z = 1, \quad 0 \le x \le 1,\ 0 \le z \le 1 \implies \overrightarrow{d\sigma_4} = \vec{j}\, dx\, dz.$$
$$S_5 : x = 0, \quad 0 \le y \le 1,\ 0 \le z \le 1 \implies \overrightarrow{d\sigma_5} = -\vec{i}\, dy\, dz.$$
$$S_6 : x = 1, \quad 0 \le y \le 1,\ 0 \le z \le 1 \implies \overrightarrow{d\sigma_6} = \vec{i}\, dy\, dz.$$

Hence

$$\begin{aligned}\iint_{\text{cube}} (x\vec{i} + y\vec{j} + z\vec{k}) \cdot \overrightarrow{d\sigma} &= \int_0^1\int_0^1 (-0)\, dx\, dy + \int_0^1\int_0^1 (1)\, dx\, dy \\ &+ \int_0^1\int_0^1 (-0)\, dx\, dz + \int_0^1\int_0^1 (1)\, dx\, dz \\ &+ \int_0^1\int_0^1 (-0)\, dy\, dz + \int_0^1\int_0^1 (1)\, dy\, dz = 3.\end{aligned}$$

(2) $\displaystyle\iint_{x^2+y^2+z^2=9} \vec{F} \cdot \overrightarrow{d\sigma} = 108\pi.$

(3) The four faces of the tetrahedron are:

$$S_1 : z = 0, \quad 0 \le x \le 2,\ 0 \le y \le 2 - x \implies \overrightarrow{d\sigma_1} = -\vec{k}\,dx\,dy.$$

$$S_2 : y = 0, \quad 0 \le x \le 2,\ 0 \le z \le -\frac{5}{2}x + 5 \implies \overrightarrow{d\sigma_2} = -\vec{j}\,dx\,dz.$$

$$S_3 : x = 0, \quad 0 \le y \le 2,\ 0 \le z \le -\frac{5}{2}x + 5 \implies \overrightarrow{d\sigma_3} = -\vec{i}\,dy\,dz.$$

$$S_4 : z = 5 - \frac{5}{2}x - \frac{5}{2}y, \quad 0 \le x \le 2, 0 \le y \le 2 - x \implies \overrightarrow{d\sigma_4} = \left(\frac{5}{2}\vec{i} + \frac{5}{2}\vec{j} + \vec{k}\right)\,dx\,dy.$$

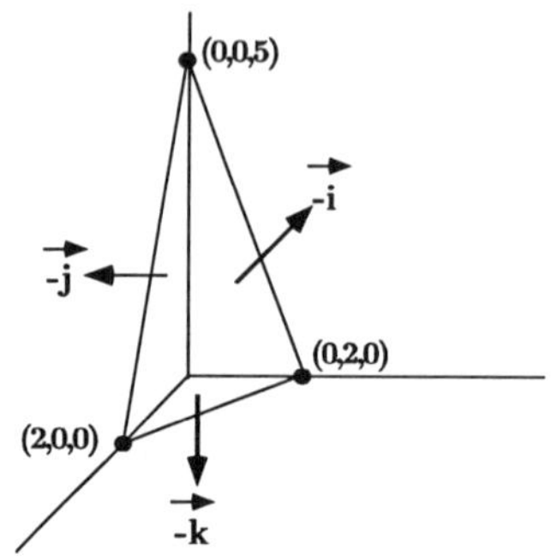

Then

$$\iint\limits_{\text{tetrahedron}} (x\vec{i} + y\vec{j} + z\vec{k}) \cdot \overrightarrow{d\sigma} = \int_0^2 \int_0^{2-x} (-0)\,dy\,dx + \int_0^2 \int_0^{-\frac{5}{2}x+5} (-0)\,dz\,dx + \int_0^2 \int_0^{-\frac{5}{2}y+2} (-0)\,dz\,dy$$
$$+ \int_0^2 \int_0^{2-x} \left[\frac{5}{2}x + \frac{5}{2}y + \left(5 - \frac{5}{2}x - \frac{5}{2}y\right)\right]\,dy\,dx = 10.$$

(4) $\iint\limits_S \vec{F} \cdot \overrightarrow{d\sigma} = 6\pi.$

(5) The four faces of the cylinder are:

$$S_1 : y = 0, \quad x^2 + z^2 \le 4 \implies \overrightarrow{d\sigma_1} = -\vec{j}\,dx\,dy.$$

$$S_2 : y = 2, \quad x^2 + z^2 \le 4 \implies \overrightarrow{d\sigma_2} = \vec{j}\,dx\,dy.$$

$$S_3 : z = \sqrt{4 - x^2}, \quad 0 \le y \le 2, -2 \le x \le 2 \implies \overrightarrow{d\sigma_3} = \left(\frac{x}{\sqrt{4 - x^2}}\vec{i} + \vec{k}\right)\,dx\,dy.$$

$$S_4 : z = -\sqrt{4 - x^2}, \quad 0 \le y \le 2, -2 \le x \le 2 \implies \overrightarrow{d\sigma_4} = \left(\frac{x}{\sqrt{4 - x^2}}\vec{i} - \vec{k}\right)\,dx\,dy.$$

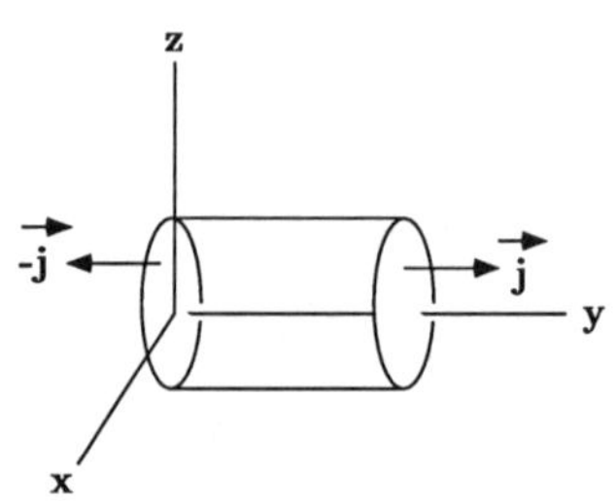

Then

$$\iint\limits_{\text{cylinder}} (x\vec{\imath} + y\vec{\jmath} + z\vec{k}) \cdot \overrightarrow{d\sigma} = \iint\limits_{x^2+z^2\leq 4} 2\,dx\,dz + 2\int_0^2 \int_{-2}^2 \left(\frac{x^2}{\sqrt{4-x^2}} + \sqrt{4-x^2}\right) dx\,dy = 24\pi.$$

(6) Note that for this vector field div $(\vec{\mathbf{F}}) = 0$.

Solutions for §28.4

(1) $\vec{\nabla} \cdot (2x^3y\vec{\imath} - z\vec{\jmath} + \vec{k}) = \left(\vec{\imath}\frac{\partial}{\partial x} + \vec{\jmath}\frac{\partial}{\partial y} + \vec{k}\frac{\partial}{\partial z}\right) \cdot (2x^3y\vec{\imath} - z\vec{\jmath} + \vec{k}) = 6x^2y.$

(2) $\vec{\nabla} \cdot (xz^2\vec{\imath} + xyz\vec{\jmath} - 3xz^4\vec{k}) = \left(\vec{\imath}\frac{\partial}{\partial x} + \vec{\jmath}\frac{\partial}{\partial y} + \vec{k}\frac{\partial}{\partial z}\right) \cdot (xz^2\vec{\imath} + xyz\vec{\jmath} - 3xz^4\vec{k}) = z^2 + xz - 12xz^3.$

(3) $\vec{\nabla} \cdot (xz\vec{\imath} + \cos(y^2z)\vec{k}) = \left(\vec{\imath}\frac{\partial}{\partial x} + \vec{\jmath}\frac{\partial}{\partial y} + \vec{k}\frac{\partial}{\partial z}\right) \cdot (xz\vec{\imath} + \cos(y^2z)\vec{k}) = z - y^2\sin(y^2z).$

(4) $\vec{\nabla} \cdot (ye^x\vec{\jmath} - \ln(x^2y)\vec{k}) = \left(\vec{\imath}\frac{\partial}{\partial x} + \vec{\jmath}\frac{\partial}{\partial y} + \vec{k}\frac{\partial}{\partial z}\right) \cdot (ye^x\vec{\jmath} - \ln(x^2y)\vec{k}) = e^x.$

(5) $\vec{\nabla} \cdot (e^{x+y}\vec{\imath} - 2\cos(yz)\vec{\jmath} + z\vec{k}) = \left(\vec{\imath}\frac{\partial}{\partial x} + \vec{\jmath}\frac{\partial}{\partial y} + \vec{k}\frac{\partial}{\partial z}\right) \cdot (e^{x+y}\vec{\imath} - 2\cos(yz)\vec{\jmath} + z\vec{k}) = e^{x+y} + 2z\sin(yz) + 1.$

(6) $\vec{\nabla} \cdot (\frac{x}{y}\vec{\imath} - \ln\left(\frac{x}{yz}\right)\vec{\jmath} + e^{z^2}\vec{k}) = \left(\vec{\imath}\frac{\partial}{\partial x} + \vec{\jmath}\frac{\partial}{\partial y} + \vec{k}\frac{\partial}{\partial z}\right) \cdot (\frac{x}{y}\vec{\imath} - \ln\left(\frac{x}{yz}\right)\vec{\jmath} + e^{z^2}\vec{k}) = \frac{2}{y} + 2ze^{z^2}.$

Let B be a solid body whose boundary ∂B is a simple closed surface. Let $\vec{F}$ be a differentiable vector field. The Divergence Theorem states that

$$\iiint\limits_B \operatorname{div}(\vec{F})\,dx\,dy\,dz = \iint\limits_{\partial B} \vec{F} \cdot \overrightarrow{d\sigma}.$$

(7) $\vec{F} = x^3\vec{\imath} + y^3\vec{\jmath} + z^3\vec{k}$ and S is the surface of a sphere of radius one. By the divergence theorem

$$\iint\limits_S \vec{F} \cdot \overrightarrow{d\sigma} = \iiint\limits_{x^2+y^2+z^2\leq 1} (3x^2 + 3y^2 + 3z^2)\,dx\,dy\,dz = \frac{12\pi}{5}.$$

(8) $\vec{F} = 3x^2y\vec{\imath} - x\vec{\jmath} + z^2\vec{k}$ and S is the surface of the unit cube $0 \leq x \leq 1$, $0 \leq y \leq 1$, $0 \leq z \leq 1$. Then

$$\iint\limits_S \vec{F} \cdot \overrightarrow{d\sigma} = \int_0^1 \int_0^1 \int_0^1 (6xy + 2z)\,dx\,dy\,dz = \frac{5}{2}.$$

(9) $\vec{F} = x^4z\vec{\imath} - 2xy^2z\vec{\jmath} + 3xz\vec{k}$ and S is the surface of the rectangular box $0 \leq x \leq 1$, $0 \leq y \leq 2$, $-1 \leq z \leq 1$. Then

$$\iint\limits_S \vec{F} \cdot \overrightarrow{d\sigma} = \int_{-1}^1 \int_0^2 \int_0^1 (4x^3z - 4xyz + 3x)\,dx\,dy\,dz = 6.$$

(10) $\vec{F} = 3x^2y\vec{\imath} - x\vec{\jmath} + z^2\vec{k}$ and S is the surface of the tetrahedron with vertices at $(0,0,0)$, $(1,0,0)$, $(0,1,0)$, $(0,0,1)$.

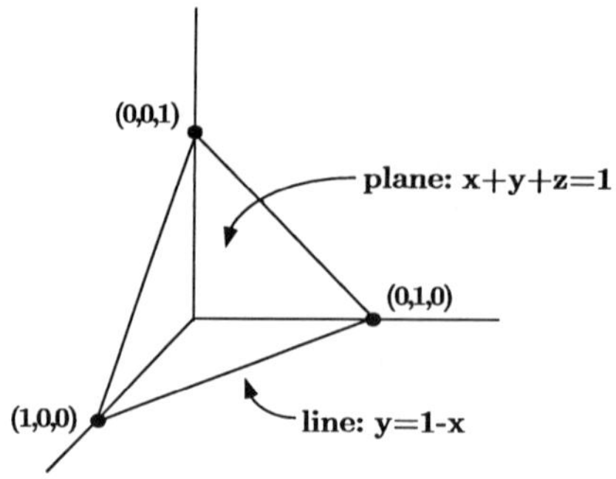

It follows from the divergence theorem that

$$\iint_S \vec{F} \cdot \overrightarrow{d\sigma} = \int_0^1 \int_0^{1-x} \int_0^{1-x-y} (6xy + 2z)\, dx\, dy\, dz = \frac{2}{15}.$$

(11) Since the divergence of $\vec{F}$ is zero we see that $\iint_S \vec{F} \cdot \overrightarrow{d\sigma} = 0$.

(12) $\vec{F} = 3x\vec{\imath} + 2\sin(x^3 z)\vec{\jmath} - z\vec{k}$ and S is the surface of the ellipsoid $x^2 + \frac{y^2}{4} + \frac{z^2}{9} \leq 4$. Then

$$\iint_S \vec{F} \cdot \overrightarrow{d\sigma} = \iiint_{x^2+\frac{y^2}{4}+\frac{z^2}{9}\leq 4} 2\, dx\, dy\, dz = 128\pi.$$

(13) $\vec{F} = x\vec{\imath} + y\vec{\jmath} + z\vec{k}$ and S is the surface of the cylinder $x^2 + y^2 \leq 9$, $-2 \leq z \leq 2$. Then

$$\iint_S \vec{F} \cdot \overrightarrow{d\sigma} = \int_{-2}^{2} \left(\iint_{x^2+y^2\leq 9} 3\, dx\, dy \right) dz = 108\pi.$$

(14), (15), (16). Use the formal definition of the divergence of a vector field.

(17) This is a vector analogue of the Product Rule. Use the Product Rule for the (partial) derivatives of a product of two real-valued functions.

(18) Check that $\left(\frac{\partial^2}{\partial x^2} + \frac{\partial^2}{\partial y^2} + \frac{\partial^2}{\partial z^2}\right)(x^2 + y^2 + z^2)^{-\frac{1}{2}} = 0$, everywhere except for the origin.

Solutions for §28.5

In problems **(1) – (9)**, *the spin of a unit disk centered at the origin in the $x\,z$, $y\,z$, $x\,y$–plane is given.*

(1) 0, 2, and 0.

(2) 0, −2, and 1.

(3) 0, 0, and 0.

(4) 0, 0, and 0.

(5) 0, 0, and −3.

(6) 0, 3, and 0.

(7) 0, 0, and 0.

(8) 0, 0, and 0.

(9) 0, 0, and 0.

Solutions for §28.6

Stoke's Theorem states that

$$\iint_S \operatorname{curl}(\vec{F}) \cdot \overrightarrow{d\sigma}^+ = \oint_{\partial S} \vec{F} \cdot \overrightarrow{dr},$$

where ∂S is the curve (counterclockwise orientation) bounding the surface S.

(1) To verify Stoke's theorem for $\vec{F} = 2z\vec{i} - 2x\vec{j} + 2y\vec{k}$ and S the surface $z = 4 - x^2 - y^2, 0 \le z \le 4$ we first compute

$$\oint_{\partial S} \vec{F} \cdot \overrightarrow{dr} = \oint_{\substack{x^2+y^2=4\\ z=0}} (2z\,dx - 2x\,dy + 2y\,dz)$$
$$= \int_0^{2\pi} (-8\cos^2(t))\,dt = -8\pi.$$

Further, $\operatorname{curl}(\vec{F}) = \vec{\nabla} \times \vec{F} = \begin{vmatrix} \vec{i} & \vec{j} & \vec{k} \\ \frac{\partial}{\partial x} & \frac{\partial}{\partial y} & \frac{\partial}{\partial z} \\ 2z & -2x & 2y \end{vmatrix} = 2\vec{i} + 2\vec{j} - 2\vec{k}$ and $\overrightarrow{d\sigma}^+ = (2\vec{i} + 2\vec{j} - 2\vec{k})\,dx\,dy$. So

$$\iint_S \operatorname{curl}(\vec{F}) \cdot \overrightarrow{d\sigma}^+ = \iint_{x^2+y^2\le 4} (2\vec{i} + 2\vec{j} - 2\vec{k}) \cdot (2x\vec{i} + 2y\vec{j} + \vec{k})\,dx\,dy$$
$$= \iint_{x^2+y^2\le 4} (4(x+y) - 2)\,dx\,dy = -8\pi.$$

(2) $\vec{F} = y\vec{i} - x\vec{j} + z\vec{k}$ and S the surface of the cone $z = \sqrt{x^2+y^2} - y^2$, $0 \le z \le 2$.

$$\oint_{\partial S} \vec{F} \cdot \overrightarrow{dr} = \oint_{x^2+y^2=4} (y\,dx - x\,dy)$$

$$= \int_0^{2\pi} \left(-4\sin^2(t) - 4\cos^2(t)\right)\,dt = -8\pi.$$

Further, $\text{curl}(\vec{F}) = \begin{vmatrix} \vec{i} & \vec{j} & \vec{k} \\ \frac{\partial}{\partial x} & \frac{\partial}{\partial y} & \frac{\partial}{\partial z} \\ y & -x & z \end{vmatrix} = -2\vec{k}$ and $\overrightarrow{d\sigma}^{+} = \left(\frac{-x}{\sqrt{x^2+y^2}}\vec{i} + \frac{-y}{\sqrt{x^2+y^2}}\vec{j} + \vec{k}\right)\,dx\,dy.$

$$\iint_S \text{curl}(\vec{F}) \cdot \overrightarrow{d\sigma}^{+} = -2 \iint_{x^2+y^2\le 4} dx\,dy = -8\pi.$$

(3) $\vec{F} = z^2\vec{i} + y^2\vec{j} + xy\vec{k}$ and S is the triangle with vertices at $(1,0,0)$, $(0,1,0)$, $(0,0,1)$.

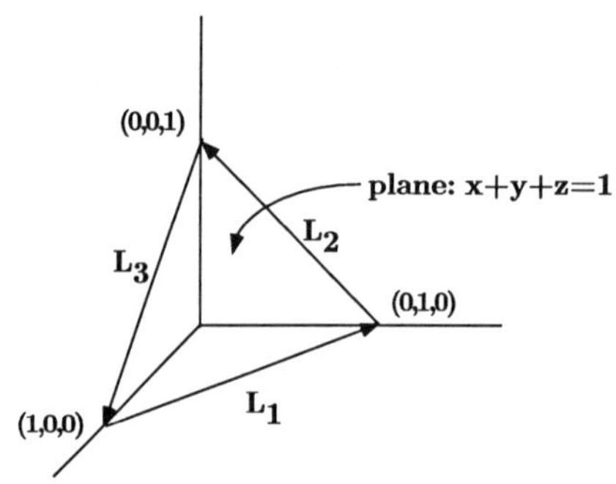

The boundary of S, ∂S consists of 3 lines, L_1, L_2, and L_3 which can be parameterized as follows:

$$L_1 = \{(1-t, t, 0) \mid 0 \le t \le 1\} \implies dx = -dt, dy = dt, dz = 0,$$
$$L_2 = \{(0, 1-t, t) \mid 0 \le t \le 1\} \implies dx = 0, dy = -dt, dz = dt,$$
$$L_3 = \{(t, 0, 1-t) \mid 0 \le t \le 1\} \implies dx = dt, dy = 0, dz = -dt,$$

Note that as t ranges from 0 to 1, the orientation is counterclockwise as required. Then

$$\oint_{\partial S} \vec{F} \cdot \overrightarrow{dr} = \oint_{\partial S} \left(z^2\,dx + y^2\,dy + xy\,dz\right)$$

$$= \int_0^1 t^2\,dt + \int_0^1 (1-t)^2\,(-dt) + \int_0^1 (1-t)^2\,dt$$

$$= \frac{1}{3}.$$

Further, $\text{curl}(\vec{F}) = \begin{vmatrix} \vec{i} & \vec{j} & \vec{k} \\ \frac{\partial}{\partial x} & \frac{\partial}{\partial y} & \frac{\partial}{\partial z} \\ z^2 & y^2 & xy \end{vmatrix} = x\vec{i} + (2 - 2x - 3y)\vec{j}$. For the surface, $z = 1 - x - y$, we have $\overrightarrow{d\sigma}^{+} = \left(\vec{i} + \vec{j} + \vec{k}\right)\,dx\,dy$. Hence,

$$\iint_S \text{curl}(\vec{F}) \cdot \overrightarrow{d\sigma}^{+} = \int_0^1 \int_0^{1-x} (2 - x - 3y)\,dx\,dy = \frac{1}{3}.$$

(4) $\vec{F} = 2x^2 z\vec{\imath} - x\vec{\jmath} + y^2\vec{k}$ and S is the hemisphere $x^2 + y^2 + z^2 = 1$, $z \geq 0$. Then ∂S is the circle $x^2 + y^2 = 1$. By Stoke's theorem:

$$\iint_S \operatorname{curl}(\vec{F}) \cdot \overrightarrow{d\sigma}^{+} = \oint_{\partial S} (2x^2 z\, dx - x\, dy + y^2\, dz) = \int_{x^2+y^2=1} -x\, dy = \int_0^{2\pi} -\cos^2(t)\, dt = -\pi.$$

(5) $\vec{F} = -y\vec{\imath} + z\vec{\jmath} - x\vec{k}$ and S is the triangle with vertices at $(1,0,0)$, $(0,-1,0)$, $(0,0,1)$.

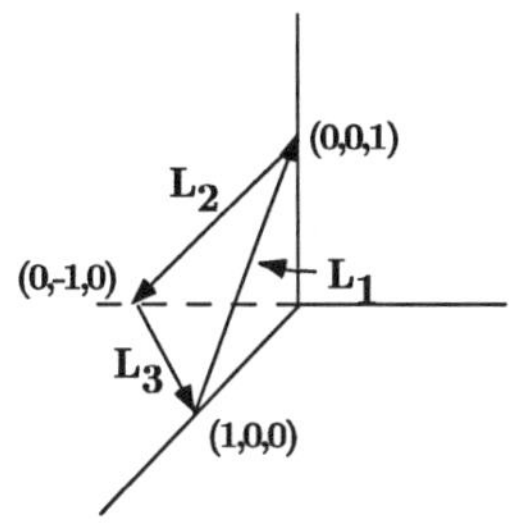

Then ∂S consists of the three lines, L_1, L_2, L_3 which can be parameterized as follows:

$$\begin{aligned} L_1 &= \{(1-t, 0, t) \mid 0 \leq t \leq 1\} \implies dx = -dt,\ dy = 0,\ dz = dt, \\ L_2 &= \{(0, -t, 1-t) \mid 0 \leq t \leq 1\} \implies dx = 0, dy = -dt, dz = -dt, \\ L_3 &= \{(t, t-1, 0) \mid 0 \leq t \leq 1\} \implies dx = dt, dy = dt, dz = 0. \end{aligned}$$

Note that as t ranges from 0 to 1, the orientation is counterclockwise as required. Then

$$\iint_S \operatorname{curl}(\vec{F}) \cdot \overrightarrow{d\sigma}^{+} = \oint_{\partial S} (-y\, dx + z\, dy - x\, dz) = \int_0^1 -(1-t)\, dt + \int_0^1 (1-t)\,(-dt) + \int_0^1 -(t-1)\, dt = -\frac{1}{2}.$$

(6) 2π.

(7) 0, because $\operatorname{curl}(\vec{\mathbf{F}}) = 0$. One could also note that $\vec{\mathbf{F}}$ is a gradient vector field, with a potential function $\phi(x, y, z) = \frac{1}{2}(x^2 + y^2 + z^2)$, and therefore the line integral along a closed curve vanishes.

(8) Evaluate the sum of the surface integrals in the left and right hand sides as an integral over the solid bounded by S_1 and S_2. Then use the identity $\operatorname{div}(\operatorname{curl}(\vec{\mathbf{F}})) = 0$.

(9) Use the identity $f\vec{\nabla} g + g\vec{\nabla} f = \vec{\nabla}(fg)$. The integral $\oint_C \vec{\nabla}(fg) \cdot \overrightarrow{dr}$ is zero as the line integral of a gradient vector field.

Solutions for Additional Exercises for Chapter 28

(1) Surface area $= \int_0^{\sqrt{2}} \int_0^{2\pi} r\sqrt{4r^2 + 1}\, d\theta\, dr = \frac{13\pi}{3}$.

(2) Surface Area $= \int_0^1 \int_x^1 \sqrt{2+4y^2}\, dy\, dx = \frac{\sqrt{3}}{2} - \frac{1}{3\sqrt{2}}.$

(3) Surface area (for cone of radius R) is given by

$$\int_0^R \int_0^{2\pi} \sqrt{2}r\; d\theta\; dr = \sqrt{2}\pi r^2.$$

Therefore, surface area $= \sqrt{2}\pi \left(6^2 - \left(\frac{1}{2}\right)^2\right) = \frac{143\sqrt{2}}{4}\pi.$

(4) Surface area $= \int_1^2 \int_0^4 \sqrt{1+4u^2}\; dv\; du = 4\sqrt{17} + \ln(4+\sqrt{17}) - 2\sqrt{5} - \ln(2+\sqrt{5}) \approx 12.671.$

(5) $\iint_S \vec{F} \cdot\, d\vec{\sigma}^+ = \int_0^{2\pi} \int_0^{\sqrt{2}} r^3 \cos(2\theta) + 2r\; dr\; d\theta = 4\pi.$

(6) $\iint_S \vec{F} \cdot\, d\vec{\sigma}^+ = \int_0^{2\pi} \int_0^{2} -r^3 \cos(2\theta)\; dr\; d\theta = 0.$

(7) $\iint_S \vec{F} \cdot\, d\vec{\sigma}^+ = \int_0^{2} \int_{-1}^{1} (-u^2v^2 - 2u^2v^3 + uv^2)\; dv\; du = -\frac{4}{9}.$

(8) $\iint_S \vec{F} \cdot d\vec{\sigma}^+ = \int_0^{\pi/2} \int_{\frac{a\sin(\theta)}{2}}^{a\sin(\theta)} a^2 dy d\theta = \frac{a^3}{2}.$

(9) $\iint_S \vec{F} \cdot\, d\vec{\sigma}^+ = \int \int_{0 \leqslant u+v \leqslant 1} 2\; du\; dv = 1.$

(10) $\vec{\nabla} \cdot \vec{F} = y^2 e^{xy^2} + \sin(y).$

(11) $\vec{\nabla} \cdot \vec{F} = \frac{2}{\sqrt{x^2+y^2+z^2}}.$

(12) $\vec{\nabla} \cdot \vec{F} = \frac{1}{x} + xze^{xyz} - \frac{x}{x^2+z^2}.$

(13) $\iint_S \vec{F} \cdot\, d\vec{\sigma}^+ = \int_0^1 \int_0^1 \int_0^1 (ye^{xy} + \sin(y))\; dz\; dy\; dx.$

(14) $\iint_S \vec{F} \cdot\, d\vec{\sigma}^+ = 0$, since $\vec{\nabla} \cdot \vec{F} = 0.$

(15) curl$(\vec{F}) = \vec{0}$. Therefore the spins in each plane are 0.

(16) curl$(\vec{F}) = \vec{0}$. Therefore the spins in each plane are 0.

(17) curl$(\vec{F}) = (0, -\infty, 0)$. Therefore the spin in xy-plane is zero and spin in yz-plane is zero.

(18) $\iint\limits_S \operatorname{curl}(\vec{F}) \cdot \, d\vec{\sigma}^+ = \int\limits_0^{2\pi} 32\cos^2(t)\sin(t)\, dt = 0.$

(19) $\iint\limits_S \operatorname{curl}(\vec{F}) \cdot \, d\vec{\sigma}^+ = \int\limits_0^{2\pi} 9\cos^2(t)\, dt = 9\pi.$

Solutions for §29.1

(1) 0. Note that the wedge product of a one-form by itself is always zero.

(2) $dx \wedge dy - z\, dy \wedge dz$.

(3) $(x - 3y) dx \wedge dy \wedge dz$.

(4) 0.

(5) $-2\, dx \wedge dy \wedge dz$.

(6) $-e^{x+y} dx \wedge dy - \cos(x^2yz)\, dz \wedge dx$.

(7) $(6z^2 - 3e^{xy})\, dx \wedge dy + (3xy - 3z^2e^y)\, dy \wedge dz + (2xy - e^{(x+1)y})\, dz \wedge dx$.

(8), **(9)** 0. Note that a form which degree is greater than the dimension of the space is identically zero.

(10) $3y^3\, dx \wedge dy \wedge dz$.

Solutions for §29.2

(1) $yz\, dx + xz\, dy + xy\, dz$.

(2) $3x^2yz^2\, dx + x^3z^2\, dy + 2x^3yz\, dz$.

(3) $-z^2\sin(xz^2)\, dx - 2xz\sin(xz^2)\, dz$.

(4) $e^x\, dx$.

(5) 0.

(6) $-x^2\, dx \wedge dy$.

(7) $-(e^x + e^y)\, dx \wedge dy$.

(8) $-xz\, dx \wedge dy - e^{x+y+z}\, dy \wedge dz + (xy + e^{x+y+z})\, dz \wedge dx$.

(9) $-x^3\, dx \wedge dy \wedge dz$.

(10) $-yz\sin(xyz)\, dx \wedge dy \wedge dz$.

(11) $dx \wedge dy \wedge dz$.

(12) 0.

(13) Use the formal definition of the operator d.

In **(ii)**, note that $\frac{\partial}{\partial x}\frac{\partial}{\partial y}f = \frac{\partial}{\partial y}\frac{\partial}{\partial x}f$, while $dx \wedge dy = -dy \wedge dx$, and the same is true for the two other pairs of variables (x, z) and (y, z). This immediately proves that for any function f, $d(df) = 0$.

Then, note that any differential form ω can be written as a function times the wedge product of dx, dy, and dz, and the operator d acts on the function only.

Solutions for Additional Exercises for Chapter 29

(1) $x^2y^2\left(x^2 - y^2\right)\, dx \wedge dy - xy(x-y)z^y\, dy \wedge dz + xy(x-y)z^y\, dx \wedge dz$.

(2) $\left(yx^2 \ln x^2 - x^2 y e^{yx}\right)\, dx \wedge dy - yx^2 \cos\left(z^y\right)\, dy \wedge dz - xy^2 \cos\left(z^y\right)\, dx \wedge dz$.

(3) $\frac{1}{x^2+y^2+z^2}\left[\left(yx^3 - x\right)\, dx \wedge dy - yx^2\, dy \wedge dz - x\, dx \wedge dz\right]$.

(4) $d\omega = \left(-2x\cos(x^2) + 2y\sin(y^2)\right)\, dx \wedge dy$.

(5) $d\omega = (x+6)\, dx \wedge dy \wedge dz$.

(6) $d\omega = (e^z - e^x)\sin(y)\, dx \wedge dy + \left[e^x\left(\tan(x) + \frac{1}{1+x^2}\right) - e^z\cos(y)\right] dx \wedge dz$.

(7) $d\omega = 0$.

(8) Closed and exact.

(9) Neither.

(10) Closed and exact.

(11) Neither.

(12) Closed and exact.

(13) Neither.

(14) Yes. The forms are independent.

Solutions for §30.1

(1) $\frac{3}{4}\sin(2\pi x) - \frac{1}{4}\sin(6\pi x)$.

(2) $\frac{7}{2} + \frac{1}{2}\sin(2\pi x) - \frac{3}{2}\cos(4\pi x) - \frac{1}{2}\sin(6\pi x)$.

(3) $\frac{5}{8}\cos(2\pi x) + \frac{5}{16}\cos(6\pi x) + \frac{1}{16}\cos(10\pi x)$.

(4) Use the trigonometric identity $\cos(2\pi mx)\cos(2\pi nx) = \frac{1}{2}(\cos(2\pi(m+n)x) + \cos(2\pi(m-n)x))$.

(5) Use the trigonometric identity $\cos(2\pi mx)\sin(2\pi nx) = \frac{1}{2}(\sin(2\pi(m+n)x) - \sin(2\pi(m-n)x))$.

(6) Use the trigonometric identity $\sin(2\pi mx)\sin(2\pi nx) = \frac{1}{2}(\cos(2\pi(m-n)x) - \cos(2\pi(m+n)x))$.

(7) $\frac{1}{2}$ for $n \neq 0$. For $n = 0$, the first integral equals 1, the second one equals 0.
Use the trigonometric identities $\cos^2(2\pi nx) = \frac{1}{2}(1 + \cos(4\pi nx))$ and $\sin^2(2\pi nx) = \frac{1}{2}(1 - \cos(4\pi nx))$.

Solutions for §30.3

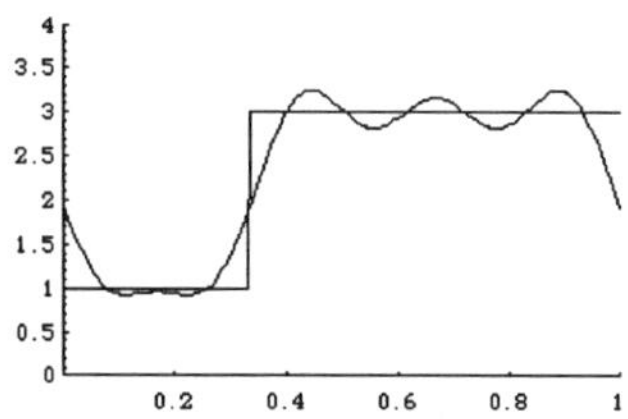

(1)

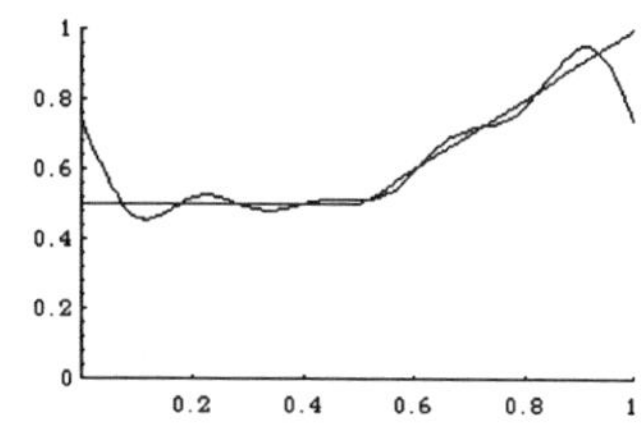

(2)

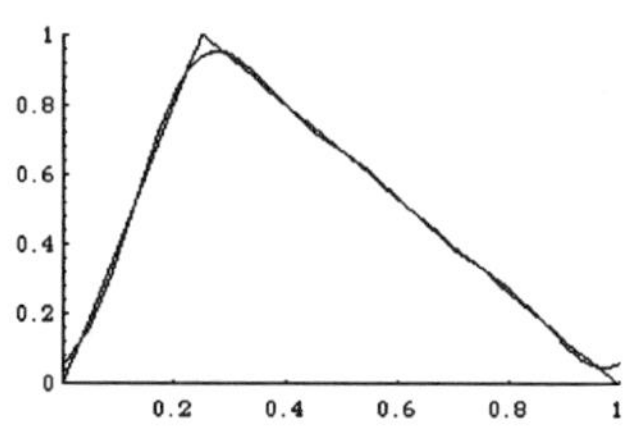

(3)

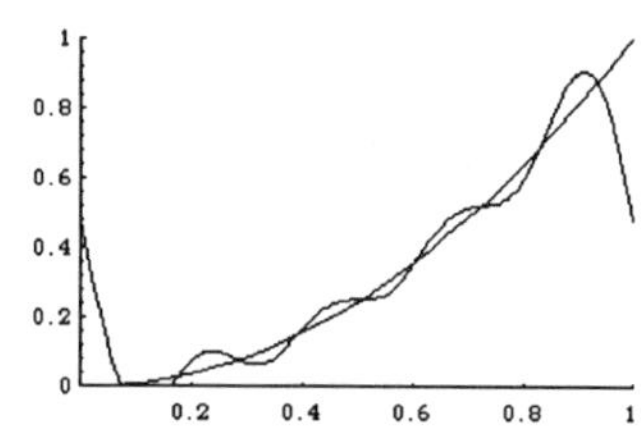

(4)

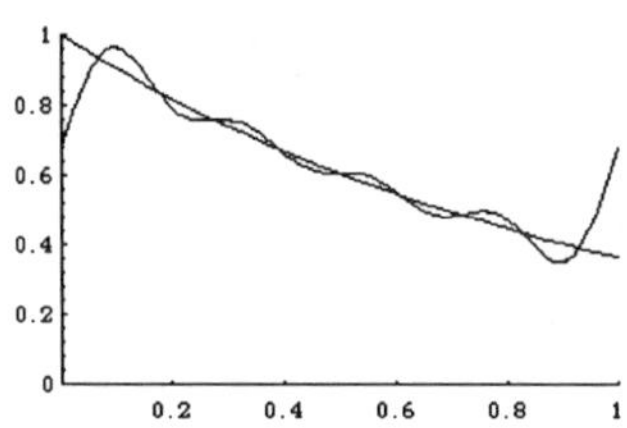

(5)

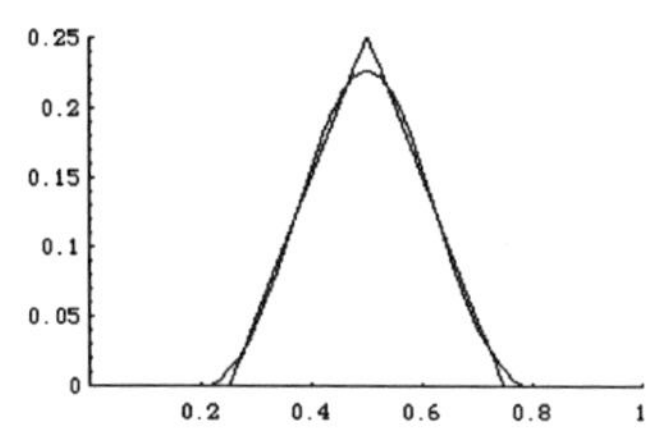

(6)

(7) Consider the function $f(x) = 2\pi^2(x^2 - x + \frac{1}{6})$. The Fourier coefficients of this function are $\hat{f}(n) = \frac{1}{n^2}$, $\hat{f}(0) = 0$, and therefore one has the equality $\int_0^1 (f(x))^2\,dx = \sum_{n=-\infty}^{\infty} |\hat{f}(n)|^2 = 2\sum_1^{\infty} \frac{1}{n^4}$.

Evaluating $\int_0^1 (f(x))^2\,dx = \frac{\pi^4}{45}$, one obtains the result $\sum_1^{\infty} \frac{1}{n^4} = \frac{\pi^4}{90}$.

(8) Fix the integer k. Integration of the expression $\hat{f}(n) = \int\limits_0^1 f(x)e^{-2\pi inx}\,dx$ by parts k times gives $\hat{f}(n) = \frac{(-1)^k}{(2\pi in)^k}\int\limits_0^1 f^{(k)}(x)e^{-2\pi inx}\,dx$. The integral $\int\limits_0^1 f^{(k)}(x)e^{-2\pi inx}\,dx$ is finite, as the function $f(x)$ is supposed to be smooth. This gives the desired bound on $\hat{f}(n)$.

(The boundary terms which appear during the integration by parts cancel out. This is because the periodic function $f(x)$ satisfy the identities $f^{(l)}(0) = f^{(l)}(1)$, for any integer l.)

Solutions for Additional Exercises for Chapter 30

(1) Fourier series expansion is $f(x) = \frac{1}{6} - \frac{1}{\pi^2}\sum\limits_{n=1}^{\infty}\frac{1}{n^2}\cos(2\pi nt)$.

(2) Fourier series expansion is $f(x) = \frac{1}{3} + \frac{4}{\pi}\sum\limits_{n=1}^{\infty}\frac{(-1)^n}{n^2}\cos(2\pi nt)$.

(3) Fourier series expansion (assuming $0 < c \leq \frac{3}{2}$) is

$$f(x) = \frac{1}{3c} + \frac{1}{\pi}\sum_{n=1}^{\infty}\frac{1}{n}\left(\sin\left(\frac{4\pi n}{3c}\right) - \sin\left(\frac{2\pi n}{3c}\right)\right)\cos(2\pi nt)$$
$$- \frac{1}{\pi}\sum_{n=1}^{\infty}\frac{1}{n}\left(\cos\left(\frac{4\pi n}{3c}\right) - \cos\left(\frac{2\pi n}{3c}\right)\right)\sin(2\pi nt).$$

(4) Fourier series expansion (assuming $c > 2$) is $f(x) = 1 - \frac{2}{c} - \frac{2}{\pi}\sum\limits_{n=1}^{\infty}\frac{1}{n}\sin\left(\frac{2\pi n}{c}\right)\cos(2\pi nt)$.

(5) Using the hint, if $f(x) = e^{a2\pi x}$,

$$\hat{f}(n) = \int_0^1 e^{2\pi ax}e^{-2\pi nix}dx$$
$$= \int_0^1 e^{2\pi(a-ni)x}dx = \frac{1}{2\pi(a-ni)}(e^{2\pi(a-ni)} - 1)$$

$$\int_0^1 e^{4\pi ax}dx = \frac{1}{4\pi^2a^2}(e^{2\pi a} - 1)^2 + 2\sum_{n=1}^{\infty}\frac{1}{4\pi^2(a^2+n^2)}(e^{2\pi a} - 1)^2$$

$$\sum_{n=1}^{\infty}\frac{1}{a^2+n^2} = \frac{1}{2a^2}\left[\pi a\frac{e^{2\pi a}+1}{e^{2\pi a}-1} - 1\right].$$

(6) Fourier series expansion is

$$f(x) = \frac{a(1-e^{-a})}{a^2+4\pi^2b^2} + (1-e^{-a})\sum_{n=1}^{\infty}\left[\frac{a}{a^2+4\pi^2(n-b)^2} + \frac{a}{a^2+4\pi^2(n+b)^2}\right]\cos(2\pi nt)$$
$$+ (1-e^{-a})\sum_{n=1}^{\infty}\left[\frac{2\pi(n-b)}{a^2+4\pi^2(n-b)^2} + \frac{2\pi(n+b)}{a^2+4\pi^2(n+b)^2}\right]\sin(2\pi nt)$$

(7) Fourier series expansion is

$$f(x) = \frac{2\pi b\,(1-e^{-a})}{a^2+4\pi^2 b^2} + (1-e^{-a})\sum_{n=1}^{\infty}\left(\frac{2\pi(b-n)}{a^2+4\pi^2(b-n)^2} + \frac{2\pi(b+n)}{a^2+4\pi^2(b+n)^2}\right)\cos(2\pi nt)$$
$$+ (1-e^{-a})\sum_{n=1}^{\infty}\left(\frac{a}{a^2+4\pi^2(b-n)^2} - \frac{a}{a^2+4\pi^2(b+n)^2}\right)\sin(2\pi nt).$$

(8) Fourier series expansion is

$$f(x) = \frac{\sinh(2\pi a)}{2\pi a} + \frac{\sinh(2\pi a)}{\pi}\sum_{n=1}^{\infty}\frac{a}{a^2+n^2}\cos(2\pi nt) - \frac{\cosh(2\pi a)-1}{\pi}\sum_{n=1}^{\infty}\frac{n}{a^2+n^2}\sin(2\pi nt).$$

(9) Fourier series expansion is

$$f(x) = \frac{\cosh(2\pi a)-1}{2\pi a} + \frac{\cosh(2\pi a)-1}{\pi}\sum_{n=1}^{\infty}\frac{a}{a^2+n^2}\cos(2\pi nt) - \frac{\sinh(2\pi a)}{\pi}\sum_{n=1}^{\infty}\frac{n}{a^2+n^2}\sin(2\pi nt)$$

(10) With $x = 2\pi u$, let $f(u) = \frac{\pi-(2\pi u)}{2} = \frac{\pi}{2} - \pi u$

$$\hat{f}(0) = \int_0^1 \left(\frac{\pi}{2} - \pi u\right)du = \frac{\pi}{2}(u-u^2)\Big|_0^1 = 0$$

and

$$\hat{f}(n) = \int_0^1 \frac{\pi}{2}e^{-2\pi inu}du - \pi\int_0^1 ue^{-2\pi inu}du$$
$$= \frac{\pi}{2}\cdot\frac{1}{2\pi in}(1-e^{-2\pi in}) + \frac{\pi}{2\pi in}e^{-2\pi in} - \frac{\pi}{4\pi^2 n^2}(1-e^{-2\pi in}) = \frac{1}{2ni}$$

$$f(u) = \frac{\pi-(2\pi u)}{2} = \sum_{n=1}^{\infty}\frac{1}{2ni}(e^{2\pi niu} - e^{-2\pi niu}) = \sum_{n=1}^{\infty}\frac{1}{2ni}\cdot 2i\sin(2\pi nu)$$

$$\frac{\pi-x}{2} = \sum_{n=1}^{\infty}\frac{\sin(nx)}{n}, \qquad \text{if } \; 0 \le x \le 2\pi$$

(11) Integrating both sides of $\frac{\pi-x}{2} = \sum\limits_{n=1}^{\infty}\frac{\sin(nx)}{n}$ yields $\frac{\pi}{2}x - \frac{x^2}{4} = -\sum\limits_{n=1}^{\infty}\frac{\cos(nx)}{n^2} + C$. Setting $x=0 \Longrightarrow$ $C = \sum\limits_{n=1}^{\infty}\frac{1}{n^2} = \frac{\pi^2}{6}$.

(12) Integrating both sides of the identity in problem (11) and setting $x = 0$ shows that the constant of integration is 0.